FESTKÖRPERPROBLEME

ADVANCES IN SOLID STATE PHYSICS 37

FESTKÖRPER PROBLEME
ADVANCES IN SOLID STATE PHYSICS 37

Edited by
Reinhard Helbig

vieweg

Die Deutsche Bibliothek – CIP-Einheitsaufnahme

Festkörperprobleme = Advances in solid state physics.

Bis Bd. 15 (1975) ausserdem im Verl. Pergamon Press,
Oxford, Elmsford, Toronto, Sydney
ISSN 0430-3393
NE: PT
ISBN 978-3-663-11946-3
NE: Deutsche Physikalische Gesellschaft / Fachausschuss
Halbleiterphysik

Editor:

Prof. Dr. Reinhard Helbig
Institut für Angewandte Physik
Universität Erlangen – Nürnberg
Staudtstr. 7
D-91058 Erlangen

Cover design: Barbara Seebohm, Braunschweig
Printed on acid-free paper

ISSN 0430-3393

ISBN 978-3-663-11946-3   ISBN 978-3-663-11944-9 (eBook)
DOI 10.1007/978-3-663-11944-9

Foreword

The volume 37 of the Advances of Solid State Physics contains contributions given as plenary talks or invited papers at the Spring Meeting of the German Physical Society (Arbeitskreis Festkörperphysik) 1997 in Münster. The number of approx. 4 000 participants of the meeting gives evidence for the attraction of the field of solid state physics especially for young physicists.

"There is no semiconductor material aside from silicon". This old statement of semiconductor people is still alive despite of the success of GaAs and the wide band gap materials in the last decade, especially in the field of optoelectronics. The volume 37 of the Advances in Solid State Physics starts with two contributions demonstrating that there is still rapid progress in the development of the material silicon: The growing of single crystals with very large diameter and the anisotropic electrochemical etching of silicon for different applications.

In optoelectronics a fast growing field is the investigation of the dynamics of semiconductor lasers, which was the topic of a symposium organized by J. Feldmann (Munich). In the "classical" optics of semiconductors an unified description seems to be achieved as shown by the contributions of Ivchenko (St. Petersburg) and Glutsch (Jena).

For the understanding of the physical properties of semiconductors it is necessary to know the lattice vibrations of the material which can now be deduced from ab initio calculations (D.Strauch (Regensburg) et al.). An important contribution of the experimental situation is given by K. Laßmann (Stuttgart).

I hope that the volume 37 of the Advances of Solid State Physics contains contributions to partial summarize the present development in solid state physics, as you may expect from a volume of this series with its long tradition.

Erlangen, July 1997   Reinhard Helbig

Contents

Silicon Wafer for the Gigabit Era

Albrecht P. Mozer

Wacker Siltronic Corporation, Burghausen, Germany

Abstract: With the increasing integration density towards the Gigabit memory, the structure size (design rule) reduces from 0.35 μm to 0.25 μm and later to 0.18 μm. The latter 0.18 μm structure is to be expected at the turn of the century. Therefore the requirements which have to be fulfilled by the base material of the integrated circuits, the silicon wafers, are constantly rising. Along with the reduction of the structure size, a change in the wafer diameter is also occurring. Currently, the semiconductor industry is getting ready for the transition from 200 mm wafer technology to 300 mm wafer technology. The new demand profile on the silicon wafer is analysed, the research and development approaches in the various fields are investigated, and the driving forces for the transition are identified.

1 Introduction

In 1996, about 40 billion silicon semiconductor elements were sold worldwide, with a total turnover of 142 billion US$. Until the turn of the century, an average volume growth rate of about 14% per year is expected. This growth is technically and technologically promoted by the DRAM (Dynamic Random Access Memory) development as well as the microprocessor development.

The evolution of the DRAM memory technology started in 1969 with the first 256 Bit DRAM and has now, in 1996, reached the point of the 64 Mbit DRAM. The next big step in the progress of this technology will be the 1 Gbit DRAM. The first prototypes of this device generation are expected around the year 2001. Therefore on average every three years a new generation of DRAM-devices has been introduced. In other words, the number of transistors per integrated circuit increases four times every three years, which is also known in literature as Moore's Law [1], published already in 1965.

The microprocessor development has taken a similarly rapid turn. In 1972 Intel introduced worldwide the first 4004 microprocessor on the market. The characteristic criteria were a 4 Bit, p-channel device with 10 μm structure size, 2 inch wafer diameter technology and a 46 instruction comprising instruction set. In 1996 the Pentium Pro-Microprocessor was introduced also by Intel, offering a 166 MHz clock frequency in 0.35 μm CMOS-technology.

Microelectronics history in short; Transistor effect demonstrated 1947. First commercial Memory Chip 1969. First commercial Microprocessor circuit 1972. First man on the moon 1969.

As already mentioned, the first commercial prototypes of the 1 Gbit DRAM memory are to be expected at the turn of the century in a technology of 0.18 μm structure size. With this increasing integration density and shrinking structure size, more and more demanding requirements emerge with respect to the base material of the integrated circuits, the silicon wafer. Simultaneously with the increasing demand profile, an enlargement of the wafer diameter from 200 mm to 300 mm takes place. The semiconductor industry is currently preparing itself for this transition. It is expected that the first 300 mm device lines will be ready for manufacturing at the turn of the century. In this paper the new silicon requirements are analyzed and the ongoing research and development activities are reviewed.

2 Technology Roadmap and Wafer-Specification Requirements

The Technology Roadmap for Semiconductors [2] was created in order to draw a consensus based picture for the future of the VLSI-Technology. The basic notion of the roadmap is to extrapolate the progress made up to the present towards the future. Fig. 1 is a part of the roadmap that describes the device generations, the minimum design rules, the wafer and chip sizes for memory and microprocessor ICs as well as the clock frequency of microprocessors as a key measure for CMOS speed performance. Remarkable is the device structure reduction which will pass from a 0.35 μm technology, which is currently in use as far as the production technology is concerned, to

Year		1995	1998	2001	2004	2007	2010
Device Parameter	Design Rule (µm)	0,35	0,25	0,18	0,13	0,10	0,07
	First DRAM shipments	64 M	256 M	1 G	4 G	16 G	64 G
	DRAM-Chip-Size (mm²)	190	280	420	640	960	1400
	Chip frequency (MHz) - cost optimization - speed optimization	150 300	200 450	300 600	400 800	500 1000	625 1100
Wafer Requirements	Microprocessor-Chip-Size (mm²)	250	300	360	430	520	620
	Wafer Diameter (mm)	200	200	300	300	400	400
	Site-Flatness (SFQD) (µm) Site-Size (mm x mm)	0,23 22x22	0,17 26x32	0,12 26x32	0,08 26x36	0,06 26x44	0,04 26x44
	Particles per Wafer Particle-Size (µm)	50 > 0,12	40 > 0,08	50 >/= 0,06	35 > 0,04	35 >/= 0,03	35 >/= 0,02
	Edge Exclusion (mm)	3	3	2	2	1	1

Figure 1 Semiconductor Industry Association Roadmap on VLSI-Technology.

0.25 μm, later to 0.18 μm and then may be reduced further to 0.13 μm and 0.10 μm. Other key observations from this roadmap for silicon wafer manufacturers are:

- Increase of wafer diameter from 200 mm to 300 mm and later to 400 mm;

- Substantial tightening of the local site flatness (Site Focal Plane Deviation, SFQD) requirements;

- Reduction of the number of particles per wafer accompanied by a significant reduction of the particle size.

Not only the VLSI device manufacturer faces great technological challenges to follow this roadmap, but also the silicon wafer supplier. The technical and technological approaches of the wafer manufacturer to achieve these challenging requirements are outlined and discussed in the following.

3   Major Technology Challenges for the Silicon Wafer Manufacturers

3.1   CZ (Czochralski) Crystal-Pulling

In Fig. 2 the wafer diameter, and the wafer area is depicted in relation to the maximum VLSI chip area used in a certain device generation. It is obvious from the consideration of Fig. 2 that larger chip areas require larger wafer diameters. An enlargement of the wafer diameter requires, for the CZ-crystal pulling method, which is exclusively applied in this diameter range, a considerable increase of the silicon melting volume. With large melting volumes we gain cylindrical crystal lengths and this is the key to be commercially efficient. In Fig. 3 the theoretical crystal ingot length as a function of crystal diameter is shown for different crucible sizes and poly silicon charges. In order to be cost competitive with 300 mm crystals, an ingot length larger than 100 cm is required, which results in a 200 kg poly silicon melting volume as a minimum. Consequently, issues like temperature fluctuations in the large melting volume, as well as the handling of the large thermal mass and crucible corrosion on the inner crucible surface are becoming severe at these melting volumes and high temperatures. The production of dislocation-free crystals is considerably more complicated at 300 mm than it is in the case of 200 mm. Major problems arise from the fact that the maximum growth rate of the ingot decreases with increasing crystal diameter. Computer simulations (Fig. 4) have shown that this is a consequence of the increasing latent heat generated at the solid/liquid interface of the crystal. This reduces the radial temperature gradient at the melt surface which destabilises the growth process and causes a significant higher concavity of the solid/liquid interface, which becomes unacceptably high for a 400 mm crystal grown at a standard 200 mm growth rate [3]. A sufficient radial temperature gradient at the melt surface and a reasonable interface deflection can only be obtained

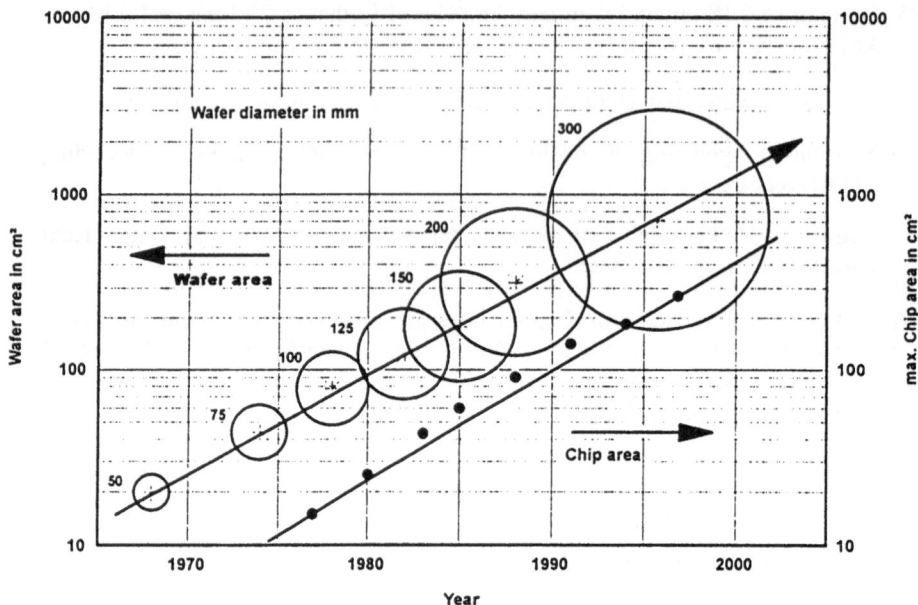

Figure 2   Wafer diameter, respectively the wafer area in relation to the maximum VLSI chip area used in certain device generations.

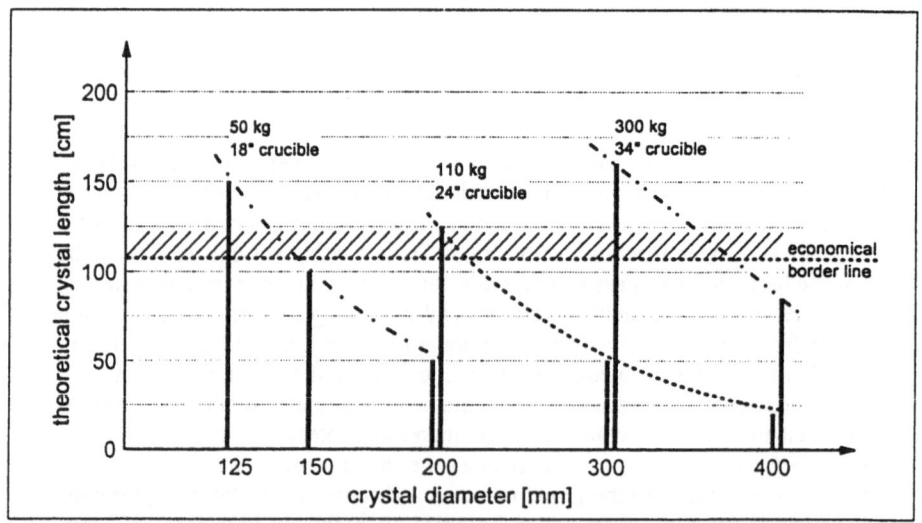

Figure 3   Theoretical crystal ingot length as function of crystal diameter, with crucible size and poly silicon charge size as parameter.

if the pull rate is lowered to about 0.4–0.5 mm/min. Unfortunately, a low pull rate not only reduces the ingot output per time, it enhances the growing time, crucible corrosion, and it increases the risk of structure loss too, which lowers the output further. Computer simulations of the entire CZ-pulling process become more and more an indispensable tool in order to reduce costly and long-lasting experiments for all of these complex analyses.

A further challenge is that if a dislocation, which is generated at the solid/liquid interface propagates back for about one length of crystal diameter into already dislocation-free grown material due to the high thermal stress. This means for the currently used

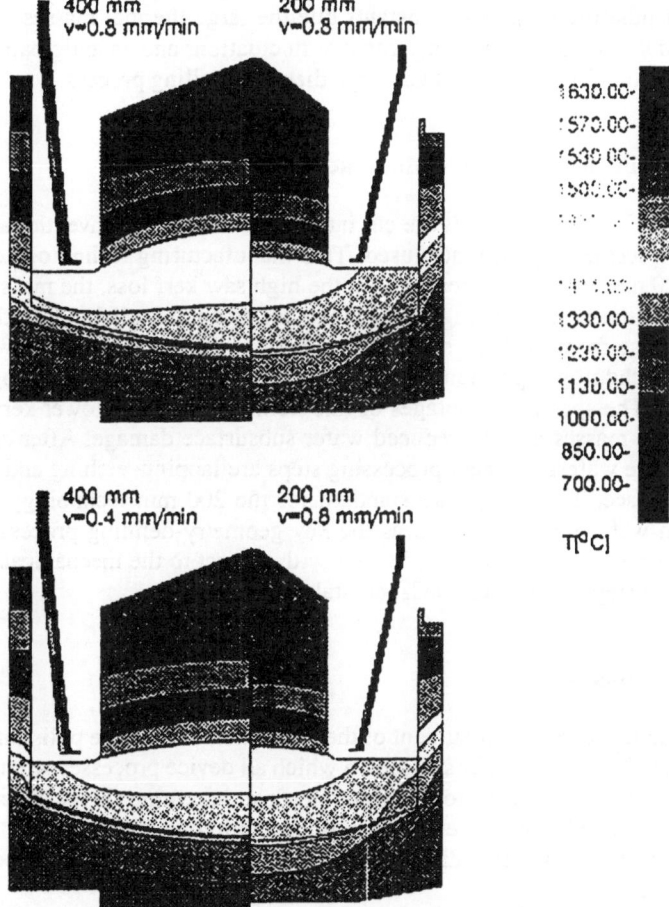

Figure 4 Comparison of isothermal temperature distribution for 200 mm and 400 mm CZ crystals at different pulling rates. The spacing between the isothermal lines is 30 °C.

300 mm pulling process at least half of the crystal ingot is lost, when a dislocation occurs. In order to obtain an economical 300 mm pulling process, comprehensive development work is absolutely necessary. This can no longer be achieved by upscaling the current 200 mm CZ-technology. Today's approaches to solve the challenges are based on the application of magnetic fields (MCZ) during the pulling process. Magnetic fields used so far range from 3 000 to 5 000 Gauss. This high fields not only raise issues as far as operational safety and work protection are concerned, but also the oxygen content in the crystals is lowered to values which no longer facilitate intrinsic gettering. The latter problem may require new innovative ideas.

On-line computer monitoring and controlling during the pulling process are prerequisites to handle the long time constants due the large thermal masses of the system. A detailed knowledge of growth parameter fluctuations and an anticipating control software is essential for a successful CZ large diameter pulling process.

3.2   Mechanical/Chemical Wafer Manufacturing Processes

After CZ-pulling, the silicon ingots are cut into raw wafers with conventional ID-saws by the 200 mm technology currently used. This manufacturing technology cannot be upscaled to 300 mm ingots. The reasons are the high saw kerf loss, the massive safety problems due to the high centrifugal force, as well as uncontrollable liquid-dynamic effects which arise from the high saw blad velocity at the larger diameter wafer. The multiwire saw technology appears to be the most promising cutting technology for the large diameters. The specific advantages of this technology are the lower kerf loss, the small transported masses and the reduced wafer subsurface damage. After cutting the ingot into separate wafers, the next processing steps are lapping, etching and polishing of the wafer surface. These steps are standard for the 200 mm technology. Grinding of the 300 mm wafer may be applied as the key geometry-defining process step. No fundamental issues have been identifed so far with respect to the mechanical/chemical wafer processing steps, compared to CZ-crystal pulling.

3.3   Polishing Process

After the mechanical/chemical treatment of the raw wafer, the surface polishing process is next. This step defines the wafer surface on which all device processes will take place later on. The atomic layers near to the wafer surface determine the electrical performance of the later VLSI circuit. Here perfection is required in all parameters in atomic layer dimensions. Currently, two different categories of requirements are specifed for the wafer surface:

- a local surface flatness on the silicon surface that allows efficient deep UV submicron lithography (geometrical requirement),

- an ideal wafer surface without any disturbances, irregularities or contamination. Requirements of less than 50 light point defects (LPDs) over the entire wafer surface of a dimension $> 0.12 \mu m$ are being specified. This parameter will be continuously tightened with the further reduction of the design rules (see Fig.1).

The geometrical parameters of the wafers must constantly be improved. Through the reduction of the structure dimensions, the exposed light wave length, applied by optical lithography, must be shortened: This means the optical depth of focus (DOF) shrinks and, consequently, the local flatness of the wafer has to be improved. Even the slightest divergence from the mathematically perfect surface geometry may lead to a failure of the corresponding VLSI circuit and thus cause lower device yield. In order to comply with these high requirements, especially with the 300 mm wafers, new polishing technologies have to be developed. Here the Double Side Polishing (DSP) or Free Floating Polishing (FFP) may be a potential solution together with advanced grinding techniques mentioned above.

However, even with highly optimized polishing processes, a certain density of Localised Light Scatterers (LLS) is still observed when wafer surfaces are inspected by highly sensitive Scanning Surface Inspection Systems (SSIS). Most of these surface scatterers on well polished wafers, however, turned out to be no particles nor other surface contaminants – as was assumed formerly – but are tiny etch pits which are called Crystal Originated Particles (COPs) [4]. The number, density and effective size of these COPs varies depending on polishing process parameters, the CZ-growing conditions and the specific ingot position of which the wafers were taken [5]. These findings indicate that COPs are mainly related to bulk crystal defects which can be influenced by pulling rate and hot-zone design [6]. Slow ingot cooling rates allow an aggregation of the defects responsible for the COP generation and an overall low density of COPs is observed. However, the average size is larger than for the ingots with faster cooling rates. Today it is commonly assumed that the origin of the COPs are vacancy clusters in the silicon ingot. Fig. 5 shows the COP distribution on 150 mm wafer (CZ, 100, p-) from different ingot growth processes, right after polishing and after an extensive SC1 (4h, 85 °C) treatment [7] which removes about 150 nm silicon. Clearly visible are the high level of COPs which are inherent to the CZ-growth process. The COPs have a negative impact on VLSI circuit performances and, in particular, on the GOI-yield (Gate Oxide Integrity) of the devices. Ingots with a high COP density show a much lower GOI-yield than ingots with a low density of COPs. An almost linear relation between the COP density and GOI defect density (GOI-DD) is observed (see Fig. 6).

A closer investigation of the COPs by an Atomic Force Microscope (AFM) reveals the shape of these etch pits. Top AFM images and AFM profiles of such crystal defects delineated by SC1 treatment, as well as by polishing are depicted in Fig. 7, showing a rectangular structure on Si(100) surfaces, which reflects the crystal orientation. The pits are sharp and well-defined features with the side walls corresponding to (111)-planes. Polishing also results in the formation of surface pits which, however, have a rather large lateral extension and a smooth transition between the pit itself and the surrounding surface due to the smoothening of the polishing process. A lot of research work has been

After polishing   After SC1 (4h, 85°C)

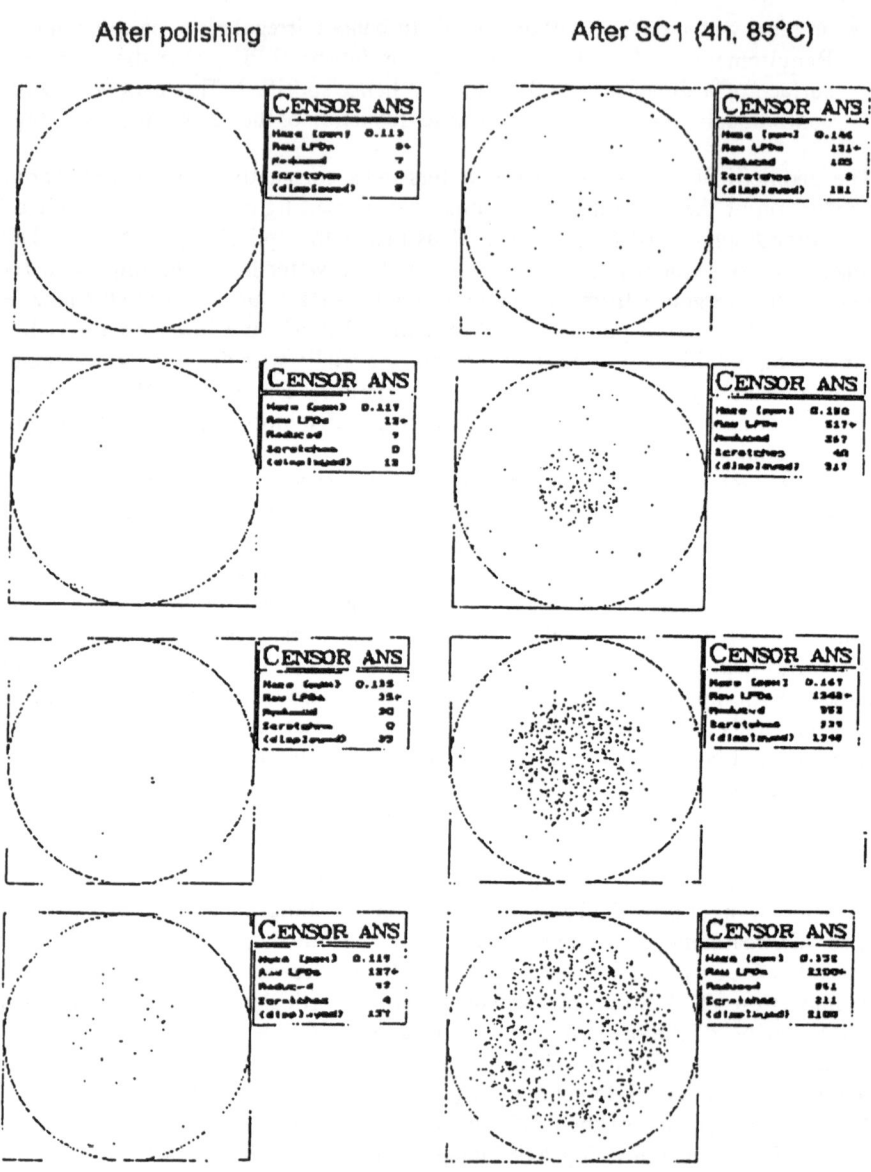

Censor ANS 100, LPD > 0,12 μm LSE
150 mm CZ 100 p-

Figure 5 COP Characterization of 150 mm Wafers from different CZ growing processes.
Left: straight after polishing; right: same Wafer after an extensive SC1 treatment.

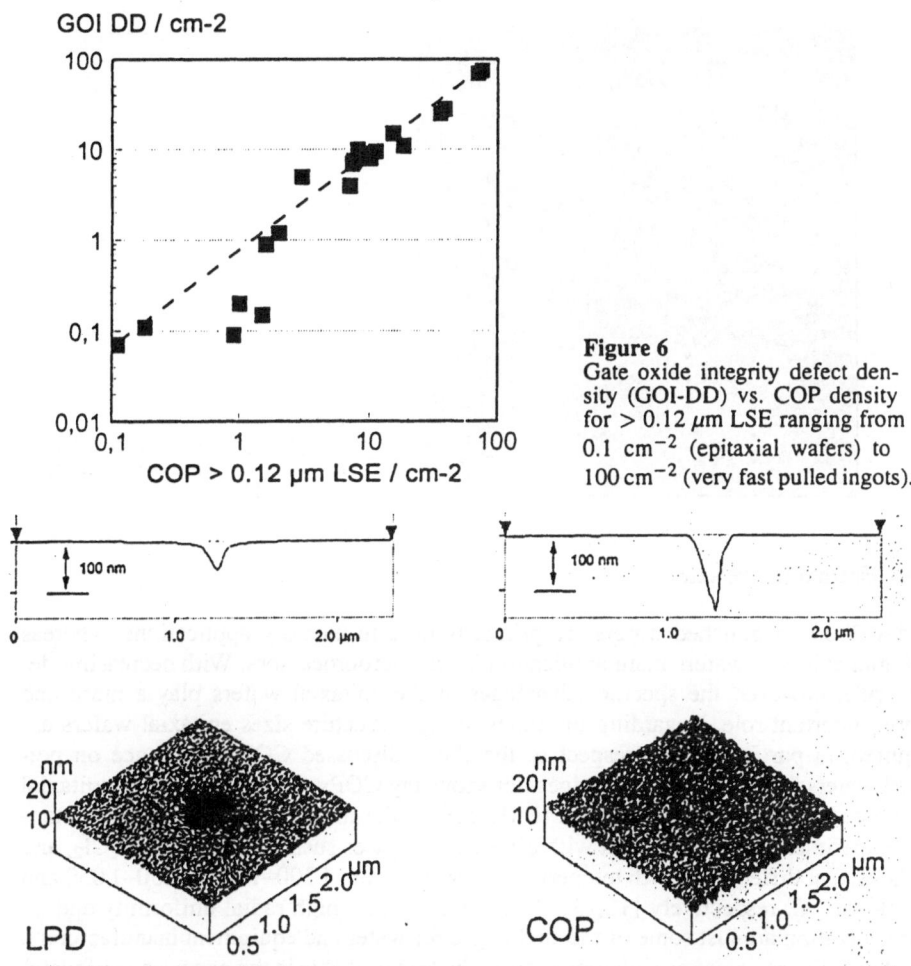

Figure 6
Gate oxide integrity defect density (GOI-DD) vs. COP density for $> 0.12\,\mu m$ LSE ranging from $0.1\ cm^{-2}$ (epitaxial wafers) to $100\ cm^{-2}$ (very fast pulled ingots).

Figure 7 Top AFM images and AFM profiles of COPs after extensive SCL treatment (right) and LPD after polishing (left)

devoted to understand whether the COPs originate from oxygen precipitates [8–10] or from vacancy aggregations (D-defects) [11]. Today it is commonly assumed that the COPs originate from octahedral vacancy clusters/voids with a morphology structure as depicted in Fig. 8 [11].

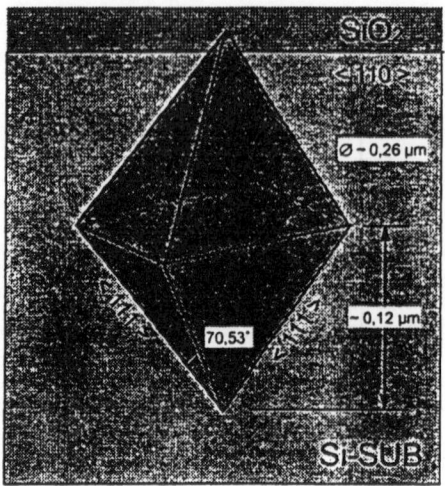

Figure 8
Morphology structure of a COP,
revealing an octahedral vacancy
cavity [11].

3.4 Epitaxial Growth

Today 200 mm polished wafers are primarily used in memory applications, whereas 200 mm epitaxial wafers main application is for microprocessors. With decreasing design rule, however, the specific advantages of the epitaxial wafers play a more and more important role. Regarding the tiniest design structure sizes epitaxial wafers are required, in particular with respect to the above discussed COP appearance on polished wafers. An epitaxial wafer does not show any COPs. The deposition of epitaxial layers on 300 mm substrates is currently one of the most challenging research topics. Experiments are carried out with different epi-gases such as $SiHCl_3$, SiH_2Cl_2, and SiH_4 used at different growth temperatures ranging from 1 200–1 100, 1 150–1 050 and 1 050–950 °C, respectively [12,13]. A sharp transition, high radial uniformity and no slip generation are just some of the challenges for wafer and equipment manufacturers. The lower the epitaxial growth temperature the less probable is the occurrence of wafer slippage. We have made substantial progress with an atmospheric pressure, $SiHCl_3$ epitaxial process below 1 100 °C and achieved preliminary encouraging results. But there is still a lot of research work to be done before 300 mm epi-wafers can be produced slip free in large quantities.

3.5 Low Thermal Budget (LTB) Wafer

Using the current employed VLSI device processing technology, the p or n well drive in processes are performed at 1 150 to 1 250 °C for about 250 to 350 min. At these temperatures and process times, different gettering phenomena can take place in the bulk and the back side of the silicon wafer. The advantage of these gettering processes are, that contaminations which find their way onto the wafer during the device processes,

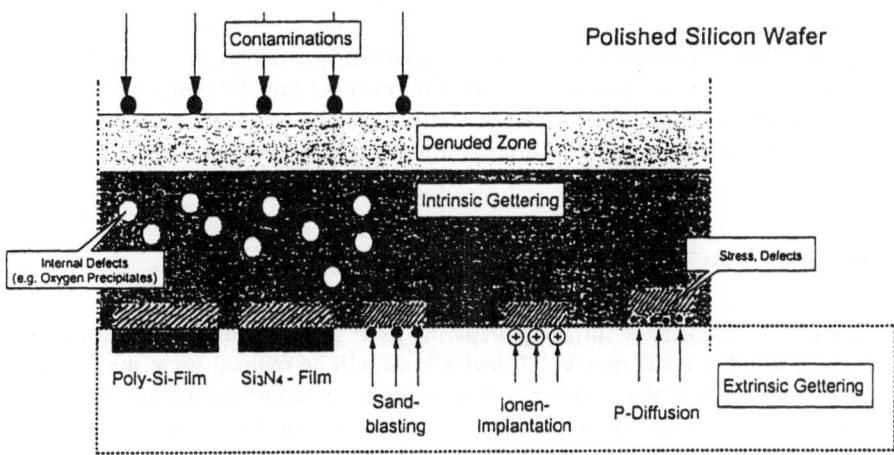

Figure 9 Summary of the different gettering technology options [14].

are trapped at the gettering centers. They cannot deteriorate the electrical performance of the VLSI-circuit. In the first process phase of the VLSI device process, the extrinsic gettering center on the wafer back side act as trapping centers in the later process phase, the intrinsic gettering via oxygen precipitates takes this role. Fig. 9 illustrates the different gettering options [14].

In the near future, the process temperature will be reduced to 850 °C to 1 000 °C for a time span of about 30 to 70 min. Furthermore the temperature ramp up and ramp down is very fast which generates a lot of thermal/mechanical stress in the silicon wafer. But more important at this low process temperature for this short period of time, the known gettering processes cannot be developed and the gettering efficiency is not sufficient to trap the contaminates. At the moment there are research and development programs in progress to preanneal/preprecipitate the silicon wafer prior to the VLSI circuit process in order to develop the gettering centers at the reduced temperatures and times. By following this road each silicon wafer would be tailored/customer engineered to the respective customer device process. Another approach would be to work without any gettering processes. This would require that during the VLSI process the wafers are not exposed to any contamination. In this case, the extrinsic gettering processes could be omitted.

4 Measuring, Characterization and Metrology

The quick, comprehensive and contact-free characterization of semiconductor wafers gains more and more importance with the increasing integration density. In addition, for monitoring and controlling of the different manufacturing process steps in a process line, more and more measuring and characterization spot checks need to be introduced.

The latter often turn into a bottleneck situation and last but not least, they are a substantial cost factor. Only a few measuring systems are currently available on the market for the characterization of 300 mm wafers at a resolution of 0.1 μm. Tremendous research work must still be carried out, before quick, non-contact and real time measuring- and evaluation systems will be available with resolutions in the submicron range.

5   Economic Considerations

The semiconductor industry is currently preparing itself for the transition into the new 300 mm wafer era. First 300 mm VLSI device lines will be ramped up at the turn of the century. Intel has had the leadership role as far as the conversion from 125 mm to 150 mm is concerned and IBM has had this role in the transition from 150 mm to 200 mm. Regarding the conversion to 300 mm, no single company will take this leading role. The industrial, production and technological challenges are too big and too fundamental to be carried out by a single company. The semiconductor industry must take this step unanimously as a "whole" industry. The international consortia – like the International 300 mm Initiative (I300I) – and the Japanese Semiconductor Leading Edge Technology (SELETE) which were founded in 1995 and 1996 respectively, play an important role in this "whole" industry approach for equipment and technology evaluation and assessment.

What are the driving forces which cause such a significant change in the semiconductor industry? It is the cost saving potential per integrated circuit. The area of a 300 mm silicon wafer is 2.25 times bigger than that of a 200 mm wafer. The total VLSI device processing costs of a 300 mm wafer are estimated to be about 1.7 times higher than that of a 200 mm wafer. The resulting conclusion is that with the conversion of the same circuit from a 200 mm wafer to a 300 mm wafer a cost saving potential of about 25% per circuit may be obtained. Consequently, the cost for the silicon base wafer will rise due to significant more difficult and demanding manufacturing processes. During the transition from a 150 mm polished wafer to a 200 mm polished wafer, the base silicon wafer costs per cm^2 increased by about 30%. For the transition from 200 mm to 300 mm wafer the silicon wafer manufacturing costs will rise many times the mentioned percentage rate due to the more complex and sophisticated wafer manufacturing processes.

6   Summary and Outlook

The requirements on silicon wafers for the Gigabit era have been analyzed and the different development attempts to master these challenges have been shown. The measures required for the transition from 200 mm wafers to 300 mm wafers have been discussed. It is evident that the CZ-crystal pulling process involves most of the challenges, followed by the polishing and epitaxy process and metrology. In these fields a lot of research work still needs to be done. Estimations show that through the transition from the 200 mm to the 300 mm wafer diameter, a cost saving potential of about 25% per integrated circuit may be obtained. This 25% saving potential per circuit is the driving force of the semiconductor industry for this diameter conversion which has to be carried out by the semiconductor industry as a whole.

Acknowledgements

The author is obliged to H. Fusstetter, P. Wagner, D. Gräf, W. v. Ammon, P. Stallhofer, T.S. Venezia and A. Pihl as well as different colleagues for the discussions and valuable comments to this article. The suggestions and recommendations from the university institutes are highly appreciated. Part of this work has been supported by the Federal Ministry of Education, Science, Reseach and Technology of Germany.

Bibliography

[1] G. Moore, "VSLI, What Does the Future Hold?", Electronic News, 42, (14), (1980).

[2] The National Technology Roadmap for Semiconductors, SIA, San Jose, CA, (1994).

[3] W. v. Ammon, Proceedings of the 2nd International Symposium on Advanced Science and Technology of Silicon Materials, Hawaii, November 25–29, p. 233, (1996).

[4] J. Ryuta, E. Morita, T. Tanaka, Y. Shimanuki, Jpn. J. Appl. Phys. 29, L1947, (1990).

[5] P. Wagner, D. Gräf, R. Schmolke, M. Suhren, Proceedings of the 2nd International Symposium on Advanced Science and Technology of Silicon Materials, Hawaii, November 25–29, p. 101, (1996).

[6] W. v. Ammon, E. Dornberger, H. Oelkrug, H. Weidner, J. Cryst. Growth 151, 273, (1995).

[7] D. Gräf, M. Suhren, U. Lambert, R. Schmolke, A. Ehlert, W. v. Ammon, P. Wagner, The Electrochem. Soc. Proc. , Vol. 96-13, 117 (1996).

[8] J.G. Park and R.A. Rozgonyi, Soild State Phenomena Vols. 47–48, 327, (1996).

[9] M. Itsumi, H. Akiya, T. Ueki, J. Appl. Phys. 78, 10, (1995).

[10] Y. Satoh, T. Shiota, Y. Murakami, T. Shingyouji, H. Furuya, J. Appl. Phys. 79, 10, (1996).

[11] J.G. Park, J.M. Park, K.C. Cho, G.S. Lee, H.K. Chung, Proceedings of the 2nd International Symposium on Advanced Science and Technology of Silicon Materials, Hawaii, November 25–29, p. 519, (1996).

[12] S.K. Grifths, R.H. Nilson, 3rd 300 mm Wafer Specification Workshop, SEMICON-West, July, (1995).

[13] S. Takasu, Proceedings of the 2nd International Symposium on Advanced Science and Technology of Silicon Materials, Hawaii, November 25–29, p. 309, (1996).

[14] F. Shimura, "Semiconductor Silicon Crystal Technology", Academic Press, Inc. p. 347, (1989).

Macroporous Silicon: Physics and Applications

V. Lehmann

Siemens AG, Dept. ZT ME 1, 81730 München, Germany

Abstract: A technique for the formation of pore arrays with high aspect ratios by electrochemical etching of n-type silicon wafers in hydrofluoric acid is presented. New devices such as silicon based capacitors or photonic bandgap materials for the infrared spectral range are fabricated based on this new technology.

1 Introduction

Porous silicon is used since decades for the manufacturing of micromechanical devices. However, porous silicon is mostly used as sacrificial layer during the manufacturing process, so far. New applications where the porous layer is used as the active part of a device are possible by an improved understanding of the pore initiation and the formation process which enables us to tailor the porous structure according to its application.

The average dimensions of porous silicon structures cover four orders of magnitude from a few nanometers to tens of micrometers. Such structures can be classified according to size in three different regimes. The microporous regime with dimensions below 2 nm is dominated by quantum size effects. All the luminescent porous silicon consists of sponge-like silicon structures with average dimensions of about 2 nm. The mesoporous regime shows average dimensions from 2 nm to 50 nm, while the dimensions of the macroporous regime are above 50 nm. The meso- and macroporous silicon formation process is dominated by space charge effects. The different pore size regimes and their dependence on the substrate resistivity are illustrated in Fig. 1.

2 Formation of Porous Silicon

Porous silicon is formed on a silicon substrate during anodization in a hydrofluoric acid electrolyte. Pore formation is only observed for anodic current densities below a critical current density J_{PS}. All electrochemical pore formation at silicon electrodes is confined to the regime indicated by a dot-pattern in the I-V-characteristic of the silicon electrode in Fig. 2.

The pore-size and wall-size distribution of an etched porous structure depends sensitively on parameters such as doping kind and density as well as illumination conditions.

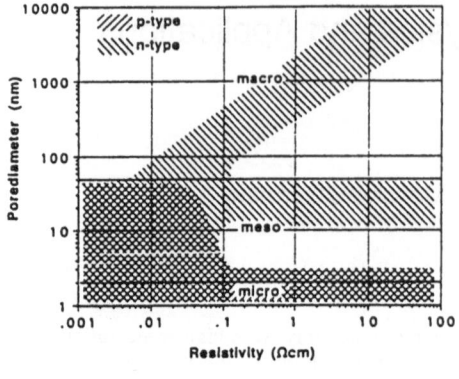

Figure 1
The different pore sizes which form on n-type and p-type silicon substrates during anodization in hydrofluoric acid are shown as a function of the substrate resistivity.

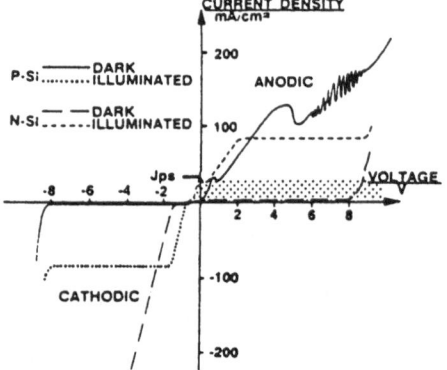

Figure 2
The I-V-characteristic of a silicon electrode in hydrofluoric acid is shown in this figure. All kinds of pore formation are confined to the anodic current densities below the critical value J_{PS}. This regime is indicated by a dot-pattern.

However, the size distribution is usually quite narrow, indicating that the structures are not fractal in character.

While mesoporous silicon is formed on degenerately doped p- or n-type substrates, microporous silicon is formed on moderately doped materials. The formation of macroporous silicon requires non-degenerate n-type substrates and illumination of the sample during anodization. Since the luminescent properties of microporous silicon are well studied, we will focus on macroporous silicon formation and its applications.

Macropore formation in silicon electrodes is characterized by a depletion of the porous region due to space charge effects [1] and a constant current density (J_{PS}) at the pore tips. This constant current density is characteristic for the steady state condition between charge transfer and mass transport at the pore tip [2]. This is in contrast to the case of microporous silicon where the depletion of the pore walls is due to quantum confinement [3] and the current density at the pore tips is always smaller than J_{PS}. Under anodic bias a space charge region (SCR) of significant extension is only present in n-type semiconductors. Consequently macropore formation in aqueous electrolytes is only observed for n-type silicon electrodes. In order to generate the electronic holes required to promote dissolution, the n-type material has to be illuminated. Illumination

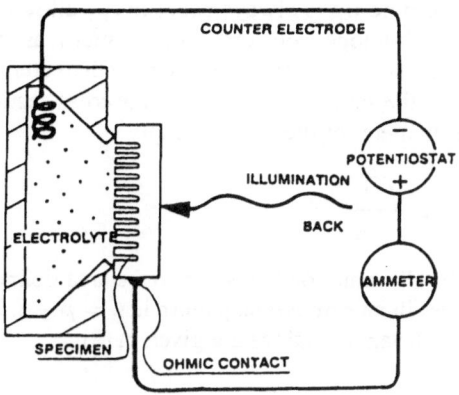

Figure 3
Sketch of the setup used for the formation of macroporous layers on silicon substrates. Note that the holes which promote the dissolution process are generated by illumination of the backside of the sample.

from the back-side of the wafer is found to be superior to frontside illumination, because in the latter case carriers are generated in the pore walls too, which leads to further dissolution in this region. Fig. 3 shows a sketch of the setup used to form macroporous layers on silicon substrates.

Pore initiation on a flat silicon surface is random and best described by an Ostwald ripening process. However, the location of the macropores on a $\langle 100 \rangle$ silicon wafer can be predetermined by etchpits, for example produced by photolithography and subsequent alkaline etching. The well known inverted pyramids generated by alkaline etching are sufficient to initiate pore growth. On the basis of the steady state condition present at the pore tip, equations can be defined which enable us to precalculate the requested pore diameter and pore depth as a function of current density and anodization time. The critical current density J_{PS} (in mA cm^{-2}) for $\langle 100 \rangle$ oriented substrates is only a function of the electrolyte concentration c (in wt% HF) and the absolute temperature T (in K) according to

$$J_{\text{PS}} = Cc^{3/2} \exp\left(-\frac{E_{\text{a}}}{k_{\text{B}}T}\right). \tag{2.1}$$

With an activation energy E_{a} of 345 meV and a constant C of 3 300 mA cm^{-2} (wt%)$^{-3/2}$. The growth rate v of pores can now be calculated according to Faraday's law, using the atomic density of silicon N_{Si} (5×10^{22} cm^{-3}) and the elementary charge e (1.602×10^{-19} C).

$$v = \frac{J_{\text{PS}}}{n(-e)N_{\text{Si}}}. \tag{2.2}$$

The number of holes consumed for the dissolution of one Si atom n, is found to be around 2.6 during macropore formation.

The fact that the current density at the pore tip is always equal to J_{PS} allows us to precalculate the pore dimensions, too. If a homogeneous orthogonal pattern, as shown in Fig. 5a, is used for pore initiation, the spacing p of the pores, the pore diameter d and the thickness w of the remaining walls are the same for all pores. If the cross section of a pore is approximated by a square, the diameter of the pore is simply

$$d = p\sqrt{\frac{J}{J_{PS}}}, \tag{2.3}$$

where J is the current density adjusted by illumination in respect to the area defined by the o-ring of the setup. The wall thickness in an orthogonal pattern is $w = p - d$. More details of the formation process and the relevant equations are given in [2].

3 Design Rules

The possibility to control pore initiation and pore geometry gives a high degree of freedom in design. However, there are a few limitations inherent to the electrochemical pore formation process which will be discussed in this section.

1. The pore length can be as large as the wafer thickness (up to 1 mm), however, this requires low electrolyte concentration, low temperature and etching times in the order of a day or more. The average growth rate of macropores is in the order of a micrometer per minute. Under standard conditions all pores have the same length.

2. The pore diameter depends on doping density and current density and can be adjusted between 0.2 μm and 20 μm. A good rule of thumb for the selection of an appropriate substrate is to square the desired pore diameter in μm and take this number as resistivity in Ωcm (for example 2 μm pores can be etched best using a 4 Ωcm n-type substrate). The pore diameter can be varied over the length of the pore by a factor up to about three for all pores simultaneously by adjusting the current density, or the illumination intensity, respectively. The taper of such pore geometries is limited to values below 30 degrees in relation to the pore axis. Note that narrow bottlenecks will significantly reduce the diffusion of chemical reactants in the pore and the formation of deep pores becomes difficult.

3. The pore cross section can vary between a circle and a four-armed star depending on the formation conditions, as shown in Fig. 4. Subsequent to the electrochemical pore formation, the cross section of the pores can be made more circular by oxidation steps or can be made more square-shaped by anisotropic chemical etching for example in aqueous HF. Branching of pores, as shown in Fig. 4e, can be suppressed by an increase of current density or a decrease of doping density, bias, or HF concentration.

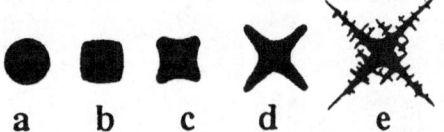

Figure 4
The pore cross section on $\langle 100 \rangle$ substrates changes from a circle (a) to a four-armed star, if doping density or bias is increased.

4. The ratio between the pitch and the pore diameter can be varied from 1.1 to 10. The number of possible pore patterns is only limited by the requirement that the relation between etched and unetched areas has to be constant on a length scale above about three times the pitch (for homogeneous backside illumination). This includes the possibility to etch a pattern with a missing pore, a missing row of pores or even two missing rows. It is also possible to enlarge or shrink the scale of the pattern across the sample surface by a factor of about three. But a pattern with an abrupt border to an unpatterned area will lead to under-etching or random pore formation in the unpatterned area. A change of the relation between etched and unetched areas would require a change in illumination intensity. However, any image projected on the backside of the wafer would generate a smoothed-out current density distribution on the frontside due to the random diffusion of the charge carriers in the bulk. This problem is reduced for thin wafers or for illumination of the frontside.

5. The pore growth direction is along the $\langle 100 \rangle$ direction and toward the source of holes. Perfect arrays of macropores can be formed on $\langle 100 \rangle$ substrates with a misorientation of up to 10 or 15 degrees, however, misorientation enhances the tendency to branching. For $\langle 110 \rangle$ or $\langle 111 \rangle$ oriented substrates there is a strong tendency of pore branching and the growth direction is determined by the $\langle 100 \rangle$ directions of the crystal as well as by the source of holes.

In conclusion it can be said that electrochemical pore etching is in some respects complementary to plasma etching: Plasma etching shows a high degree of freedom in the design in x- and y-direction (the wafer surface), however, the etching depth is limited and the variability in z-direction is limited to cylindrical and conical shapes. Electrochemical pore designs are limited in x- and y-direction, but the flexibility in z-direction and the possibility to etch trough the wafer offer new applications.

4   Applications

Out of the various applications of macroporous silicon substrates, two will be discussed in detail. One is the manufacturing of photonic crystals, which requires to control the pore wall thickness and the periodicity within a few percent. The second is a new capacitor concept, which visualizes in an exemplary way the application of basic semiconductor manufacturing steps such as doping, oxidation and chemical vapor deposition to a macroporous material.

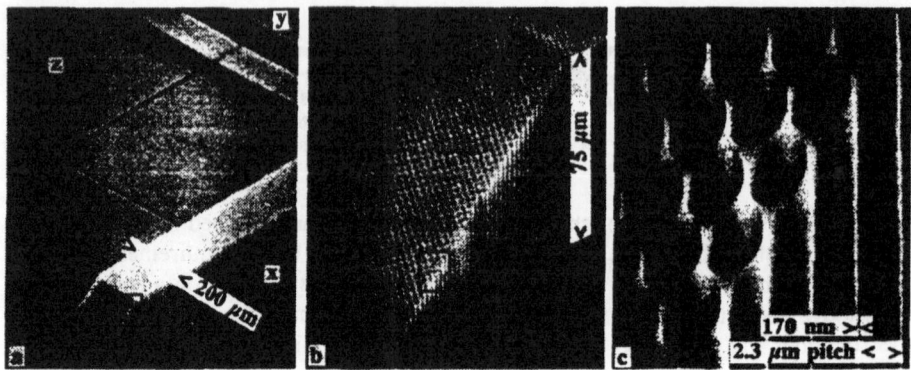

Figure 5 SEM micrograph of a patterned layer of macroporous silicon which exhibits a full two-dimensional photonic band gap. The pore diameter and the walls thickness of this two-dimensional triangular pore lattice are 2.13 μm and 170 nm, respectively.

A photonic crystal is a particularly structured metallic or dielectric material with a periodicity on a wavelength scale. In such regular structures propagation of radiation can be described analogous to the case of electrons in the crystal lattice of a semiconductor. Photonic crystals show a forbidden band gap, i.e. photons of a specific frequency interval are not allowed to propagate in the material.

The triangular pore lattice shown in Fig. 5 exhibits a full two-dimensional photonic band gap for a wavelength around 5 μm [6].

For the transmission measurements the porous layer was structured in 200 μm wide bars using a micro-machining technique described in [5]. The experimentally determined band gap agrees well with the calculated band structure as shown in Fig. 6. The controlled electrochemical pore formation processes open a new door for the experimental investigation of the fascinating optical properties of photonic crystals for the IR spectral range.

Conventional capacitors suffer from a few drawbacks: either a superior dielectric like polystyrene or SiO_2 is selected which leads to low surface to volume ratios of the electrodes or the electrode area is increased by etching or sintering techniques which entails the disadvantages of anodically formed oxides as in the case of aluminum or tantalum capacitors. Ceramic dielectrics were developed to solve this problem by increasing the dielectric constant instead of the electrode area. A drawback of this approach is the strong dependence on temperature and applied electric field of the dielectric constant in these materials.

A solution of these problems is offered by a capacitor concept based on macroporous silicon which combines an electrochemically enlarged electrode surface with a superior dielectric. A macroporous silicon substrate with pores of about one micrometer in diameter and a pore depth of a few tenths of a millimeter offers a surface area which is enlarged by about two or three orders of magnitude as compared to an unetched substrate surface. If standard microelectronic processes, like the deposition of an ONO dielectric (SiO_2-Si_3N_4-SiO_2) and a polysilicon electrode are applied to such substrates,

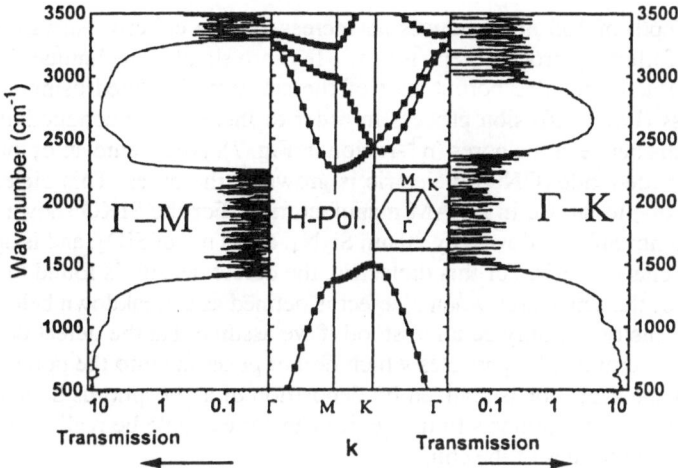

Figure 6 In the center part, the calculated photonic band structure of the two-dimensional triangular lattice shown in Fig. 5 is plotted. On both sides the transmission spectra of the sample for different directions are shown between 500 and 3 500 cm^{-1}. Note that the transmission is plotted on a logarithmic scale.

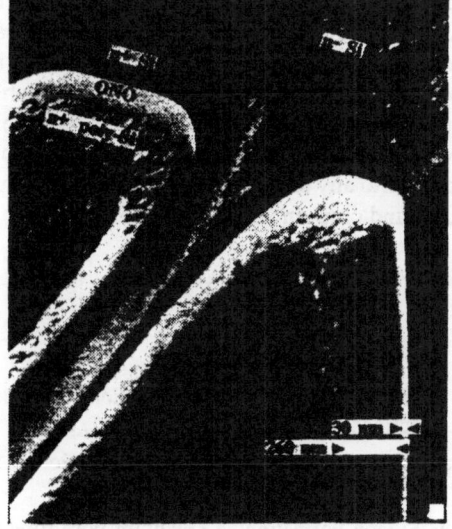

Figure 7
This figure shows a bevelled sample similar to the one in Fig. 5 at high magnification. The heart of the capacitor consists of a highly doped substrate and a highly doped poly-silicon layer with a thin ONO-dielectric in between.

a capacitor of superior properties can be fabricated. This silicon based capacitor is designated SIKO which is an abbreviation of the German words "Silizium Kondensator" (silicon capacitor) [4]. An example of a macroporous substrate used for the fabrication of a SIKO is shown in Fig. 5.

A pore depth of 165 μm produces an increase of the effective surface in the order of 100. All following process steps for the SIKO are standard techniques in microelectronic manufacturing. The porous part of the substrate is doped using a phosphorus silicate glass (PSG) diffusion process in order to increase the conductivity of the remaining walls between the pores (n^+-region in Fig. 7). After removal of the PSG layer, an oxide-nitride-oxide (ONO) dielectric is grown in the pores. This dielectric is well understood due to its use in DRAM manufacturing. For the SIKO shown in Fig. 7, it consists of 5 nm SiO_2, followed by 20 nm Si_3N_4, and 5 nm of SiO_2, and is optimized for a operating bias of 6.3 V. For this dielectric, the defect density is found to be less than 10^{-3} cm^{-2} of the active area when a defect is defined as a breakdown below 15 V. This low defect density can only be understood if we assume that the defect density of planar films is determined by particles which do not penetrate into the pores. The second electrode of the capacitor is realized by deposition of a phosphorus doped polysilicon film (Fig. 7). The two contacts to the device can for example be realized by sputtering aluminum on both sides of the chip.

The electrical characteristics of the SIKO surpass most of the electrical data of other standard capacitor technologies. The remarkable invariability of the capacitance of the SIKO under variation of bias, temperature, frequency, and time of operation is a consequence of the superior properties of its ONO dielectric. The maximum operating temperature of the SIKO chip is 200 °C. The electrochemically etched array of straight pores produces a comb-like electrode structure which allows us to realize an electrical series resistance for the SIKO chip which is lower than for any of the other capacitor technologies. The specific capacitance of the SIKO is close to 4 μFV mm^{-3}, but preliminary data of a shrinked design show that values up to 12 μFV mm^{-3} are achievable. The dissipation factor of the SIKO at room temperature is below 10^{-4}. Due to its perfect compatibility with standard silicon device technology the SIKO chip can be easily incorporated by wire or flip chip bonding into any silicon chip assembly, like multi chip modules (MCMs), chip on board (COB) and chip on smart and chip cards.

Bibliography

[1] M. J. J. Theunissen, "Etch Channel Formation during Anodic Dissolution of n-Type Silicon in Aqueous Hydro-fluoric Acid", J. Electrochem. Soc., Vol. 119, pp. 351, 1972.

[2] V. Lehmann, "The Physics of Macropore Formation in Low Doped n-Type Porous Silicon", J. Electrochem. Soc., Vol. 140, pp. 2836-2843, 1993.

[3] V. Lehmann and U. Gösele, "Porous Silicon Formation: A Quantum Wire Effect", Appl. Phys. Lett., Vol. 58, pp. 856, 1991.

[4] V. Lehmann, W. Hönlein, H. Reisinger, A. Spitzer, H. Wendt, and J. Willer, "A new capacitor technology based on porous silicon", Solid State Technol., Vol. 38, pp. 99, 1995.

[5] S. Ottow, V. Lehmann, and H. Föll, "Processing of three-dimensional microstructures using macroporous n-type silicon", J. Electrochem. Soc., Vol. 143, pp. 385, 1996.

[6] U. Grüning, V. Lehmann, S. Ottow, and K. Busch, "Macroporous silicon with a complete two-dimensional photonic band gap centered at 5 μm", Appl. Phys. Lett., Vol. 68, pp. 747, 1996.

$Si_{1-y}C_y$ and $Si_{1-x-y}Ge_xC_y$ Alloy Layers on Si Substrate

Karl Eberl

Max-Planck-Institut für Festkörperforschung,
Heisenbergstr. 1, D-70569 Stuttgart, (Germany);
email: eberl@servix.mpi-stuttgart.mpg.de

Abstract: $Si_{1-y}C_y$ and $Si_{1-x-y}Ge_xC_y$ alloy layers with a carbon concentration of a few percent are prepared by solid source molecular beam epitaxy. Near band-edge photoluminescence (PL) is observed from $Si/Si_{1-y}C_y$ multiple quantum well (MQW) structures. The band gap in the pseudomorphic films is reduced by about 65 meV per percent C. The PL indicate a type I heterostructure with the band offset being mainly in the conduction band. In $Si_{1-x-y}Ge_xC_y$ MQWs compressive strain caused by Ge is partially compensated by C alloying and the band gap increases with y. PL measurements from closely spaced $Si_{1-y}C_y/Si_{1-x}Ge_x$ layers show a lower transition energy than separate $Si_{1-y}C_y$ and $Si_{1-x}Ge_x$ reference samples. This is attributed to spatially indirect PL transitions between the electrons confined in the $Si_{1-y}C_y$ layers and the heavy holes located in the $Si_{1-x}Ge_x$ layers.

Electron mobility enhancement is measured for strained n-type doped $Si_{0.996}C_{0.004}$ layers below 180 K, which is attributed to the splitting of the Δ valleys in the conduction band. In p-type modulation doped 8 nm thick $Si_{0.49}Ge_{0.49}C_{0.02}$ quantum wells we observe an improved hole mobility at room temperature and 77 K compared to corresponding samples without C, which is a consequence of the reduced strain in the layer due to substitutional C. The realisation of a $Si_{0.54}Ge_{0.45}C_{0.012}$ p-channel MODFET on Si is presented. A room temperature transconductance of $g_{me} = 57$ mS/mm and a saturation current of $I_D = 40$ mA/mm is demonstrated for a 0.75 μm gate-length non-recessed device.

1 Introduction

Si/SiGe heterostructures are very interesting for device applications like heterobipolar transistors and photodetectors [1]. Significant flexibility in the design of device structures is achieved by using thick relaxed SiGe buffer layers as a virtual substrate, which allows one to adjust the strain. This concept opened the possibility to prepare high mobility electron and hole channels in tensily strained Si and compressively strained Ge-rich SiGe quantum wells, respectively [2]. The main problems of strain relaxed buffer layers, however, are the several μm total film thickness, the still very high defect density and the characteristic surface roughness.

A different concept towards strain adjustment was suggested by adding C into the Si/SiGe material system. Pseudomorphic $Si_{1-y}C_y$ layers with tensile strain and SiGeC films with reduced compressive strain directly on Si were reported in 1991 [3–5]. Before that, Posthill et al. published the first epitaxial growth of $Si_{1-y}C_y$ alloy layers on Si, but the films where not free of SiC precipitates since they used relatively high substrate temperatures during growth [6]. Since that time several groups started to prepare Si and SiGe films with substitutional C by MBE [7,11] and CVD [12–17] technique. A recent overview on structural and phonon properties was given by Jain et al. [18].

Pseudomorphic $Si_{1-y}C_y$ and $Si_{1-x-y}Ge_xC_y$ alloy layers on Si have a good crystal quality and are essentially free of extended lattice defects when prepared under appropriate conditions. Demkov and Sankey discussed the influence of local strain around C atoms in the Si matrix effect on the basis of a first principles model calculation [19] and predicted that the fundamental band pap in unstrained $Si_{1-y}C_y$ is reduced for a small amount of C in the Si lattice. As a consequence of that the question arises what the effects on the electronic and optical properties are, and how the carrier mobilities are affected. In this contribution we discuss photoluminescence measurements and electrical transport measurements of pseudomorphic $Si_{1-y}C_y$ and $Si_{1-x-y}Ge_xC_y$ alloy layers on Si. The influence of substitutional C on the fundamental band gap and the band alignments is discussed.

2 Experimental Conditions

The samples discussed in the following are prepared by solid source MBE on undoped Si(100) substrate with a resistivity of above 5 000 Wcm. A directly heated pyrolytic graphite filament is used for C sublimation with a typical growth rate of up to 0.002 nm/s. The hot filament is completely surrounded by graphite shieldings. In addition the shutter is lined with a graphite shielding to reduce problems with metal, CO and CO_2 contamination. Si is evaporated in an electron beam evaporator, which is feedback controlled by a mass-spectrometer. For Ge we use a special PBN crucible effusion cell with a Si orifice plate to reduce B background doping. Elemental B and a GaP decomposition source [20] are used for p- and n-type doping, respectively. Typically the active multilayer structures are grown on a 300 to 400 nm thick Si buffer layer. Optimised sample quality with respect to photoluminescence (PL) intensity for $Si/Si_{1-y}C_y$ QW structures is achieved at a substrate temperature of about 550 °C [21].

The structural properties of the samples were determined by double-crystal high resolution X-ray diffraction (XRD) and by comparing the data with dynamical simulation results, which are based on Vegard's law and linearly interpolated elastic constants. The reproducibility for layer thickness, Ge and C concentration from sample to sample is about 5%. The C content measured by XRD agrees well with the C nominally deposited and calibrated by secondary-ion mass spectroscopy (SIMS) analysis. However, the absolute uncertainty in SIMS is about a factor of two due to the missing absolute reference samples for SIMS in the high C concentration regime. A pronounced dependence of the substitutional C incorporation on growth temperature is not observed within the temperature range from 400 to 600 °C for C concentrations below 3%.

For the PL studies the samples were excited by a 476 nm Kr$^+$ laser beam at a power density of about 0.2 W/cm^2. The temperature of the sample during measurement was about 8 K. PL is analysed by a liquid-nitrogen cooled Ge detector using standard lock-in techniques.

3 $Si_{1-y}C_y$ Alloy Layers

Figure 1 shows the band gap versus lattice constant for Si, Ge, cubic SiC and diamond. The solid curve between Si and Ge indicates the band gap for unstrained SiGe alloys with the characteristic kink at about 85% Ge, which is due to the crossover from the Si like 6-fold D minima to the Ge like 4-fold L minima in the conduction band (CB). The dashed line towards Ge shows the reduction of the band gap for pseudomorphic SiGe on Si, which is determined by the strain-degenerate 4-fold Δ minima in the CB and the heavy holes (hh) in the valance band (VB). The lattice mismatch between Si and Ge is only 4% therefore the fundamental properties can be interpolated for SiGe alloys to a large degree. The intrinsic band gap of $Si_{1-y}C_y$ alloys was expected to homogeneously increase with the C content, because the band structures of Si, cubic SiC and diamond are very similar except the energy value of the band gap [23,24]. The PL data obtained recently indicate a strong band gap reduction for pseudomorphic $Si_{1-y}C_y$ alloy films as shown by the dashed line starting from Si towards SiC. Subtracting the effect of tensile

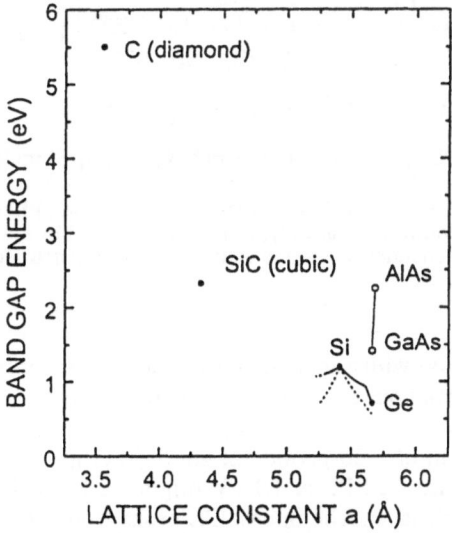

Figure 1
Fundamental band gap versus lattice constant for Si, Ge, SiC and C. The dashed lines show the band gap for pseudomorphic SiGe and $Si_{1-y}C_y$ alloy layers on Si.

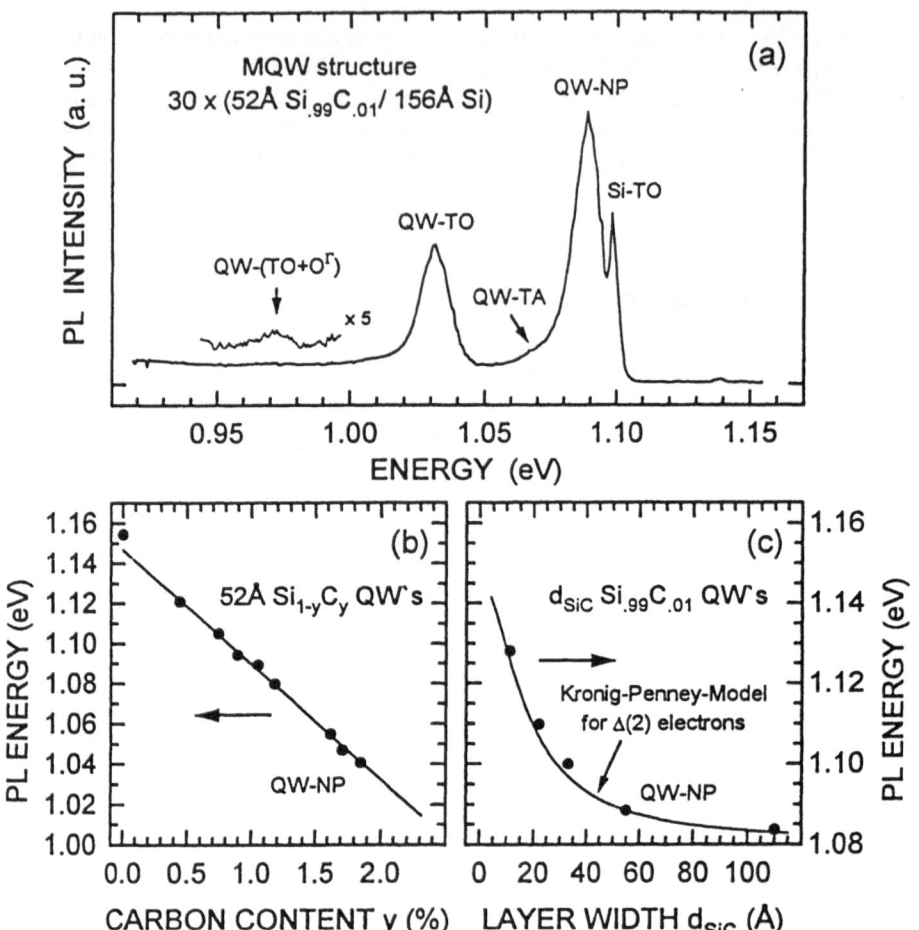

Figure 2 a) Low temperature (8K) PL spectrum of a 30 period $Si_{0.99}C_{0.01}$/Si multiple quantum well structure. The Si and $Si_{1-y}C_y$ layers are 15.6 nm and 5.2 nm thick, respectively. b) Energy of the no-phonon PL peak as a function of the carbon content. c) Energy of the no-phonon PL peak as a function of the $Si_{0.99}C_{0.01}$ layers thickness. For smaller thickness the PL peak shifts to higher energies due to the quantum confinement effect in the $Si_{1-y}C_y$ quantum wells.

strain reveals that unstrained $Si_{1-y}C_y$ alloys with a small C content have a smaller band gap than Si which is indicated by the continuous curve. Further details are discussed in the following.

Figure 2a shows a typical PL spectrum of a 30-period 5.2 nm $Si_{0.99}C_{0.01}$/15.6 nm Si MQW structure. The Si-TO phonon assisted PL at 1.1 eV originates from the Si substrate, buffer layer and 200 nm Si cap layer. The intense PL peaks from excitons confined in the QWs are the Si–Si TO phonon replica at 1.032 eV and the no-phonon

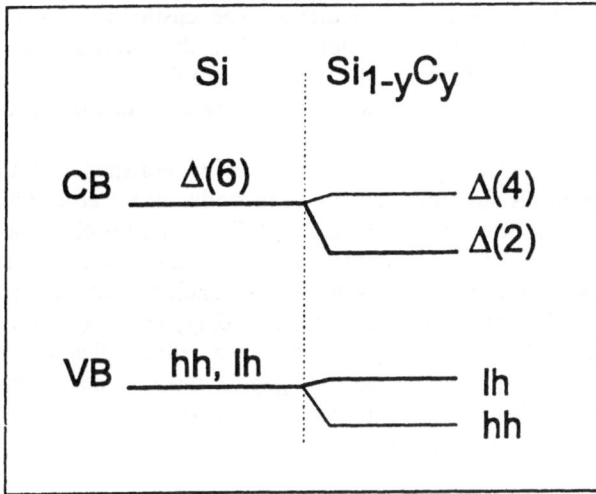

Figure 3
Schematic diagram of the conduction (CB) and valence-band-edge (VB) for pseudomorphic Si$_{1-y}$C$_y$ on Si(100). It is based on a deformation potential calculation taking the tensile strain within the Si$_{1-y}$C$_y$ layer into account and neglecting intrinsic effects of C on the band offset.

(NP) transition at 1.089 eV. The linewidth of these peaks is about 12 meV. The weak PL shoulders labelled as QW-TA and QW-(TO+O$^\Gamma$) are probably TA phonon and TO plus Γ point optical phonon assisted recombinations of carriers confined in the Si$_{1-y}$C$_y$ quantum wells. All lines labelled with QW shift as a function of the C content y in the Si$_{1-y}$C$_y$ layers as shown for the no-phonon line in Figure 2b. The PL energy of the NP line shifts linearly by $\Delta E^{NP} = -y \cdot (5.7 \text{ eV})$ reflecting the decrease of the fundamental band gap in the MQW structure. For the layer thickness of 5.2 nm and a C content of about 2% or more we observe an increasing broad emission background at low energy, which is similar to the broad PL reported earlier [25]. In thinner Si$_{1-y}$C$_y$ MQWs we observe band-edge PL for a C content up to about $y = 6.4\%$ [21].

Figure 2c shows the energy dependence of the NP PL peak of 30-period Si$_{0.99}$C$_{0.01}$/ 15.6 nm Si MQWs as a function of the Si$_{0.99}$C$_{0.01}$ alloy layer thickness d_{SiC}. The peak is shifting to higher energy by reducing the QW width d_{SiC}. This proves, that the PL lines originate from quantum confined levels and not from defect states related to C incorporation. In other words, it proves that Si$_{1-y}$C$_y$ alloy layers are a reasonable semiconductor material with a well-defined band gap. The measured data follow the curve, which represents a Kronig-Penney-Model calculation for the twofold degenerate conduction band-edge $\Delta(2)$ and thus using $m = 0.92 m_0$ as the effective mass for the electrons and neglecting any band offset in the valence band. This and more detailed PL studies on MQW structures with the Si$_{1-y}$C$_y$ layer thickness constant and varying the Si layer thickness indicate that the band offset in this heterostructure is mainly in the conduction band [22], which is complementary to the situation for pseudomorphic Si/SiGe heterostructures, where the band offset is mainly in the valence band.

This model of the band alignment is also supported by just looking at the strain induced splitting of the electron and hole valleys. Figure 3 shows a schematic diagram of the conduction (CB) and valence band-edge (VB) for pseudomorphic Si$_{1-y}$C$_y$ layers on

Si(100). It is based on a deformation potential calculation. The tensile biaxial strain introduced by substitutional C splits the electron valleys and thus, the twofold degenerate conduction band-edge $\Delta(2)$ is shifted down in energy by about $\Delta E^{\Delta(2)} = -y \cdot (4.6\,\text{eV})$. In the valence band the light hole states are expected to increase only slightly in energy with increasing tensile strain $\varepsilon = 0.35 \cdot y$.

Summarising the data for $Si_{1-y}C_y$ alloys we find the following: For small C contents of at least up to 2.5% these alloys are a high quality semiconductor material with the optical band gap shrinking by about $\Delta E(y) = -y \cdot (6.5\,\text{eV})$. This takes the electron confinement into account, which is about 8 meV for the 5.2 nm QWs discussed in Figure 2. Since the tensile strain accounts for about $-y \cdot (4.6\,\text{eV})$ we conclude that the intrinsic band gap of unstrained $Si_{1-y}C_y$ will decrease roughly by $\Delta E_g(y) = -y \cdot (1.9\,\text{eV})$ for small C concentrations. Exciton binding of electrons and holes with a typical energy of 10 meV is neglected. The tentative change of the band gap for unstrained and strained $Si_{1-y}C_y$ alloys is schematically illustrated in Figure 1 by the continuous and the dashed curves starting from Si toward SiC, respectively.

4   SiGeC Alloy Layers

Figure 4 shows low temperature PL spectra of a $Si_{0.84}Ge_{0.16}$ and two $Si_{0.84-y}Ge_{0.16}C_y$ MQW structures with $y = 0.008$ and $y = 0.01$. The samples have 25 periods and the thickness of the $Si_{0.84}Ge_{0.16}$ and $Si_{0.84-y}Ge_{0.16}C_y$ layers are 4 nm. The QWs are separated by 17 nm thick Si layers. The main PL originating from the QWs are no-phonon and TO phonon lines of confined carriers. The intense peak at $E = 1.1$ eV is the TO phonon replica from the thick Si layers. For compressively strained $Si_{1-x}Ge_x$ on Si the band offset is mainly in the valence band with the heavy hole (hh) shift being about $\Delta E_{hh} = x \cdot (0.72\,\text{eV})$ [27]. In the conduction band there is a small energy lowering of the $\Delta(4)$ electron valleys [28]. The band alignment for Si/SiGe is schematically shown in the left part of Figure 5. The PL from $Si_{0.842}Ge_{0.16}C_{0.008}$ MQWs is shifted to about 20 meV higher energy compared to the $Si_{0.84}Ge_{0.16}$ as shown in Figure 4. The substitutional incorporation of 0.8% C into $Si_{0.84}Ge_{0.16}$ reduces the compressive strain from $\varepsilon = -0.64\%$ to -0.36%. The PL blueshift is attributed to the strain-induced increase and a minor intrinsic lowering of the fundamental band gap [29]. For C concentrations in $Si_{0.84-y}Ge_{0.16}C_y$ which are higher than 0.8% we find a downward shift of the PL energies as shown by the topmost spectrum in Figure 4 with $y = 1\%$ and more precisely in the insert of Figure 4.

The band gap increase in $Si_{1-x-y}Ge_xC_y$ is schematically illustrated in the right part of Figure 5. Due to partial relief of compressive strain the $\Delta(2)$ to $\Delta(4)$ and the hh to lh splittings are reduced and would be brought back to degeneracy in case of perfect strain compensation, which is achieved for about $y = x/8.2$. Additionally, there must be an intrinsic effect of C on the band gap for relaxed $Si_{1-x-y}Ge_xC_y$ like for $Si_{1-y}C_y$ alloys as discussed in the last section.

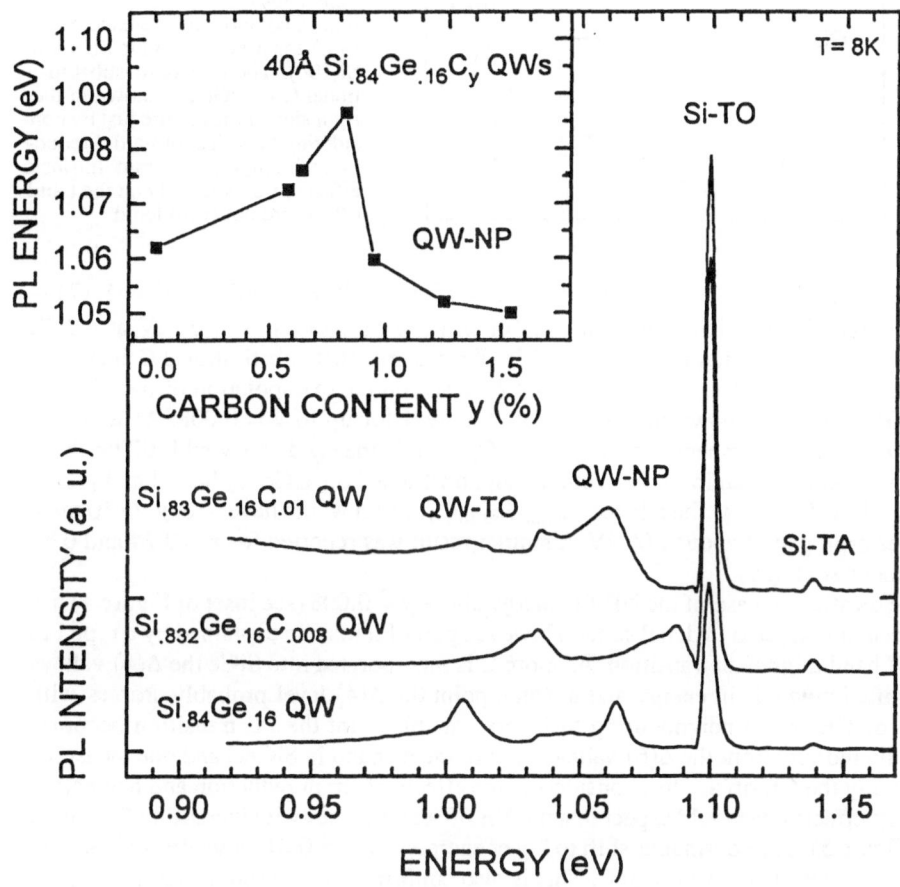

Figure 4 PL spectra from a 25 period $Si/Si_{0.84}Ge_{0.16}$ and two $Si/Si_{0.84-y}Ge_{0.16}C_y$ multiple quantum well structures with $y = 0.008$ and $y = 0.01$. The Si and $Si_{1-x-y}Ge_xC_y$ layers are 20 nm and 4 nm thick, respectively. The TO phonon replica and the no-phonon (NP) PL peaks labeled with QW originate from the quantum wells. The low intensity shoulder close to $E = 1.03$ eV is the TO+O(Γ) phonon assisted transition from the Si substrate and buffer layers. The insert shows the energy shift of the no-phonon PL energy as a function of the carbon concentration.

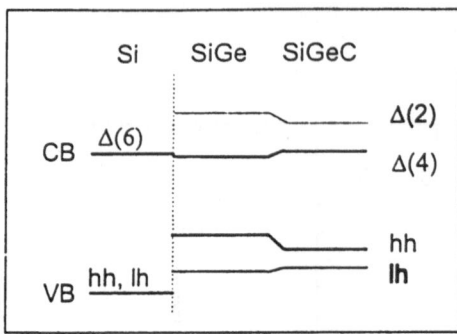

Figure 5
Schematic diagram of the conduction (CB) and valence-band edge (VB) for pseudomorphic SiGe on Si(100). The effects of substitutional C in SiGe are shown in the right side of the picture just by considering the effect of partial reduction of compressive strain. Intrinsic effects of C especially on the band offsets are not considered.

The band gap in pseudomorphic $Si_{1-x}Ge_x$ with $x < 25\%$ is [30]: $E_g(x) = 1.171 - 1.01x + 0.835x^2$ eV. Thus, the band gap of strained $Si_{0.84}Ge_{0.16}$ is 1.03 eV at 4.2 K, which is 140 meV smaller than that of Si. We assume, that substitutional C increases the energy gap by about $\Delta E = y \cdot (0.24$ eV) based on the extrapolation of the increasing NP-PL energy shown in the insert of Figure 4 for up to $y = 0.008$. As a consequence, for strain compensated $Si_{0.82}Ge_{0.16}C_{0.02}$ on Si the band gap would still be about 100 meV smaller than in Si. For comparison, unstrained $Si_{0.84}Ge_{0.16}$ has a band gap of about 1.11 eV, whereas the relaxed $Si_{0.82}Ge_{0.16}C_{0.02}$ film would have a roughly 30 meV smaller band gap of about 1.07 eV. A similar result was reported for $x = 0.24$ and 0.38 by Amour et al. [29].

The drastic decrease of the NP-PL energy above $y = 0.008$ (see inset of Figure 4) has not been investigated in detail so far. However, part of it may be explained by a type I to type II band alignment transition. As more C is incorporated into SiGe the $\Delta(4)$ valleys are shifted upwards in energy and at some point the $\Delta(4)$ level probably crosses with the $\Delta(6)$ CB in the neighbouring Si layers. At this point the PL transition becomes spatially indirect from the $\Delta(6)$ valleys in Si to the hh band in SiGeC and one measures essentially the CB offset. Just considering the effects of strain reduction and neglecting intrinsic effects one would expect that the VB offset decreases with increasing C content (see Figure 5). The continuous shift to lower energies for $y > 0.01$, however indicates an increasing VB offset in Si/SiGeC. This is also confirmed in measurements on coupled quantum wells discussed in Figure 6 and 7.

In summary for SiGeC alloys we find a very similar downward bowing of the intrinsic energy gap like for $Si_{1-y}C_y$ discussed in Figure 1. When starting from a certain unstrained SiGe alloy composition and adding substitutional C into the layer, the energy gap goes down by about 10–20 meV per percent C for unstrained alloys at least for small C concentrations. For pseudomorphic layers, C reduces the strain in the films and increases the band gap by about $\Delta E = y \cdot (0.24$ eV), but this increase is only part of the one which would be obtained by reducing the Ge content to achieve similar strain reduction [29]. The experimental data suggest that SiGeC layers can be prepared on Si, which have small and adjustable lateral strain, a smaller band gap and a substantial band offset in the VB.

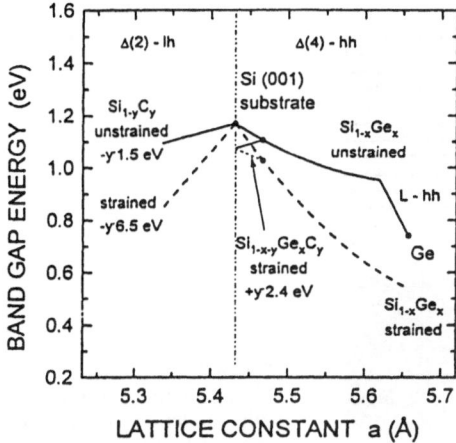

Figure 6
Band gap in Si$_{1-y}$C$_y$, Si$_{1-x}$Ge$_x$ and Si$_{1-x-y}$Ge$_x$C$_y$ versus intrinsic lattice constant. The PL results from strained Si$_{1-y}$C$_y$ and Si$_{1-x-y}$Ge$_x$C$_y$ layers on Si are indicated by a dashed lines. The band gap in unstrained Si$_{1-y}$C$_y$ and Si$_{1-x-y}$Ge$_x$C$_y$ is deduced from PL data considering deformation potential calculations (solid line). The Si$_{1-x}$Ge$_x$ data are taken from literature.

Figure 6 summarises the data discussed in the last two sections. It shows the band gap in Si$_{1-y}$C$_y$, Si$_{1-x}$Ge$_x$ and Si$_{1-x-y}$Ge$_x$C$_y$ versus intrinsic lattice constant. The PL results from strained Si$_{1-y}$C$_y$ and Si$_{1-x-y}$Ge$_x$C$_y$ layers on Si are indicated by a dashed lines. The band gap in unstrained Si$_{1-y}$C$_y$ and Si$_{1-x-y}$Ge$_x$C$_y$ is deduced from PL data considering deformation potential calculations (solid line). The data for strained and unstrained Si$_{1-x}$Ge$_x$ are taken from literature.

5 Coupled Si$_{1-x-y}$Ge$_x$C$_y$/Si$_{1-z}$C$_z$ Layers

The starting idea of the experiments discussed in the following was to put a SiGe layer right next to a Si$_{1-z}$C$_z$ quantum well in order to achieve strong confinement for both holes and electrons in the corresponding layers directly on Si. Similar concepts have been realised on SiGe buffer layers with a Si layer between compressively strained Ge quantum wells [31]. Figure 7 shows cross-sectional TEM micrographs with moderate (a) and with high resolution (b) of a 25-period 4 nm Si$_{0.84}$Ge$_{0.16}$/3.3 nm Si$_{0.988}$C$_{0.012}$/17 nm Si double quantum well structure grown on Si(100). The layers are free of extended defects and symmetric in strain. That means, the Si$_{0.84}$Ge$_{0.16}$ layers are compressively strained, whereas the Si$_{0.988}$C$_{0.012}$ films are under tensile strain to match the Si substrate. PL spectra of two coupled Si$_{1-x-y}$Ge$_x$C$_y$/Si$_{1-z}$C$_z$ quantum well structures with $x = 0.16$, $y = 0$ or 0.01, and $z = 0.012$ are shown in Figure 8. The Si$_{1-x-y}$Ge$_x$C$_y$ and the Si$_{1-z}$C$_z$ layers are again 4 nm and 3.3 nm thick, respectively. The arrows between the Si-TO line from the substrate and the QW-NP PL line from the DQWs mark the energetic position of the reference samples with only Si$_{0.84}$Ge$_{0.16}$ and Si$_{0.988}$C$_{0.012}$ QWs. The DQWs reveal PL lines at lower energy than the corresponding single QWs. They are attributed to spatially indirect transitions of $\Delta(2)$ electron and

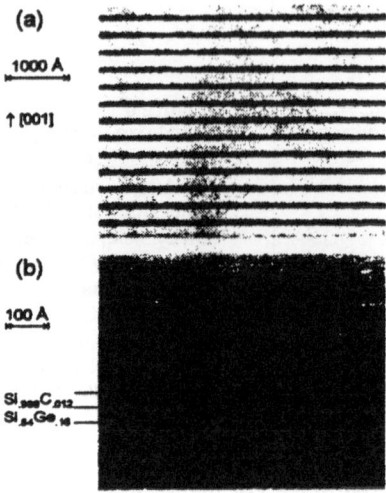

Figure 7
Cross-sectional TEM micrographs with moderate (a) and with high resolution (b) of a 25-period 4 nm $Si_{0.84}Ge_{0.16}$/3.3 nm $Si_{0.988}C_{0.012}$/17 nm Si double quantum well structure grown on Si(100).

heavy hole states localised within the neighbouring $Si_{0.988}C_{0.012}$ and $Si_{0.84}Ge_{0.16}$ layers, respectively. The band alignment is schematically illustrated in upper left corner of Figure 9. The no-phonon line is strongly enhanced compared to the TO phonon line. The intensity gain observed for the DQW with respect to the corresponding single QWs is considerable.

Figure 9 shows PL spectra from $Si_{0.84}Ge_{0.16}$/$Si_{0.988}C_{0.012}$ DQW structures with varied layer width $d_{SiGe} = 1.2 d_{SiC}$ in Si. The ratio of 1.2 between the individual layer thicknesses is chosen to account for the heavier effective mass and thus the stronger confinement of the electrons in the $\Delta(2)$ states compared to the heavy hole (hh) states in the valence band. The optimised QW thicknesses with respect to maximum overlap of the wave functions for electrons in $Si_{0.988}C_{0.012}$ and holes in $Si_{0.84}Ge_{0.16}$ is achieved for $d_{SiGe} = 2.3$ nm. This sample shows a three orders of magnitude intensity increase. For a further discussion of these data see also ref. [26].

A strong dependence of the DQW PL on the spatial separation of carriers is demonstrated by a 40% intensity decrease observed for a DQW structure with a 0.4 nm thick Si layer in-between the $Si_{0.84}Ge_{0.16}$ and $Si_{0.988}C_{0.012}$ layers. Adding C into the $Si_{0.84}Ge_{0.16}$ layers results in a further PL intensity enhancement and a slight downwards shift in energy.

The insert of Figure 8 shows the energy of the no-phonon PL peak as a function of the carbon content in $Si_{0.84}Ge_{0.16}$/$Si_{1-z}C_z$ layers with y varied from zero to 2% (upper curve), The lower two curves in the insert show the energy shift for $Si_{0.84-y}Ge_{0.16}C_y$/$Si_{1-z}C_z$ double quantum wells for different C concentrations in the both layers. The behaviour of the upper curve is interpreted as follows: For $y = 0$ we have a spatially

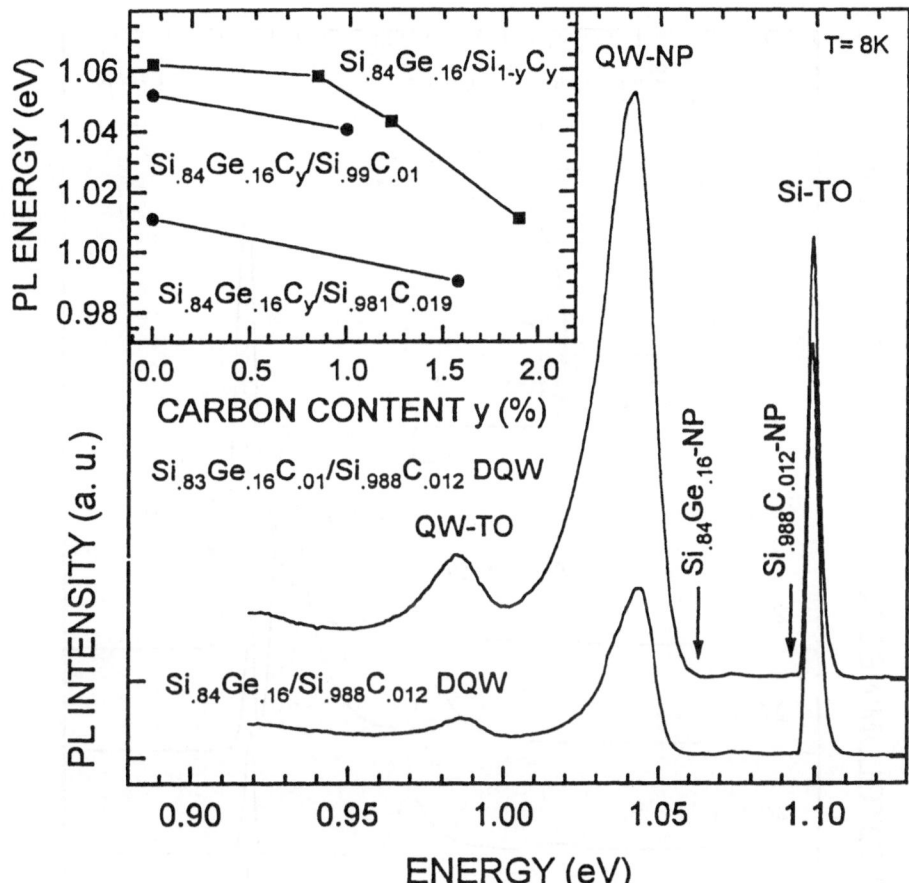

Figure 8 PL spectra of coupled Si$_{1-x-y}$Ge$_x$C$_y$ / Si$_{1-z}$C$_z$ quantum well structures with $x = 0.16$, $y = 0$ or 0.01, and $z = 0.012$. The Si$_{1-x-y}$Ge$_x$C$_y$ and the Si$1-z$Cz layers are 4 nm and 3.3 nm thick, respectively. The coupled quantum wells are repeated 25 times with thick Si layers in between. The insert shows the energy of the no-phonon PL peak as a function of the carbon content z in Si$_{1-x}$Ge$_x$/Si$_{1-z}$C$_z$ layers (upper curve). The lower two curves in the insert show the energy shift for Si$_{1-x-y}$Ge$_x$C$_y$/Si$_{1-z}$C$_z$ double quantum wells for different C concentrations in the individual layers.

direct, type I transition from the $\Delta(4)$ CB to the hh VB within the Si$_{0.84}$Ge$_{0.16}$ layer. Up to about $z = 0.8\%$ there is hardly any shift of the NP-PL energy. As more and more C is incorporated into the adjacent Si$_{1-z}$C$_z$ layer, the $\Delta(2)$ CB shifts downwards and a crossing of the $\Delta(4)$ in SiGe and the D(2) level in Si$_{1-z}$C$_z$ takes place. For $z \geq 0.008$ optically excited electrons relax into the $\Delta(2)$ CB states in Si$_{1-z}$C$_z$ and thus the PL energy decreases as observed from Si$_{1-z}$C$_z$ single QW structures.

A small PL redshift of about -10 meV/%C is observed for the Si$_{0.84-y}$Ge$_{0.16}$C$_y$/

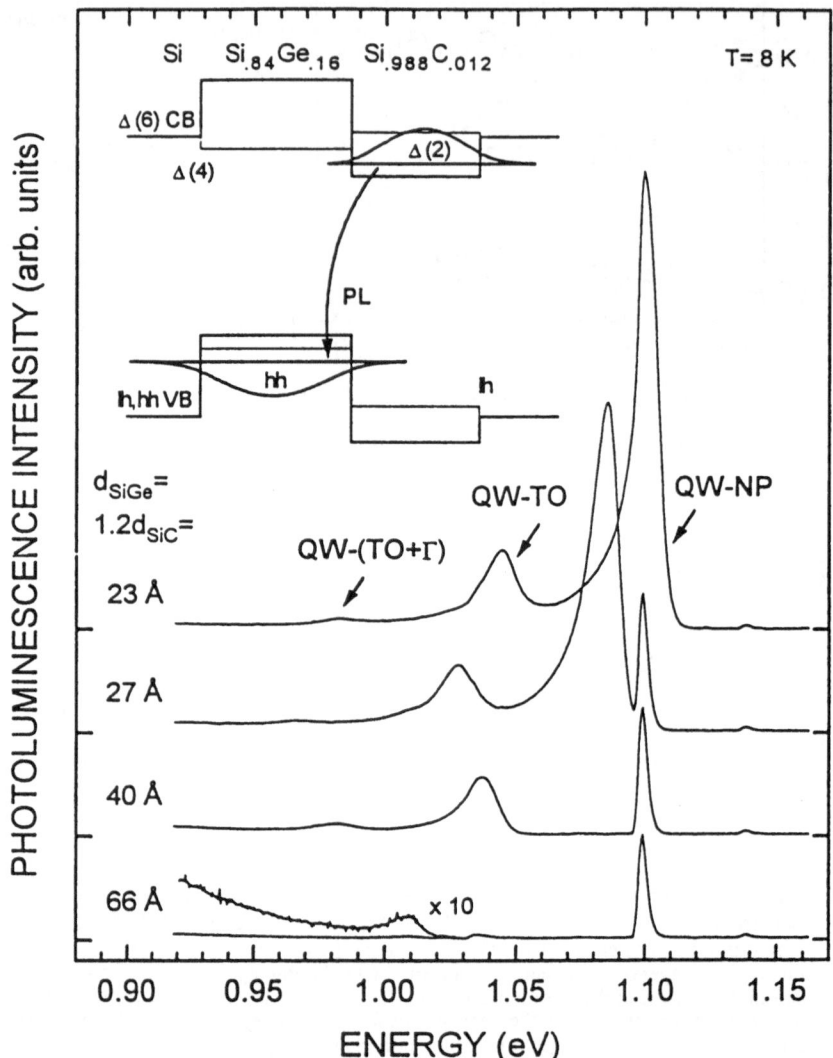

Figure 9 PL spectra from 25-period $Si_{0.84}Ge_{0.16}/Si_{0.988}C_{0.012}/17$ nm Si double quantum well structure with varied layer width $d_{SiGe} = d_{SiC}$ on Si(100). The spectra are plotted in the same intensity scale for direct comparison. A maximum in the overlap of the wave functions and the maximum intensity of the quantum well no-phonon PL peak is achieved for $d_{SiGe} = 2.3$ nm and $d_{SiC} = 1.9$ nm. The insert in the upper part shows a schematic diagram of the conduction (CB) and valence-band edge (VB). The thicker line indicates the two-fold $\Delta(2)$ minima and the heavy hole bandedge in the CB and the VB, respectively. The thinner lines indicate the four-fold $\Delta(4)$ minima and the light hole bandedge. The arrow indicates the PL transition from the $\Delta(2)$ CB states in the $Si_{0.988}C_{0.012}$ layer to the hh VB states in the $Si_{0.84}Ge_{0.16}$ layer.

$Si_{1-z}C_z$ double quantum wells with respect to corresponding $Si_{0.84}Ge_{0.16}/Si_{1-z}C_z$ DQW structures. This can be assigned to an increase of the hh VB offset in a 4 nm $Si_{0.84-y}Ge_{0.16}C_y$ quantum well by C incorporation relative to Si. This VB shift is driven by intrinsic properties and not by the effect of strain reduction, which would rather decrease the VB offset as discussed above and shown in Figure 5. However, more detailed studies are needed to get a reliable picture of the quantitative change of the VB offset, since other groups found a slight decrease of the VB offset in capacitance-voltage measurements from Si/SiGeC heterostructures [16]. One has to keep in mind that confinement effects, a modified excitonic binding energy in DQWs, inhomogenities within the layers or impurities may affect the quantitative PL shifts with C content given in Figure 8.

In conclusion, we can describe the energy shifts of the PL lines for $Si_{0.84}Ge_{0.16}/Si_{1-z}C_z$ and $Si_{0.84-y}Ge_{0.16}C_y/Si_{1-z}C_z$ DQWs considering strain and intrinsic effects of C on the alloy layers within the picture of the CB and VB alignments shown in Figure 8. An improved PL efficiency of the DQWs embedded in Si is observed compared to single QWs. Further enhanced no-phonon transitions and even more efficient capture of excited carriers may be expected for larger Ge and C contents. It is also interesting that the DQW structures presented here are nearly symmetric in strain, which means that there are no strong overall thickness limitations for the superlattice. This makes this concept interesting for future applications in waveguides, infrared detectors, modulators and electroluminescence devices grown directly on Si.

6   Electrical Transport Properties

Ershov and Ryzhii [32] published a theoretical study of electron transport in strained $Si_{1-y}C_y$ and predicted an increase of the mobility for reasonably small alloy scattering potentials. First experimental results from Sb doped $Si_{1-y}C_y$ alloys were reported by Faschinger et al. [9]. We have prepared 0.5 μm thick $Si_{0.996}C_{0.004}$ and Si reference layers with nominally the same Phosphorus doping of about $3 \cdot 10^{17}$ cm^{-3}. The $Si_{0.996}C_{0.004}$ is fully strained as measured by XRD. The critical thickness for pseudomorphic growth is significantly beyond 0.5 μm for the small C content used in this case. The $Si_{0.996}C_{0.004}$ layers were grown on n-Si(100) substrate with a resistivity above 5 000 Wcm. A 200 nm thick undoped Si buffer layer is deposited at 700 °C before growing the doped layer at 550 °C. A careful chemical precleaning is performed to keep the Boron doping at the substrate interface below $5 \cdot 10^{10}$ cm^{-2}.

The electron density and Hall mobility from the samples described above is shown in Figure 10. Over the whole temperature range from 50 to 300 K the electron density is essentially the same in the $Si_{0.996}C_{0.004}$ and the reference sample. No electron capture by C or by any C-related defects is observed. In undoped $Si_{0.996}C_{0.004}$ layers we find a high resistivity with no ohmic contacts to the layer. The electron mobility of the $Si_{0.996}C_{0.004}$ sample and the reference sample are also closely matching down to about 180 K. For lower temperatures the mobility in the $Si_{0.996}C_{0.004}$ sample is getting

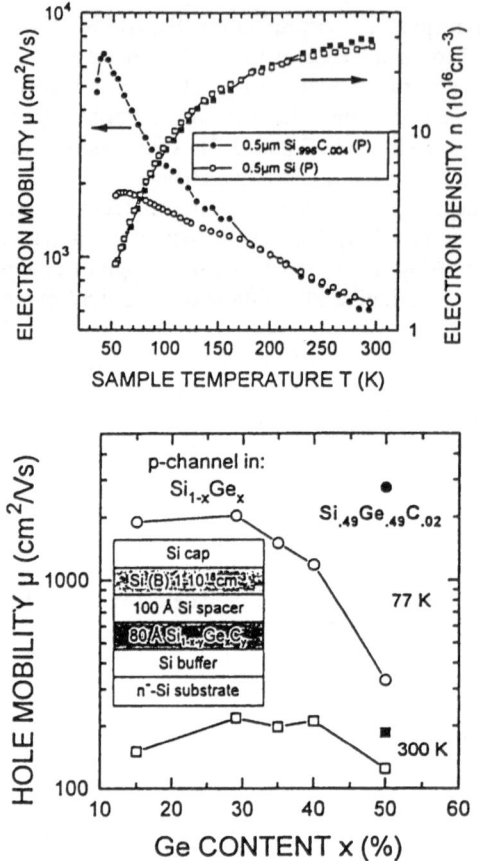

Figure 10
Hall mobility and electron density (corrected for surface depletion) measured as a function of temperature in a Van-der-Pauw geometry from an n-type (Phosphorus) $3 \cdot 10^{17}$ cm^{-3} doped 0.5 μm thick Si$_{0.996}$C$_{0.004}$ layer and a reference layer without C. The layers were grown on (n-) 5 000 Wcm Si substrates with a 200 nm thick undoped Si buffer layer.

Figure 11
Hall mobilities measured at room temperature and 77 K in a Van-der-Pauw geometry from p-channel Si$_{1-x-y}$Ge$_x$C$_y$ modulation doped quantum well structures. The layer sequence is schematically shown in the inset.

significantly higher as compared to the Si sample with the same electron density. This can be explained by the tensile strain in the Si$_{0.996}$C$_{0.004}$ layer, which induces a splitting of the six-fold Δ-valleys. The $\Delta(2)$ valleys become the CB minima with the small in-plane effective mass for the electron. The splitting of the $\Delta(4)$ and $\Delta(2)$ levels is only about 20 meV for the small C concentration used in this sample, therefore a mobility enhancement is expected only for low temperatures. The results on the thick homogeneously doped Si$_{1-y}$C$_y$ alloy layers demonstrate that the samples show reasonable doping behaviour and that the mobilities do not degrade. For very small C concentrations we observe a significant mobility enhancement, which can be explained by the smaller effective mass and the reduced intervalley scattering due to the tensile strain within the Si$_{1-y}$C$_y$ layer.

Figure 11 shows Hall-mobilities measured at room temperature and 77 K in a Van-der-Pauw geometry from several p-channel Si$_{1-x}$Ge$_x$ and one Si$_{0.49}$Ge$_{0.49}$C$_{0.02}$ modulation doped quantum well structure represented by open and full symbols, respectively. The layer sequence is 400 nm undoped Si, 8 nm Si$_{1-x-y}$Ge$_x$C$_y$, i0 nm Si spacer,

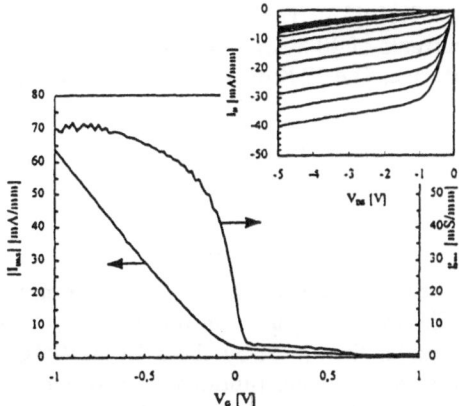

Figure 12
Measured DC transconductance gme and drain current I_D in the saturation range at $V_{DS} = -3$ V as a function of the gate bias voltage for a $0.75+\mu m$ gate-length p-channel $Si_{0.54}Ge_{0.45}C_{0.012}$ heterostructure field-effect-transistor (FET) at 77 K. The room temperature output characteristic of the device is shown in the right part of the figure. The gate voltage sweep is $V_G = -0.7$ V to $+1.0$ V with a step width of 0.1 V.

a Boron doped Si layer and a 2.5 nm Si cap layer (see also the insert of Figure 9.). Looking at the hole mobilities in the $Si_{1-x}Ge_x$ QWs we observe a maximum of about 220 cm²/Vs at 300 K and about 2 000 cm²/Vs at 77 K for $x = 30\%$. For higher Ge concentrations the mobilities go down. This is attributed to strain induced interface roughness and misfit dislocations. The 8 nm quantum well thickness is above the equilibrium critical thickness for pseudomorphic $Si_{0.5}Ge_{0.5}$ on Si. In the $Si_{0.49}Ge_{0.49}C_{0.02}$ sample the compressive strain is reduced by about 30%. In other words, the strain is similar to a virtual Ge concentration of about 35%. The mobilities for the $Si_{0.49}Ge_{0.49}C_{0.02}$ QW are 185 cm²/Vs at 300 K and 2 750 cm²/Vs at 77 K, the later being significantly improved against the corresponding sample without C. The higher hole mobility for the $Si_{0.49}Ge_{0.49}C_{0.02}$ QW is a consequence of the reduced strain in the layer due to substitutional C. For the $Si_{0.5}Ge_{0.5}$ sample we measure a hole density of $5 \cdot 10^{11}$ cm^{-2} at 300 K and $1.5 \cdot 10^{11}$ cm^{-2} at 77 K. The hole density in the C alloyed QW is also $5 \cdot 10^{11}$ cm^{-2} at 300 K but only about $0.5 \cdot 10^{11}$ cm^{-2} at 77 K. So far, the samples are not optimised for maximum carrier transfer from the doping layer to the QW. The reduction of carrier density at low temperature may be a consequence of alloy inhomogenities and local strain fields around the C atoms in the $Si_{0.5}Ge_{0.5}$ matrix. This, however, seems to be not a significant problem for room temperature devices, because the mobilities for electrons and holes at 300 K are not degrading with C alloying.

Figure 12 shows the DC transconductance (g_{me}) and drain current (I_{DSS}) in the saturation range at $V_{DS} = -3$ V as a function of the gate bias voltage for a 0.75 μm gate-length p-channel $Si_{0.538}Ge_{0.45}C_{0.012}$ modulation doped field-effect-transistor (MODFET) at 77 K. The room temperature output characteristic of the device is shown in the upper right part of the figure. The layer sequence of the sample consists of a 300 nm thick Si buffer layer, an 8 nm thick $Si_{0.54}Ge_{0.45}C_{0.012}$ layer, which serves as the two-dimensional hole gas channel, a 5 nm thick undoped Si spacer, a 3 nm thick $1 \cdot 10^{19}$ cm^{-3} Boron doped Si layer and a 15 nm thick Si cap layer. The carrier density and hole mobility for this sample was $1.4 \cdot 10^{12}$ cm^{-2} and 113 cm²/Vs, respectively. For details of the device fabrication and characteristics see Ref. [33]. The room temperature transconductance of 57 mS/mm and saturation currents of 40 mA/mm are demonstrated for this first 0.75 μm gate-length, non-recessed device. The 77 K data are $g_{me} = 70$ mS/mm and $I_{DSS} = 63$ mA/mm.

7 Summary

In pseudomorphic $Si_{1-y}C_y$ layers on Si the band gap is reduced by about 65 meV/%C. Considering deformation potentials we find the band gap of unstrained $Si_{1-y}C_y$ alloys with small carbon concentration to be smaller than in Si by about 15 meV/%C. The band gap in pseudomorphic SiGeC is increased by substitutional C incorporation by about 24 meV/%C. The band alignment in $Si/Si_{1-y}C_y$ is most probably type I with the offset mainly in the conduction band. A strong electron and hole confinement is achieved directly on Si in coupled $Si_{1-x}Ge_x/Si_{1-y}C_y$ quantum wells, which results in a strong no-phonon PL intensity.

Enhanced electron mobilities are observed in thick pseudomorphic n-type doped $Si_{1-y}C_y$ layers at low temperatures, which is attributed to the splitting of the D valleys with the $\Delta(2)$ valleys being the CB minima. In a modulation doped p-type $Si_{0.49}Ge_{0.49}C_{0.02}$ QW we observe an improved hole mobility at room temperature and 77 K compared to a corresponding sample without C, which is a consequence of the reduced strain in the layer due to substitutional C.

The PL investigations and the Hall measurements presented provide a picture of the influence of substitutional C on the band gap and the band alignment. The results demonstrate that $Si_{1-y}C_y$ and $Si_{1-x-y}Ge_xC_y$ alloys offer more flexibility in the design of device structures directly on Si. We have shown that non-optimised MODFETs with promising device characteristics are realised from simple Si/SiGeC layer sequences without any thick relaxed SiGe buffer layers. A significant performance improvement is expected for layers with improved hole mobilities, higher hole densities and after down-scaling of the gate-length.

Acknowledgement

It is a pleasure to thank K. Brunner, Y. Jin-Phillipp and W. Winter for their excellent work on this project. We are also greatful to M. Glück and U. König for the good collaboration and especially for processing and measuring the FETs. The work has been supported financially by the Bundesministerium für Bildung und Forschung within the "Si Nanoelektronik" project. The author gratefully acknowledges the continuous support of K. von Klitzing and H. G. Grimmeiss.

Bibliography

[1] Properties of Strained and Relaxed Silicon Germanium, edited by E. Kasper, EMIS Datareviews Series No 12, (1995).

[2] K. Ismail, F.K. LeGoues, K.L. Saenger, M. Arafa, J.O. Chu, P.M. Mooney and B. S. Meyerson, Phys. Rev. Lett., 73, 3447 (1994).

[3] S.S. Iyer, K. Eberl, M. Goorsky, F.K. LeGoues, J.C. Tsang, and F. Cardone, Appl. Phys. Lett. 60, 356 (1992).

[4] K. Eberl, S.S. Iyer, J.C. Tsang, M.S. Goorsky and F.K. LeGoues, J. Vac. Sci. Technol. B. 10 934 (1992).

[5] K. Eberl , S.S. Iyer, S. Zollner, J.C. Tsang, and F.K. LeGoues, Appl. Phys. Lett. 60, 3033 (1992).

[6] J.B. Posthill, R.A. Rudder, S.V. Hattanggady, C.G. Fountain and R.J. Markunas, Appl. Phys. Lett. 56, 734 (1990).

[7] H.J. Osten, E. Bugiel and P. Zaumseil, Appl. Phys. Lett. 64, 3440 (1994). And H. Rücker, M. Methfessel, E. Bugiel and H.J. Osten, Phys. Rev. Lett. 72, 3578 (1994).

[8] J. Kolodzey, P.R. Berger, B.A. Omer, D. Hits, F. Chen, A. Khan, S. Shao, M.M. Waite, S. Ismat Shah, C.P. Swann and K.M. Unruh, J. Crystal Growth 157, 386 (1995).

[9] W. Faschinger, S. Zerlauth, G. Bauer and L. Palmetshofer, Appl. Phys. Lett. 67, 3933 (1995).

[10] P.O. Pettersson, C.C. Ahn, T.C. McGill, E.T. Croke and A. T. Hunter, Appl. Phys. Lett. 67, 2530 (1995).

[11] G. He, M. D. Savellano and H. A. Atwater, Appl. Phys. Lett. 65, 1159 (1995).

[12] J.W. Strane, H.J. Stein, L.R. Lee, B.L. Doyle, S.T. Picraux and J.W. Mayer, Appl. Phys. Lett. 63, 2786 (1993).

[13] J.L. Regolini, F. Gisbert, G. Dolino and P. Boucaud, Mater. Lett. 18, 57 (1993).

[14] Z. Atzomon, A.E. Bair, E.J. Jaquez, J.W. Mayer, D. Chandrasekhar, D.J. Smith, R.L. Hervig and McD. Robinson, Appl. Phys. Lett. 65, 2559 (1994).

[15] L.D. Lanzerotti, A. S. Amour, C.W. Liu and J. C. Sturm, Tech. Dig. IEDM, 930, (1994)

[16] K. Rim, S. Takagi, J.J. Welser, J.L. Hoyt and J.F. Gibbons, Mat. Res. Soc. Symp. Proc. Vol. 397, 327 (1995)

[17] J. Mi, P. Watten, P. Letourneau, M. Judelewicx, M. Gailhanou, M. Dutoit, C. Dubois and J.C. Dupuy, Appl. Phys. Lett, 67, 259 (1995).

[18] S.C. Jain, H.J. Osten, B. Dietrich and H. Rücker, Semicond. Sci. Technol. 10, 1289 (1995).

[19] A.A. Demkov and O.F. Sankey, Phys. Rev. B 48, 2207 (1993).

[20] G. Lippert, H.J. Osten, D. Krüger, P. Gaworzewski and K. Eberl, Appl. Phys. Lett. 66, 3197 (1995).

[21] K. Brunner, K. Eberl and W. Winter, Appl. Phys. Lett. 69 ,91 (1996).

[22] K. Brunner, K. Eberl and W. Winter, Phys. Rev. Lett. 76, 303 (1996).

[23] R.A. Soref, J. Appl. Phys. 70, 2470 (1991).

[24] A.R. Powell, K. Eberl, B.A. Ek and S.S. Iyer, J. Cryst. Growth 127, 425 (1993).

[25] P. Boucaud, C. Francis, A. Larré, F.H. Julien, J.M. Lourtioz, D. Bouchier, S. Bonar and J.T. Regolini, Appl. Phys. Lett. 66, 70 (1995).

[26] K. Brunner, W. Winter and K. Eberl, Appl. Phys. Lett. 69, 1279 (1996).

[27] J.C. Sturm, H. Manoharan, L.C. Lenchyshyn, M.L.W. Htewalt, N.L. Rowell, J.-P. Noel and D.C.Houghton, Phys. Rev. Lett. 66, 1362 (1991).

[28] D.C. Houghton, G.C. Aers, S.-R. E. Yang, E. Wang, and N.L. Rowell, Phys. Rev. Lett. 75, 866 (1995).

[29] A. S. Amour, C.W. Liu, J. C. Sturm, Y. Lacroix and M. L. W. Thewalt, Appl. Phys. Lett. 67, 3915 (1995).

[30] D. Dutartre, G. Bremond, A. Souifi, T. Benyattou, Phys. Rev. B 44, 11525, (1991).

[31] M. Gail, G. Abstreiter, J. Olajos, J. Engvall, H. Grimmeis, H. Kibbel and H. Presting, Appl. Phys. Lett. 66, 2978 (1995).

[32] M. Ershov and V. Ryzhii, J. Appl. Phys. 76, 1924 (1994).

[33] M. Glück, U. König, K. Brunner, W. Winter and K. Eberl, Physica B, Condensed Matter (1998) in print.

Beryllium-containing II-VI Compounds: Properties and Applications

A.Waag[2], Th.Litz [1], F. Fischer[1], H.-J. Lugauer[1], T. Baron[1],
K. Schüll[1], U. Zehnder[1], T. Gerhard[1], U. Lunz[1], M. Keim[1],
G. Reuscher[1], G. Landwehr[1]

[1] Physikalisches Institut, Universität Würzburg,
Am Hubland, 97074 Würzburg, Germany
[2] School of Electrical and Computer Engineering,
Purdue University, W. Lafayette, USA

1 Introduction

During the last years a substantial progress in the development of ZnSe based light emitting devices has been made [1]. By optimizing the device design as well as reducing the defect densities, laser diodes with low threshold current densities and high optical output power could be fabricated, emitting in the blue and green spectral range. The lifetime of such devices now reaches 100 h at room temperature under CW operation [2], being close to a possible commercialization of ZnSe laser diodes for first applications where pulsed laser sources can be used.

An additional, yet untried approach to further improve the reliability of ZnSe based devices is to use beryllium containing II-VI compounds. BeS, BeSe and BeTe are characterized by a considerable amount of covalent bonding and a high bond energy. This distinguishes these materials from the conventional ionic wide gap II-VI semiconductors like ZnSe, ZnTe or CdTe. In addition, quaternary Be-containing selenides and tellurides can be lattice matched to GaAs. The incorporation of beryllium into ZnSe based materials is expected to increase the covalency, hardness and bond energy of the material and therefore reduce the degradation of optoelectronic devices.

Recently, thin film structures using Be-compounds have been fabricated and characterized for the first time at the university of Würzburg. It became clear that – besides the application aspects – these materials are also very interesting from a fundamental point of view. Using Be-containing II-VI compounds, ionic and covalent lattice matched II-VI materials can be combined in quantum well structures. The type II band alignment of BeTe and ZnSe gives additional freedom in the band gap engineering, and it is possible to grow lattice matched quaternaries of low polarity onto silicon. These aspects in combination with the possibility to fabricate device quality structures with very high band gaps gives plenty of room for interesting fundamental research.

In this review, the properties of Be containing II-VI compounds – as far as they are known so far – will be described, and the potential of these novel materials both for basic research as well as applications will be discussed.

Since the first demonstration of ZnSe based blue-green emitters by M. A. Haase et al. in 1991 [3], the development of ZnSe laser diodes has essentially been based on three main achievements: The incorporation of plasma activated nitrogen doping, allowing the fabrication of p-type ZnSe layers [4,5], was the initial breakthrough, which allowed to fabricate LED's and laser diodes. Doping by activated nitrogen is still the only viable way to date to obtain p-type ZnSe. All attempts to fabricate highly doped p-ZnSe with alternative dopants [6] or other growth techniques like MOCVD were unsuccessful so far. A second important improvement was the introduction of magnesium for the use of ZnMgSSe as high band gap cladding layers in laser diodes [7]. This solved the problem of a too small confinement energy in the first laser diodes, which had been based on a ZnSSe-ZnSe-ZnCdSe structure. By using ZnMgSSe, the band gaps could now be engineered, though still being lattice matched. In this way, a sufficiently large optical confinement (difference in index of refraction) and electrical confinement (difference in band edges) could be realized in separate confinement double heterostructure quantum well laser diodes. A third improvement, which finally allowed to achieve CW operation at room temperature was the introduction of ZnSe-ZnTe pseudogradings for ohmic contact layers to p-type ZnSe [8]. The low lying valence band of ZnSe is a problem for the fabrication of ohmic contacts with a simple metal contact. No metal has a work function which would be high enough for this wide band gap material. In addition, ZnSe can not be degenerately doped p-type, which leads to wide depletion layers and therefore high ohmic losses across the metal contact. One solution is to grade the valence band from the low lying ZnSe up to ZnTe, where the valence band position is approximately 0.6 eV higher than compared to ZnSe. In addition, ZnTe can easily be degenerately doped p-type, and therefore efficient tunneling contacts can be fabricated on ZnTe.

All three main developments led to the fabrication of lasers which could be operated continuously at room temperature. As the threshold current densities as well as operating voltages could be decreased, it became clear that degradation due to grown in defects mainly originating at the GaAs-ZnSe interface and their propagation were responsible for the limited device lifetime of the laser diodes. Subsequently, the epitaxy of the II-VI growth on GaAs was improved resulting finally in defect densities as low as 10^3 cm^{-2} [9]. Such low defect densities clearly demonstrate that the heteroepitaxy of II-VI on III-V substrates can well be under control. Meanwhile, ZnSe lasers have already been used for a prototype demonstration in high density CD players [10].

Despite the considerable improvements described above, the lifetime of ZnSe laser diodes is still not high enough for commercialization. An alternative approach in order to solve the existing problems is the incorporation of beryllium.

Recently, C. Verie has studied the elastic constants of beryllium chalcogenides theoretically. He concluded that beryllium chalcogenides are promising in respect to lattice hardening due to their pronounced covalent bonding [11], which is supposed to have an important impact on the defect generation and propagation and therefore the device lifetime. For the estimation of the experimentally unknown shear moduli, a semi-empirical law has been considered, linking the shear modulus with the covalency of a compound, making use of W. A. Harrisons calculations based on the linear combination of atomic orbitals (LCAO) [12]. It is predicted that the shear modulus of BeS is higher than that of GaAs and that the shear moduli of BeTe and BeSe are considerably higher than that of ZnSe.

Only very few data on the basic properties of Be compounds are available in the literature. Nevertheless one can judge already from the existing publications that beryllium chalcogenides – due to their band gaps and lattice constants – are interesting materials for light emitting diodes (LED's) and laser diodes. Therefore, we started to fabricate Be containing II-VI compounds by molecular beam epitaxy and systematically investigated their properties.

In the following sections, the fabrication of thin film structures will be described first, followed by a discussion of the properties of selenides and tellurides known so far. In the last two paragraphs, interesting new aspects of quantum well structures will be discussed, both from a fundamental and application point of view.

2 Growth of Be-Containing Thin Film Structures

All groups which have taken up the investigation of Be compounds after the first realization of the epitaxy of such materials in summer 1995 are using molecular beam epitaxy as a growth technique, providing a good flexibility in terms of growth temperature and dopant species. Beryllium has a very low vapor pressure, therefore the beryllium effusion cell has to be operated at a temperature of approximately 1 000 °C, which is very unusual for a II-VI MBE system. All other effusion cells are operated at much lower temperatures of 150 °C to 400 °C. In addition, low substrate temperatures between 230 °C and 350 °C are used for the MBE growth of most of the wide gap II-VI semiconductors. At such low substrate temperatures, a pronounced heating of the substrate due to the hot Be effusion cell is inevitable. Despite of this, the Be flux can easily be controlled and a high reproducibility of the Be composition can be achieved, as will be demonstrated subsequently. For the n-type doping, a ZnI_2 source has been used, and the p-type doping can be achieved by employing plasma-activated nitrogen from a rf plasma source. The thin films and quantum well structures are grown on (100) GaAs substrates, which is advantageous due to the close lattice matching of ZnSe and GaAs. To improve the surface quality of the GaAs substrates, a GaAs epilayer is grown on top of the epiready substrate in a separate GaAs MBE chamber, which is connected to the II-VI growth chamber via a UHV transfer system. BeMgZnSe compounds were in general grown at a substrate temperature of 300 °C. For BeTe, the substrate temperatures varied between 250 °C and 550 °C. It should be noted that the growth window of Be containing selenides and tellurides obviously overlaps with the one of ZnSe, which simplifies the epitaxy of Be compounds considerably. This is in contrast to the competitive wide gap material GaN-InN, where the growth regimes of InN and GaN are incompatible, leading to substantial difficulties for the growth of the ternary compound InGaN used as the active layer in nitride based LED's and laser diodes. Furthermore, elemental Be, Se, Te, Mg, and Zn effusion cells can be employed.

3   Covalency and Bond Energy

In Figure 1, the extrapolated cohesive energy per bond (bond energy) for various compound semiconductors, has been plotted vs. hybrid covalency, following the discussions in [12]. It is an empirical rule that the behavior of bond energy vs. covalency is quite linear in an isoelectronic series, this is a series of binary compounds with the same total number of electrons. In this way, the plot in Fig. 1 can be used to estimate the bond energies of beryllium chalcogenides employing their theoretically calculated covalencies [12]. This was done for BeSe and BeS, as well as for the (zincblende) nitride compounds. Other values for the bond energies in Figure 1 are experimental ones taken from Ref. [12]. As can be seen from the figure, the bond energies for beryllium chalcogenides are in general much higher as compared to conventional II-VI-compounds like CdTe and ZnSe, and can even reach values comparable with GaN. It should also be mentioned that BeO is a high temperature ceramic, which has very unusual properties for a II-VI-compound. This underlines the special bond situation in Be compounds as compared to other conventional much more ionic II-VI compounds.

4   BeMgZnSe

In Figure 2 band gap and lattice constant for a variety of different II-VI binaries which are relevant for blue-green laser technology are shown, including the two novel binaries BeSe and BeTe. Reflection measurements on thin BeSe films indicate a band gap of as high as 5.6 eV at room temperature [13]. At present it is not clear whether the binary BeSe is an direct or indirect semiconductor. Transmission measurements are in preparation to clarify this point.

As is shown in Fig. 2, the quaternary BeMgZnSe can be grown lattice matched onto GaAs as well as silicon, tuning the band gap from 2.8 eV up to above 4 eV. The sticking coefficient of the 3 metals is high and relatively independent of the substrate temperature, resulting in a good control of the lattice parameter for quaternary BeMgZnSe. This good control of the lattice parameter is reflected by the width of the rocking curves of quaternary layers, where a drift of composition in time would lead to a broadening of

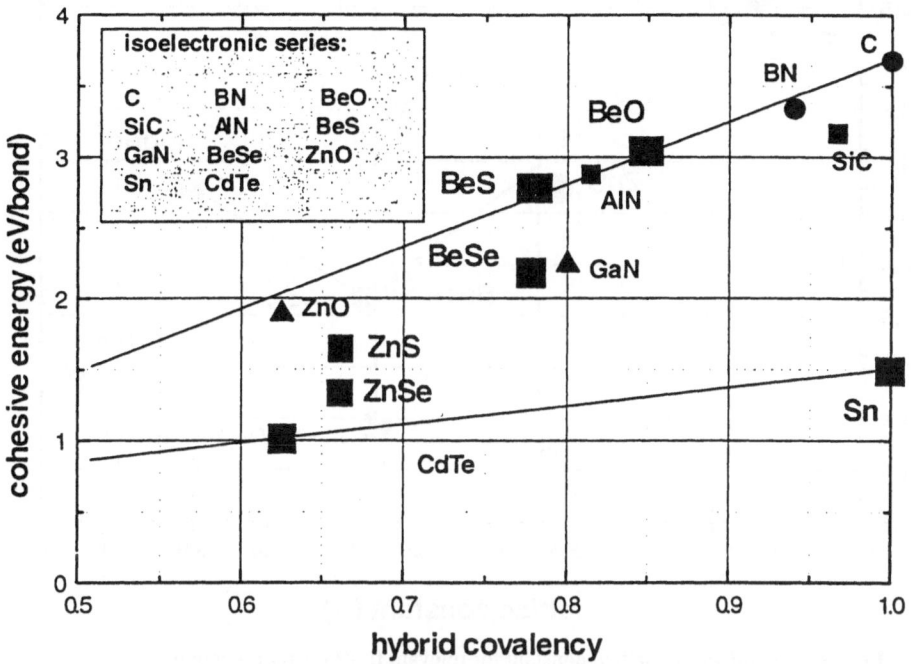

Figure 1 Cohesive energy per bond vs. hybride covalency. Values for BeSe and BeS as well as the nitrides are derived from the covalency, all other values are experimentally determined [12].

the rocking curves. We could reproducibly achieve values below 25 arcsec (004, Cu-K$_\alpha$) with a minimum value of 17 arcsec for BeMgZnSe layers with a band gap of 2.9 eV and a layer thickness of 2 μm. Rocking curves of a 2 μm thick BeMgZnSe layer on GaAs substrate are shown in Fig. 3, indicating the high structural quality which can be reached in these epilayers.

For the ternary BeMgSe, a Be-concentration of 2.8% has to be used for lattice matching to GaAs. Again, rocking curves with a width of 20 arcsec. have been obtained for such ternary films.

In the case of conventional ZnMgSSe, the incorporation of sulfur is problematic because the sulfur sticking coefficient is very sensitive to the substrate temperature [14]. The surface temperature itself is in general changing with growth time due to the changing surface emissivity, and as a consequence it is not straight forward to control the lattice parameter precisely. In addition, especially sulfur as well as selenium is known to have a strong trend towards a chemical reaction with the GaAs surface, and thereby deteriorating the structural quality. Therefore care has to be taken not to expose the GaAs surface with sulfur or selenium prior to the growth of a II-VI film. Usually, a ZnSe

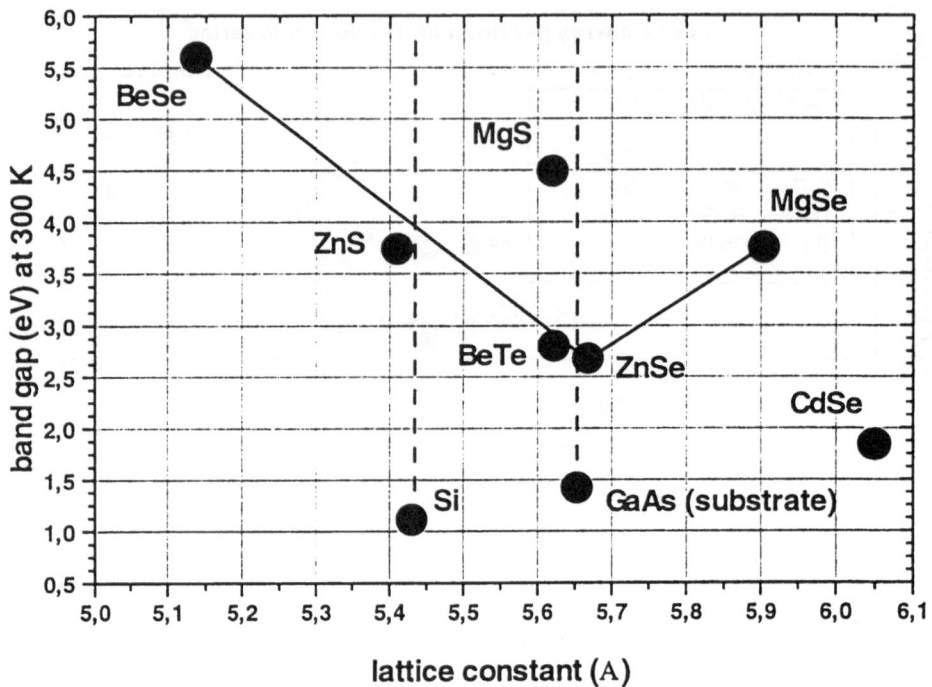

Figure 2 Band gap vs. lattice constant for relevant II-VI binary compounds.

buffer is grown with a zinc rich growth start in order to prevent such an unintentional surface reaction. This is another advantage of the BeMgZnSe system, where no sulfur is used.

In Fig. 4, the band gap of BeMgZnSe is shown as a function of Be and Mg content. The Be and Mg concentrations necessary for a lattice matching to GaAs are also indicated. Due to the large band gap of BeSe, only small concentrations of Be and Mg have to be employed for a band gap of 2.9 eV typical for claddings of blue green ZnSe laser diodes.

With increasing beryllium and magnesium content the band gap of quaternary Be-MgZnSe shifts to higher energies. In photoluminescence at low temperatures, no additional defect bands occur with the incorporation of beryllium, demonstrating that the high temperature Be effusion cell does not necessarily produce any impurity problems.

A variety of ZnSe quantum wells embedded in BeZnSe or BeMgZnSe have been fabricated and investigated. The PL linewidth of the quantum well peak increases from 1 meV to 7 meV if the confinement energy is increased from 20 meV to 100 meV, respectively, which is expected due to the fact that interface fluctuations get more important in thinner quantum wells.

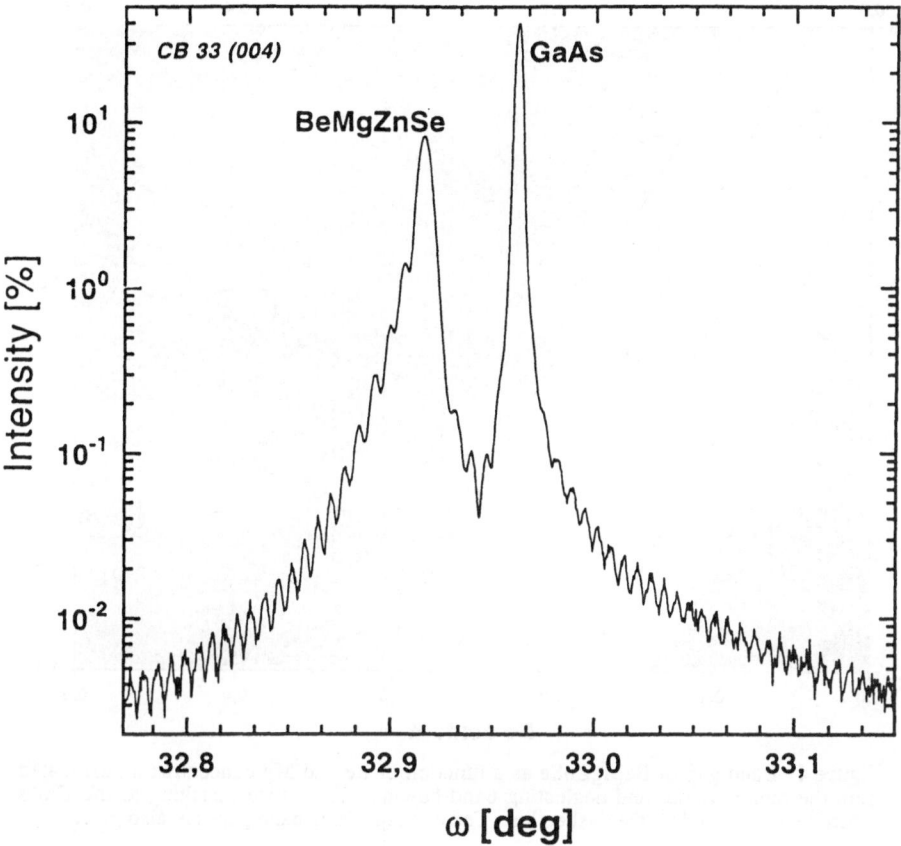

Figure 3 Rocking curve of a BeMgZnSe layer. The FWHM of 17 arcsec is at the theoretical limit for this apparatus function and layer thickness.

An important parameter for the design of laser devices is the change of the index of refraction as a function of the quaternary composition. We determined the index of refraction for various layers by fitting the reflection data. For that the film thickness was measured after the epitaxy on a cleaved sample by scanning electron microscopy. The change of index of refraction with band gap is very similar for the two different quaternary materials BeMgZnSe and ZnMgSSe, as is shown in Fig. 5, where the index of refraction for ZnSe, BeMgZnSe and ZnMgSSe is plotted as function of energy.

As mentioned above, one important advantage of beryllium containing compounds is bond strengthening. To get informations on the rigidity of the lattice, we did perform nano-indentation experiments. The micro hardness of some ternaries and quaternaries have been measured by nano-indentation. A larger solid solution hardening effect can be obtained by alloying ZnSe and beryllium instead of alloying ZnSe with sulfur, for

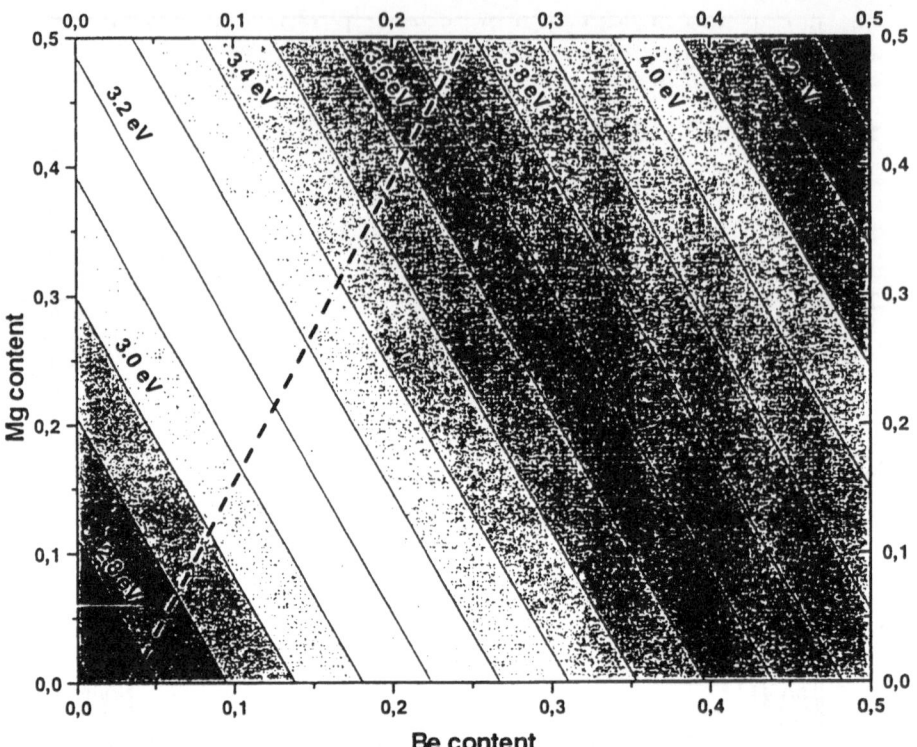

Figure 4 Band gap of BeMgZnSe as a function of Be and Mg concentration, estimated from the binary values and neglecting band bowing.. The lattice matching to the GaAs substrate is indicated by the dashed line. The corresponding band gaps are also given.

comparable composition [15]. Hardness of a material is governed by two processes: dislocation formation, which is related to the bond strength, and dislocation propagation, which is strongly influenced by the size and elastic modulus difference between the elements of an alloy [16]. Here, the pronounced lattice hardening is believed to be mainly due to the small atomic radius of the beryllium atoms and can not be taken as a direct proof for the higher covalency of these compounds.

BeMgZnSe can be degenerately doped n-type, which has been demonstrated for band gaps of up to 3.1 eV. In contrast to that a sufficiently large p-type doping in terms of light emitting devices can only be accomplished for band gaps below 2.9 eV. The limit of the p-type doping has not yet been systematically investigated. This is very similar to the case of ZnMgSSe, where a compensation mechanism becomes more and more pronounced with increasing band gap of the quaternary [17], a behavior which is quite typical also for other wide gap II-VI semiconductors.

In summary, the composition and therefore the lattice constant of BeMgZnSe can precisely be controlled and the material can be used as a high quality cladding in quantum well structures.

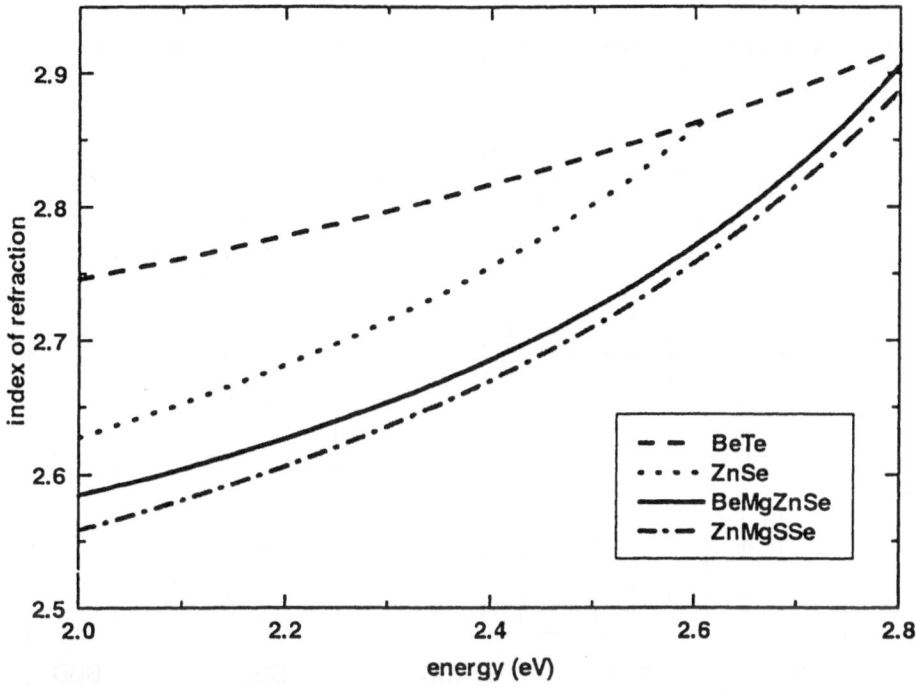

Figure 5 Index of refraction for ZnSe, BeMgZnSe and ZnMgSSe is plotted as function of energy as obtained from reflection measurements (from [15]).

5   BeTe

The close lattice matching of BeTe and GaAs (0.6%) opens the possibility for a pseudomorphic combination of tellurides (BeTe, high p-type doping) and selenides (BeMgZnSe, high n-type doping).

It is indicated by reflection as well as ellipsometry measurements that the band gap of BeTe is indirect, which is consistent with other findings [18]. An indirect gap of 2.8 eV has been derived from reflection measurements (compare also Fig. 2). The indirect nature of the band gap in BeTe is uncritical if BeTe is used in contacts, the first practical application of this material (see below). BeTe can even be introduced into the waveguide or be part of the active layer regions as long as the wavelength is no shorter than 445 nm. The direct transition of BeTe has an energy of 4.1 eV at room temperature.

The following experiment gives an indication of the higher bond energy of beryllium chalcogenides: The desorption of a compound in vacuum can be measured by the RHEED (High Energy Electron Diffraction) oscillation technique measuring the growth rate of the compound as a function of substrate temperature [19], using the RHEED oscillation technique. In the case of BeTe – due to the relatively good lattice matching of

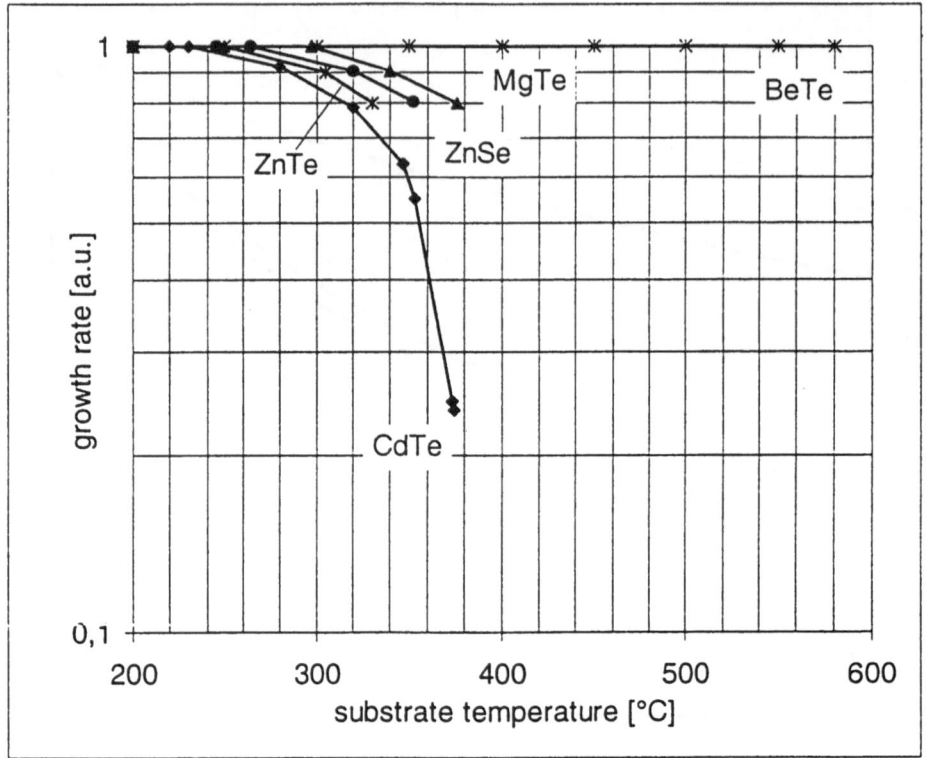

Figure 6   Growth rate as a function of substrate temperature.

BeTe to the GaAs substrate (0.6%) – well pronounced and long lasting RHEED oscillations can in general be observed during growth. The onset of desorption with increasing substrate temperature can usually be followed by the decrease of the growth rate. The decrease of the growth rate then corresponds to a first approximation to the desorption rate – neglecting temperature dependent sticking coefficients for the moment – and experimental values of the growth rate as a function of substrate temperature are shown in Fig. 6 for various II-VI compounds.

In contrast to other II-VI compounds like e.g. ZnSe, the growth rate of BeTe stays constant up to a substrate temperature of 550 °C. This indicates that in the case of BeTe desorption does not play a role even for such high substrate temperatures. The molecular beam epitaxial growth of BeTe is compatible even with the one of GaAs, and BeTe-GaAs heterostructures can be fabricated at the substrate temperatures typically used for the MBE growth of GaAs. This result demonstrates that the bond energy of BeTe is indeed high as compared to the conventional II-VI-compounds CdTe, ZnSe and ZnTe.

One severe problem of ZnSe based laser diodes is the fabrication of good ohmic contacts to p-type ZnSe. As described before, ZnTe-ZnSe pseudogradings are used today in order to tune the valence band from the low lying ZnSe related position up to the ZnTe position, taking additional advantage of the high p-type dopability of ZnTe for a subsequent metal contact. The main disadvantage of this approach is the high lattice mismatch between these two materials, resulting in an extremely high defect density. High defect densities are known to promote diffusion of both dopant atoms as well as matrix materials [20].

By replacing ZnTe by BeTe the problem of lattice mismatched graded contacts can be solved. Assuming a high lying valence band for BeTe, the possibility for the fabrication of BeTe-ZnSe graded contacts has already been discussed on a theoretical basis [21]. Recently, we have determined the valence band offset between BeTe and ZnSe (0.9 eV) experimentally in the following way.

Due to a type II band offset, this value can directly be measured by utilizing the optical transition between the conduction band of ZnSe and the valence band of BeTe in ZnSe-BeTe quantum well structures. The spatially indirect transition is usually low in intensity, but detectable. In Fig. 7, the PL of such a structure with two different BeTe quantum wells is shown. The recombination at 1.8 eV and 2 eV results from the spatially indirect transitions into the hole subband of the two BeTe quantum wells with different thickness, and the energetic difference between the ZnSe band gap PL and the indirect transition line for the wide quantum well (confinement energy small) directly gives the BeTe-ZnSe valence band offset. This assumes, however, that the residual doping level in both the ZnSe and BeTe is low enough, so that band bending effects can be neglected (flat band conditions). Recent data based on photoelectron spectroscopy indicate, however, that the band offset can vary by more than 0.5 eV due to the actual configuration of the interface between BeTe and ZnSe [22].

The valence band offset between ZnSe and GaAs is also around 1 eV [23] in the case of Zn rich growth conditions. As a consequence there should be a quite close valence band matching between GaAs and BeTe provided that the transitivity rule for band offsets holds. In addition, BeTe, ZnSe and GaAs are relatively well lattice matched (see Figure 1), which allows to grow a BeTe-ZnSe pseudograding between GaAs and the ZnSe based device structure, and thereby opening the way to a new type of device design utilizing a p-type substrate. For such a pseudograding, however, the BeTe has to be heavily doped p-type. We checked this by using plasma activated nitrogen doping, and free hole concentrations have been measured by both van der Pauw as well as infrared reflection. P-type doping levels above $1 \cdot 10^{20}$ cm^{-3} could be achieved.

The possibility of a high p-type doping makes BeTe interesting for the p-type cladding, too. In ZnSeTe and ZnMgSeTe, it has been demonstrated that the incorporation of Te enhances the possibility for heavy p-type doping of the material drastically [24,25]. In both cases, however, the lattice matching to GaAs is sacrificed due to the large lattice parameter of ZnTe. Using beryllium, the lattice constant can be corrected, and lattice matched quaternary BeZnSeTe can be fabricated which is expected to allow heavy p-type doping. This aspect could also be interesting for the growth of laser diodes by metal organic chemical vapor deposition, where the p-type doping of ZnSe is still an

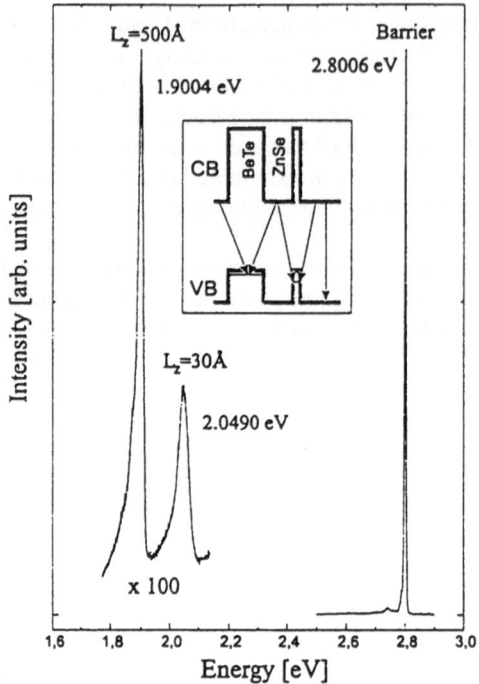

Figure 7
Photoluminescence of a BeTe-ZnSe quantum well structure for band offset measurements.

unresolved problem. Metal-organic precursors for beryllium are available. One should mention that due to the type II band offset between BeTe and ZnSe a pronounced band bowing is expected.

The IIA-VI compounds like MgSe, CaSe, MgTe and CaTe are known to be very hygroscopic. In the case of BeTe the hygroscopic character is much less pronounced. One has to be careful, however, not to heat BeTe at normal atmosphere above 100 °C. In this case the surface will deteriorate due to some surface reaction.

BeTe seems to be an ideal candidate for the growth start on GaAs. As described above, the valence bands are nearly aligned. Secondly, Te has a much lower chemical reactivity with the GaAs surface as compared to selenium and sulfur [26]. Recent surface investigations showed that tellurium can act like a surfactant for the growth of ZnSe on GaAs. Dislocation densities as low as 10^4 cm^{-2} have been reached so far by using a BeTe buffer.

In order to assess the covalency of BeTe, FIR measurements have been performed and compared with simulated spectra. The parameters for this simulation are directly connected with the optical polarity of the compound, and values found by a best fit to the experiment are consistent with the values from LCAO calculations in ref. [12]. Details of this analysis have been described elsewhere [27].

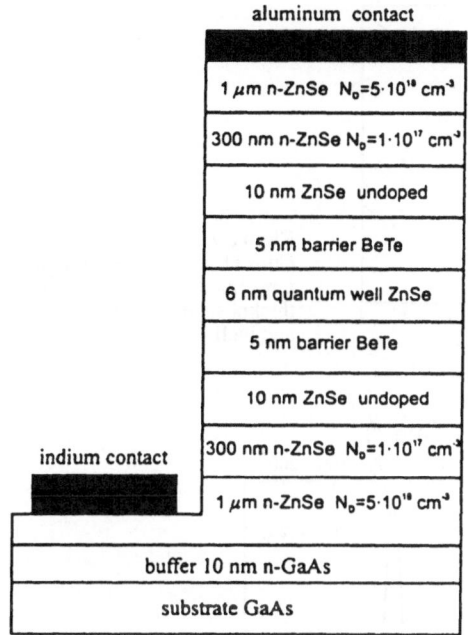

aluminum contact

1 μm n-ZnSe $N_o = 5 \cdot 10^{18}$ cm^{-3}
300 nm n-ZnSe $N_o = 1 \cdot 10^{17}$ cm^{-3}
10 nm ZnSe undoped
5 nm barrier BeTe
6 nm quantum well ZnSe
5 nm barrier BeTe
10 nm ZnSe undoped
300 nm n-ZnSe $N_o = 1 \cdot 10^{17}$ cm^{-3}
1 μm n-ZnSe $N_o = 5 \cdot 10^{18}$ cm^{-3}

indium contact

buffer 10 nm n-GaAs
substrate GaAs

Figure 8
Structure of a double barrier resonant tunneling device.

6   Resonant Tunneling Structures

ZnSe-BeTe structures have potential not only as low resistance ohmic contacts for light emitting devices, they are also interesting from the band gap engineering point of view. Due to the good lattice matching between GaAs, BeTe and BeMgZnSe, a variety of combinations of a type I and type II band offset can be chosen.

BeTe-ZnSe can e.g. be used for the fabrication of resonant tunneling structures. Vertical transport through semiconductor heterostructures has been intensely investigated since the pioneering work of Chang et al. in 1974 [28]. Resonant tunneling in double barrier structures in high magnetic fields allows to study the Landau quantization as well as the energy versus wave vector relation of electrons and holes. So far work on resonant tunneling has almost exclusively been confined to III-V materials. Tunneling data on II-VI wide gap semiconductors were obtained already on CdTe/CdMgTe double barrier structures [29]. The type II band alignment between BeTe and ZnSe with its large barrier above 2 eV has opened new possibilities because of the large energetic distance of bound states, the high quality of the interface and the possibility to introduce manganese into the quantum well in order to investigate resonant tunneling through semimagnetic compounds. A typical structure of such a device is shown in Figure 8.

The current voltage characteristics of a resonant tunneling diode with a thickness of 4 nm for both barrier and well are shown in Fig. 9. Four peaks in the I-V characteristics can be resolved. Details of the first resonance are shown in Figure 10. The structure

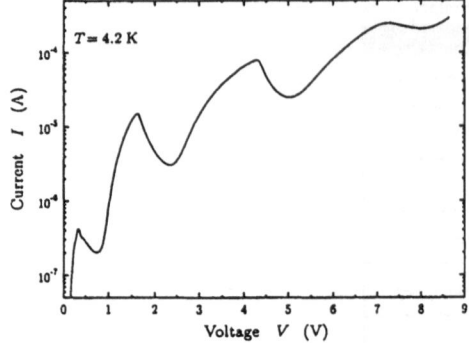

Figure 9
Current voltage characteristics of
a resonant tunneling diode with a
thickness of 4 nm for both barrier
and well.

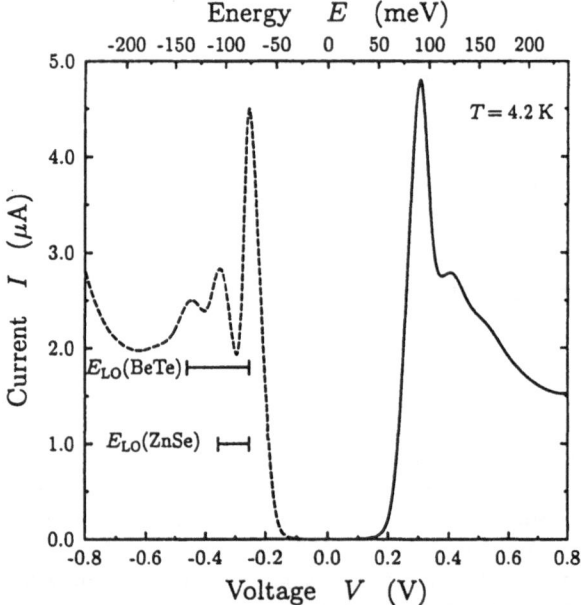

Figure 10
Details of the first
resonance. (Cur-
rent vs. voltage).
The additional
peaks are attributed
to a phonon as-
sisted tunneling
process.

around 0.4 eV can be attributed to LO-phonon assisted tunneling. Application of a magnetic field parallel to the tunneling current reveals well resolved quantum oscillations,
shown in Figure 11.

These can be attributed to the 2-dimensional nature of an electron accumulation layer
in the n-type ZnSe emitter layer. In addition, the carrier concentration in the 2D emitter can be changed with voltage, as is demonstrated in Fig. 11 by the change in oscillation period. The resonant tunneling has been analyzed in detail as a function of
temperature and magnetic field. For further details see Ref. [30]. In the future it will
be interesting to investigate semimagnetic resonant tunneling structures, especially in

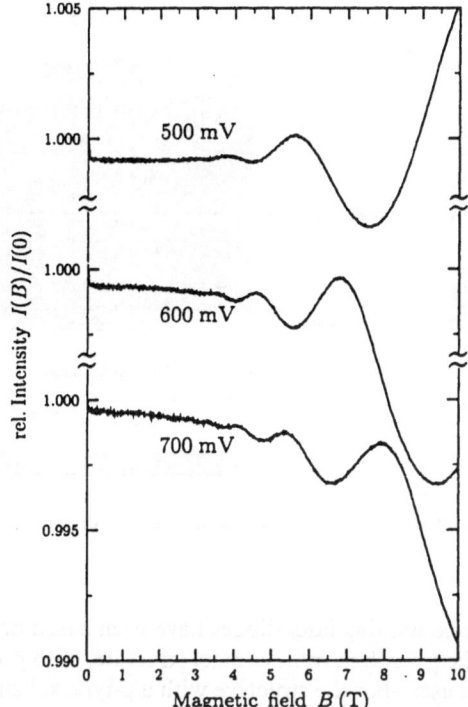

Figure 11
Shubnikov de Haas oscillations
of the tunneling current through a
resonant tunneling device, due to
2 dimensional emitter states.

reduced dimensions. For that the quaternary BeZnMnSe can be used to fabricate lattice matched semimagnetic quantum well structures. Resonant tunneling can then be a means to e.g. perform an electrical subband spectroscopy of semimagnetic material in reduced dimensions.

7   Light Emitting Devices

The degradation mechanisms of II-VI laser diodes based on sulfur containing II-VI materials have been studied in detail over the last years. It turned out that – depending on defect densities – either extended defects originating mainly at the interface between GaAs and ZnSe [31], or point defects in the active region are responsible for the early degradation of conventional laser diodes. The degradation has been identified to be enhanced by defect recombination of electron hole pairs [32]. In order to strengthen the active ZnCdSe region in laser diodes for an enhancement of the laser lifetimes, the additional incorporation of Be into the quantum well suggests itself resulting in a quaternary BeZnCdSe active region. In addition, the ternary BeZnSe seems to be a natural, easy to grow and lattice matched waveguide material with a suitable band gap.

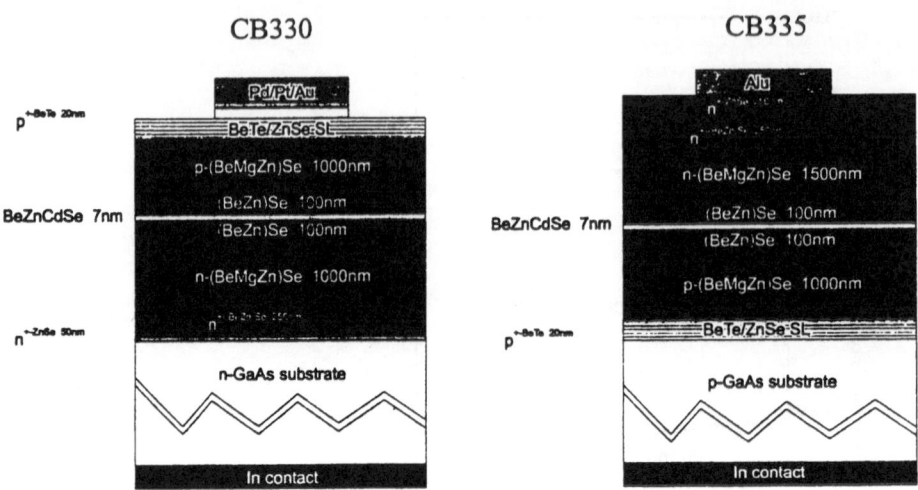

Figure 12 Typical structures of the first Be-containing laser diodes, grown on both p-type and n-type GaAs.

Therefore, first attempts to fabricate Be-containing laser diodes have been based on the structures shown in Figure 12. Due to the pseudomorphic contact structure, both p-type and n-type GaAs substrates can now be used. For the structure with a p-type substrate, there are some potential advantageous: a current spreading through the p-type cladding would lead to a lower current density across the contact at the bottom of the structure towards the GaAs substrate. In addition – during processing – a n-type contact can then be fabricated on a degenerately doped n-type top ZnSe layer, which is much easier from a technological point of view.

The emission of such a laser diode below, at and above threshold is shown in Fig. 13. It is interesting to note that first laser diodes grown on p-type and n-type GaAs had similar I-V characteristics. These first devices had a threshold current density of 230–300 A/cm^2 at 80 K. Meanwhile, Be-containing laser diodes with a lifetime of 57 h under CW operation at room temperature have been fabricated [33], the lifetime being close to the maximum lifetime of 100 h reported up to now for the best ZnMgSSe devices [2]. In view of the short development time for Be-lasers, this is certainly an exciting result. In addition, it should be noted that Be-laser diodes up to now use only ZnCdSe or BeZnCdSe (2.8% Be) quantum wells. This is due to the fact that it is advantageous to keep the Be temperature constant throughout the epitaxy of the waveguide and active region, avoiding a growth interruption at sensitive interfaces. This implies that waveguide and active region then have the same low Be concentration of 2.8%, corresponding to a lattice matched ternary BeZnSe. Clearly, the Be concentration of the active region should be increased in the future to bring the larger covalency of Be compounds into play. Up to now only the good reproducibility of the lattice constant as well as the pseudomorphic and non-absorbing contact structure have been exploited.

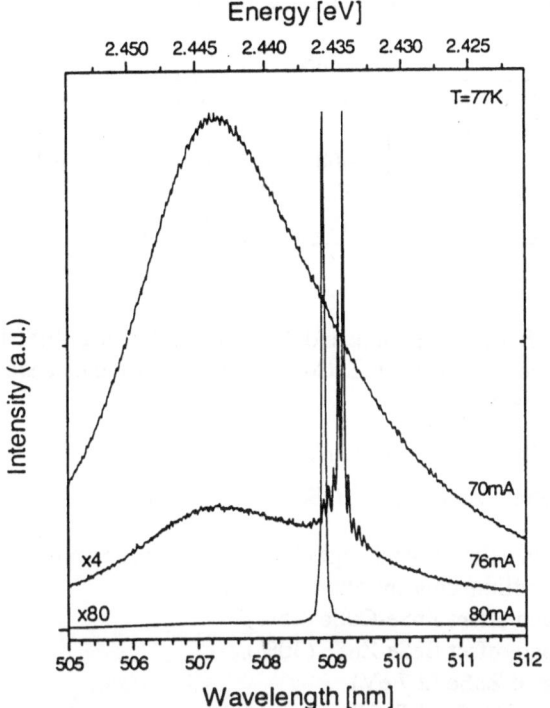

Figure 13
Emission of a laser diode.

One approach for increasing the Be concentration in the quantum wells is to increase both Be and Cd concentration of the BeZnCdSe, keeping the lattice parameter constant. The novel quaternary compound BeZnCdSe as the active region should therefore be considered in some more detail. It turns out that the amount of Be which can be introduced in this way is relatively limited. First of all, the band gap of the quaternary BeZnCdSe increases much faster with Be as it decreases with Cd concentration, resulting in a larger band gap of the BeZnCdSe as compared to the ZnCdSe compound (provided the lattice parameter is kept constant).

In addition, the band offset between the BeZnCdSe and the electrical confinement layer of a laser diode has to be considered. The change of the valence band edge of BeZnCdSe as a function of Be and Cd concentration has been estimated from the ZnCdSe/ZnSe and BeZnSe/ZnSe valence band offsets. The latter could experimentally be assessed using BeTe-BeMgZnSe quantum well structures. The type II band alignment allows a direct determination of the valence band offset via PL, as described above for BeTe-ZnSe structures. A value of approximately 30% for the valence band offset could be obtained for the BeMgZnSe-ZnSe heterostructure, though the data are still preliminary. The valence band offset for the ZnCdSe-ZnSe system has been reported in the literature by various groups, the results varying between 10% and 30% [34,35].

Table 1 Parameters used for the calculation of BeTe/ZnSe superlattices.

	ZnSe		BeTe	
Direct (indirect) bandgap @ @ RT (eV)	2.67		4.1 (2.8–2.9)	[4]
Valence band offset to ZnSe (eV)	–.–		0.9	
lattice constant a_0 (Å)	5.668		5.622	[4]
Effective electron mass m_e^*/m_0	0.17	[8]	0.17	
Effective hole mass m_h^*/m_0	0.65	[8]	0.65	

Assuming a value of 30% for the BeMgZnSe-ZnSe and 25% for the ZnCdSe-ZnSe heterostructure, the band offset for the quaternary BeZnCdSe can be estimated by a simple linear interpolation in the following way:

$$\Delta E_{VB} = x \cdot VBO(ZnSe - CdSe) \cdot (E_G(ZnSe) - E_G(CdSe)) - $$
$$y \cdot VBO(BeSe - ZnSe) \cdot (E_G(BeSe) - E_G(ZnSe)) \tag{7.1}$$

ΔE_{VB}	difference in valence band edges between BeZnCdSe and ZnSe
x,y	Cd and Beryllium concentration
VBO(ZnSe-CdSe)	valence band offset ZnSe-CdSe (25%)
VBO(BeSe-ZnSe)	valence band offset BeSe-ZnSe (30%)
$E_G(ZnSe)$	energy gap of ZnSe (2.7 eV)
$E_G(CdSe)$	energy gap of CdSe (1.7 eV)
$E_G(BeSe)$	energy gap of BeSe (5.6 eV)

This implies that no bowing occurs. Results of this estimation are shown in Fig. 14, where the position of the valence band of ZnSe is the reference energy. It turns out that – starting with a ZnCdSe active region with 30% Cd – the Be concentration can only be increased to 8% before the valence band of the BeZnCdSe is shifted below the one of ZnSe, resulting in a type II band alignment between ZnSe and the BeZnCdSe quantum well. This means that the band gap of the waveguide has to become considerably larger than the one of ZnSe, if Be concentrations above 8% in the quantum well in conjunction with a type I confinement should be obtained.

Another approach to increase the Be concentration in the quantum well is to use BeTe-ZnSe superlattices. If the period is small enough, minibands are formed allowing an efficient vertical transport and recombination despite the type II band alignment between BeTe and ZnSe. In Fig. 15 the miniband gap for a BeTe-ZnSe superlattice as a function of superlattice period is shown. For the calculations, parameters given in Table 1 have been used.

At a BeTe-ZnSe interface, two types of interface configurations can be expected: A) Te-Be-Se-Zn, or B) Be-Te-Zn-Se (or, of course, a mixture of both). In case A the intermediate layer consists of BeSe with a lattice mismatch of −10%, in case B the intermediate layer is ZnTe, with a lattice mismatch of +7%. Nevertheless, BeTe-ZnSe superlattices can be fabricated with a high structural quality, as is demonstrated in Fig. 16,

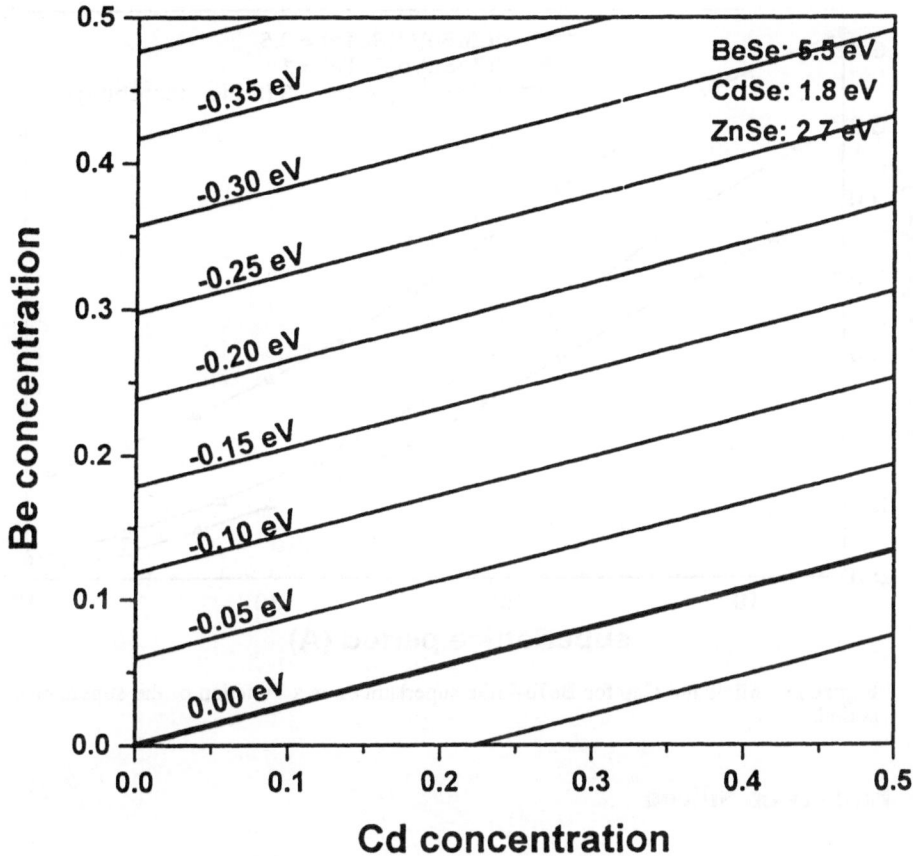

Figure 14 Estimated BeZnCdSe valence band position as a function of Be and Cd content. Values given are in eV relative to the ZnSe valence band.

where a (004) rocking curve for such a BeTe-ZnSe superlattice with 10 nm BeTe and 20 nm ZnSe is plotted.

This indicates that BeTe-ZnSe superlattices could indeed be used as active layer material for blue green laser diodes. The different interface configurations (A or B) give rise to a considerable interface strain, which influences the mean lattice constant of the superlattice. This has recently been demonstrated by detailed TEM investigations [36]. Up to now, no results on laser diodes using this approach have been reported.

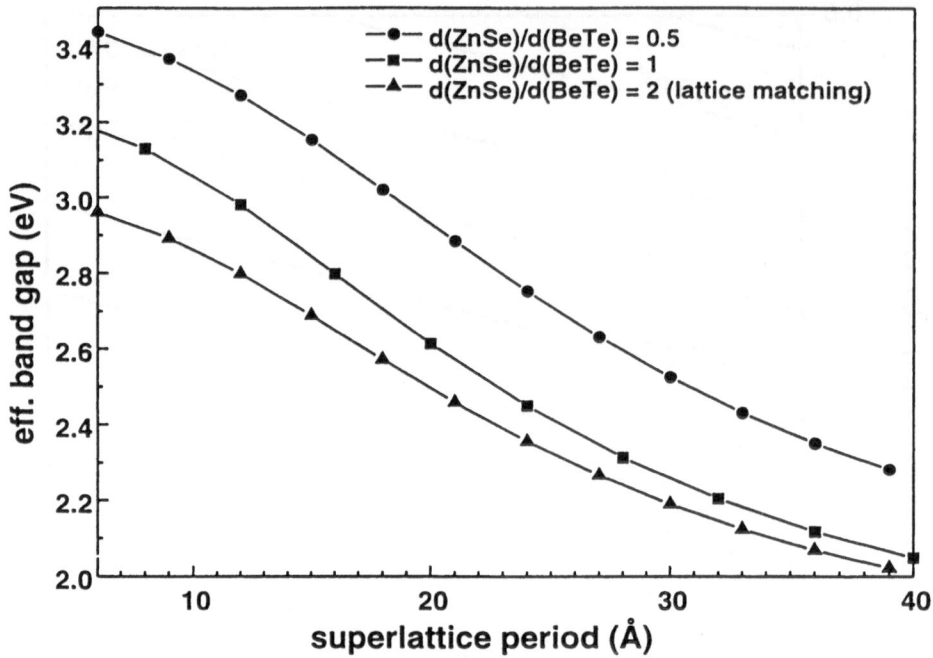

Figure 15 Miniband-Gap for BeTe-ZnSe superlattices as a Function of the superlattice period.

8   Epitaxy on Silicon

Another interesting aspect of Be compounds has not yet been discussed: the possibility to grow lattice matched thin films onto silicon substrates. The ternary BeZnSe is lattice matched to silicon with approximately 40% Be. Based on that, the band gap can be increased by using the quaternary BeMgZnSe. Even if BeSe is an indirect semiconductor, the ternary BeZnSe is expected to be still direct at the Be concentrations necessary for a lattice matching to silicon.

A lattice matched direct band gap semiconductor like BeMgZnSe on silicon would give way to fabricate optoelectronic devices on silicon, with a possible integration of II-VI optoelectronic devices and silicon electronics, e.g. for UV solar blind detectors. Due to the high Be concentration, a pronounced increase in II-VI lattice rigidity can be expected. The lattice matching should result in high quality layers with low defect densities and therefore e.g. an improved modulation frequency of detectors in comparison to GaN devices [37].

The first attempt to grow Be compounds onto silicon has already been reported in [38]. BeTe has been grown onto silicon. It has been pointed out that the quality of the epitaxy of BeTe on silicon is high as compared to other compound semiconductors with similarly large lattice mismatch, and even in comparison to ZnS with a lattice mismatch

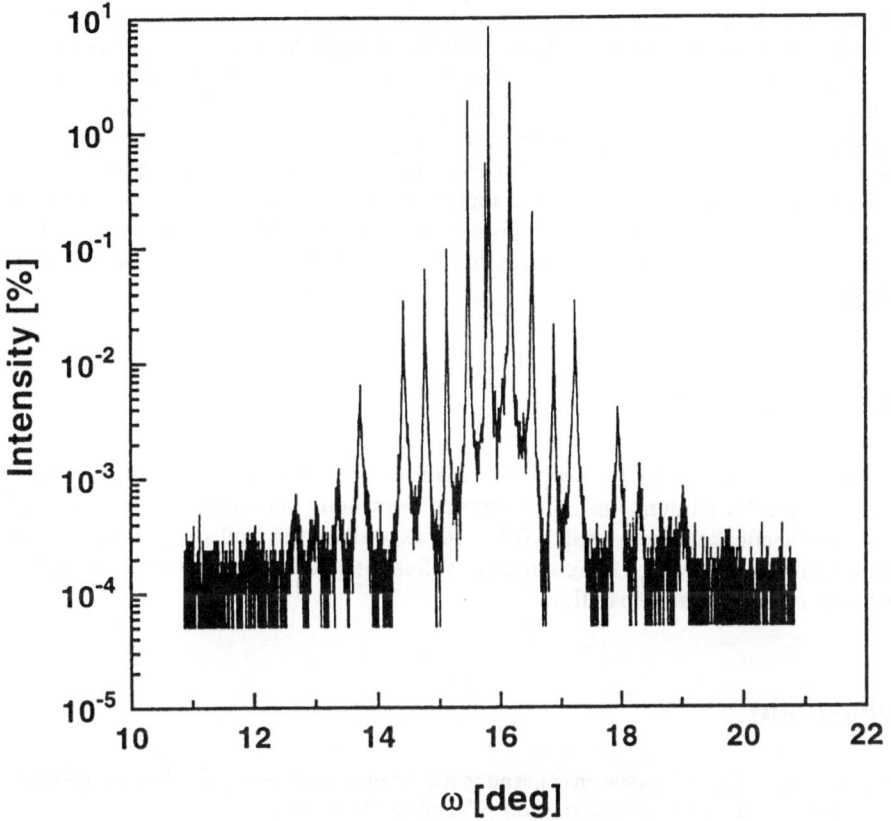

Figure 16 XRD spectrum of a 10 nm BeTe/20 nm ZnSe superlattice.

as small as 0.4% to silicon. This behavior can be explained by the large energy of formation of extended defects in BeTe (and also BeSe and BeS) in comparison to other II-VI and III-V semiconductors. This can lead to a localization of defects at the interface between BeTe and silicon, and only relatively few extended defects can penetrate into the BeTe (for details see [38]). It is unclear, however, whether device quality material can be grown under such highly lattice mismatched conditions.

9   Summary

A review on the potential of beryllium chalcogenides for the design of blue green laser diodes and other quantum well structures has been given. In addition to the device aspects, beryllium compounds provide substantial flexibility for the band gap engineering. This is due to the close lattice matching of BeTe, BeMgZnSe and GaAs, in a combination of both a type I and a type II band offset situation, as well as the potential combination with semimagnetic manganese containing compounds like BeZnMnSe. In order to bring the covalency of Be-compounds into play for an improvement of device performance it will be necessary to increase the Be concentration of the active region of ZnSe based laser diodes. Two of the possible approaches have been discussed here. Some of the basic properties of beryllium compounds are still speculative or totally unknown. This is expected to change in the future due to the increasing world wide activities in the field.

Acknowledgements

We gratefully acknowledge the support of the Bayrische Forschungsstiftung, the Bundesministerium für Bildung und Wissenschaft, Forschung und Technologie, Bonn, the Deutsche Forschungsgemeinschaft (SFB 410) as well as the Volkswagenstiftung. We want to thank also P. Wolf, Th. Schumann, A. Schönteich and R. Brauner for technical assistance and substrate preparation.

Bibliography

[1] For a comprehensive review on conventional II-VI matierials see e.g. II-VI semiconductors, R. L. Gunshor and A. Nurmikko (Ed.), Willardson Beer 1997.

[2] S. Taniguchi, T. Hino, S. Itoh, K. Nakano, N. Nakayama, A. Ishibashi, M. Ikeda, Electronic Letters 32 (1996) 552.

[3] M. A. Haase, J. Qui, J. M. dePuydt, H. Cheng, Appl. Phys. Lett. 59 (1991) 1272.

[4] R. M. Park, M. B. Troffer, C. M. Rouleau, J. M. dePuydt, M. A. Haase, Appl. Phys. Lett. 57 (1990) 2127.

[5] K. Ohkawa, T. Karasawa, T. Mitsuyu, Jpn. J. Appl. Phys. 30 (1991) L152.

[6] H.-J. Lugauer, Th. Litz, F. Fischer, A. Waag, T. Gerhard, U. Zehnder, W. Ossau, G. Landwehr Int. Conf. on Molecular Beam Epitaxy, Proceedings to be published in Journal of Crystal Growth, E. Tournier et al, European Materials Research Society Meeting, Strasbourg, EMRS 1996, Proceedings

[7] H. Okuyama, T. Miyajima, Y. Morinaga, F. Hiei, M. Ozawa, K. Akimoto, Elec. Lett. 28 (1992) 1798.

[8] Y. Fan, J. Han, L. He, J. Saraie, R. L. Gunshor, M. Hagerott, H. Jeon, V. Nurmikko, G. C. Hua, N. Otsuka, Appl. Phys. Lett. 61 (1992) 3160; A. Ishibashi, Y. Mori, J. Cryst. Growth 138 (1994) 677.

[9] J. Petruzzello, T. Marshall, S. Herko, K. W. Haberern, M. Buijs, K. K. Law, T. J. Miller, G. M. Haugen, Int. Symp. on Blue Laser and Light Emitting Diodes, Chiba, Japan 1996, Proceedings, p. 230

[10] K. W. Haberern et al., SPIE, Physics and Simulation of Optoelectronic Devices, San Jose 1997, proceedings.

[11] C. Verie, in Semiconductor Heteroepitaxy, B. Gil and R. -L. Aulombard (Ed.), World Scientific (1995) p. 73

[12] W. Harrison, in Electronic Structure and the Properties of Solids, Dover Publications, p. 176.

[13] M. Stutzmann, priv. comm. as has been pointed out in earlier reports, a band gap of 5.0 eV was the lower estimate due to the limitation of the equipment used then. .

[14] M. Ringle, D. C. Grillo, J. Han, R. L. Gunshor, G. C. Hua, A. V. Nurmikko, Inst. Phys. Conf. Series 141 (1995) 513 IOP Publishing

[15] A. Waag, F. Fischer, H.-J. Lugauer, T. Litz, T. Gerhardt, J. Nürnberger, U. Lunz, U. Zehnder, W. Ossau, G. Landwehr, EMRS spring meeting 1996, Strasbourg (invited), Symposium C, to be published.

[16] B. Roos, C. J. Santana, C. R. Abernathy, K. S. Jones, Thin Films: Stresses and Mechanical Properties, ed. by W. W. Gerberich et al. Pittsburgh, PA, MRS Symp. Proc. 436 (1996)

[17] J. Han, R. -L. Gunshor, A. Nurmikko, phys. stat. sol. (b) 187 (1995) 285.

[18] R. G. Dandrea, C. B. Duke, Appl. Phys. Lett. 64 (1994) 2145.

[19] Th. Behr, T. Litz, A. Waag, G. Landwehr, J. Cryst. Growth 156 (1995) 206.

[20] Th. Baron, thesis, University of Grenoble 1996.

[21] P. M. Mensz, Appl. Phys. Lett. 64 (1994) 2148.

[22] M. Nagelstraßer, H. Dröge, H.-P. Steinrück, F. Fischer, T. Litz, A. Waag, G. Landwehr BESSY Jahresbericht 1996.

[23] A. Franciosi, L. Vanzetti, L. Sorba, A. Bonnani, R. Cingolani, M. Lomascolo, D. Greco, Mater. Science Forum vol. 182-184 (1995) 17.

[24] W. Faschinger, S. Ferreira, H. Sitter, R. Krump, G. Brunthaler, Materials Science Forum Vols. 182-184 (1995) 29.

[25] R. Krump, S. O. Ferreira, W. Faschinger, G. Brunthaler, H. Sitter, Mat. Science Forum 182-184 (1995) 349.

[26] W. Spahn, H. R. Reß, C. Fischer, R. Ebel, W. Faschinger, M. Ehinger, G. Landwehr, Physics and Simulation of Optoelectronic Devices IV, SPIE 1996.

[27] J. Geurts et al., Int. Conf. on the Physics of Semiconductors, Berlin 1996

[28] L. L. Chang, L. Esaki, R. Tsu, Appl. Phys. Lett. 24 (1974) 593.

[29] G. Reuscher, M. Keim, F. Fischer, A. Waag, G. Landwehr, Phys. Rev. B53 (1996) 16414.

[30] U. Lunz, M. Keim, G. Reuscher, F. Fischer, K. Schüll, A. Waag, G. Landwehr, J. Appl. Physics 80 (1996) 6329.

[31] S. Guha, J. M. dePuydt, M. A. Haase, J. Qiu, H. Cheng, Appl. Phys. Lett. 63 (1993) 3107.

[32] A. Ishibashi, M. Ukito, S. Tomiya, Int. Conf. Physics of Semiconductors, Berlin 1996, Proceedings

[33] M. A. Haase et al (3M), announced at the SPIE meeting San Jose 1997.

[34] V. Pelligrini, R. Atanasov, A. Tredicucci, F. Beltram, C, Amzulini, L. Sorba, L. Vanzetti, A. Franciosi, Phys. Rev. B 51 (1995) 5171

[35] J. Ding, N. Pelekanos, A. V. Nurmikko, H. Luo, N. Samarth, J. K. Furdyna, Appl. Phys. Lett. 57, 2885 (1990) .

[36] Th. Walter, A. Rosenauer, D. Gerhtsen, F. Fischer, R. Gall, Th. Litz, A. Waag, G. Landwehr, Microscopy of Semiconducting Materials, Oxford 1997, Proceedings.

[37] B. Goldenberg, J. D. Zook, R. J. Ulmer, Topical workshop on III-V nitrides, Nagoya, Japan 1995, proceedings.

[38] X. Zhou, Sh. Jiang, W. P. Kirk, 9th Int. Conf. MBE, Malibu 1996, proceedings.

White Light Emitting Diodes

J. Baur, P. Schlotter and J. Schneider

Fraunhofer-Institut für Angewandte Festkörperphysik,
Tullastr. 72, D-79108 Freiburg

Abstract: Using blue-emitting GaN LEDs on SiC substrate chips as primary light sources, we have fabricated green, yellow, red and white light emitting diodes (LUCOLEDs). The generation of mixed colors, as turquoise and magenta, is also demonstrated. The underlying physical principle is that of luminescence down-conversion (Stokes shift), as typical for organic dye molecules and many inorganic phosphors. For white light generation via the LUCOLED principle, the phosphor $Y_3Al_5O_{12}:Ce^{3+}(4f^1)$ is ideally suited. The optical characteristics of $Ce^{3+}(4f^1)$ in $Y_3Al_5O_{12}$ (YAG) are discussed in detail. Possibilities to "tune" the white color by various substitutions in the garnet lattice are shortly outlined.

1 Introduction

Different blue light emitting diodes (LEDs) based on gallium nitride (GaN) are now commercially available: The Nichia [1] SQW diode, grown on sapphire (α-Al_2O_3) substrate, available with different wavelengths between $\lambda_{max} = 450$ nm and $\lambda_{max} = 520$ nm, and the Cree [2] "blue chip", grown on silicon carbide (6H-SiC) substrate, $\lambda_{max} = 430$ nm. These primary light sources can be used as efficient pumps to excite organic and inorganic luminescent materials for subsequent photon emission at lower energies. The name LUCOLED (Luminescence Conversion LED) has been coined for such a device [3].

If luminescent organic dyes are embedded in organic matrices, as epoxy resin or polymethylmetacrylate (PMMA) the luminescent species is, in general, incorporated in molecular form. The spectral shift of the luminescence toward longer wavelength is a manifestation of the well known Stokes-shift (found in 1853). As a typical example Fig. 1 shows the absorption and emission spectrum of a commercial green-emitting perylene-based dye embedded in epoxy resin. Also shown in Fig. 1 is the electroluminescence spectrum of a Cree LED, used as a pump for our LUCOLEDs. A strong spectral overlap of the blue LED-emission with the absorption spectrum of the green-emitting dye converter is apparent. This enables the fabrication of green-emitting LUCOLEDs.

A slightly different situation arises, if inorganic phosphors are used for LUCOLEDs. Here, the luminescent powders will, in general, not dissolve in the organic matrix and

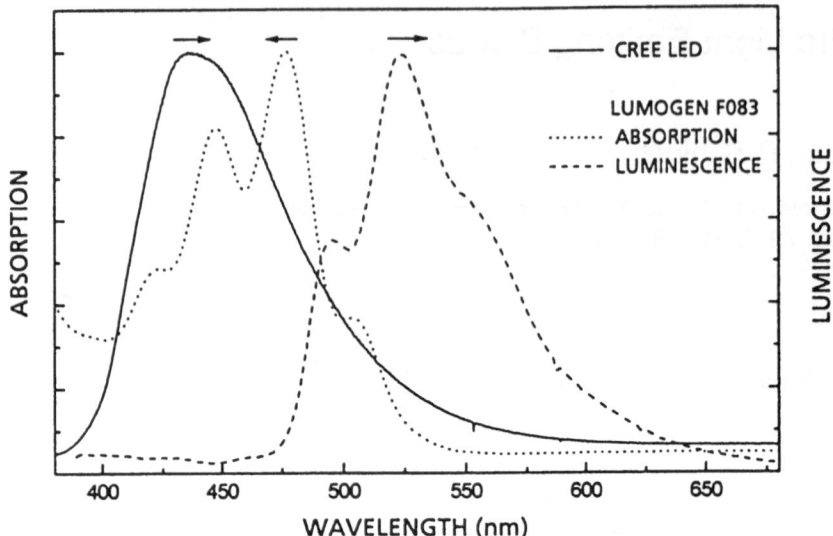

Figure 1 Absorption and photoluminescence spectra of a perylene-based dye (BASF, Lumogen F083) dissolved in epoxy resin, contrasted to the emission spectrum of the dye exciting blue Cree LED.

the luminescent properties of the composite are essentially those of the luminescent rare-earth or transition metal ion in the respective inorganic single crystal host. The situation encountered here is basically the same as found in fluorescent tubes, where the efficient 254 nm line of mercury is used to excite visible luminescence in various inorganic phosphor powders cladded on the inner side of the tube.

2 LUCOLED Fabrication

The cross-sectional view of a LUCOLED is shown in Fig. 2. Here, we used a blue emitting Cree LED-chip (GaN/SiC) as the primary light source. It is mounted in a reflector cup and embedded in an epoxy resin matrix by standard LED technology. Since the 6H-SiC substrate is semiconducting, in contrast to sapphire, only one bond-wire is required in this case.

The luminescence converting material, either organic or inorganic, is mixed in tiny quantity ($\approx \mu$g) to the epoxy resin immediately on top of the semiconductor chip. Alternatively, also the entire plastic dome can be filled with highly diluted dyes or phosphors. This, of course, results in a more diffuse emission characteristic of the LUCOLED.

Typical examples for green and reddish emitting LUCOLEDs, using perylene-based dye molecules as converter, are shown in Fig. 3. White emitting diodes can also be realized, if a green emitting and red-emitting dye are simultaneously added, see Fig. 3.

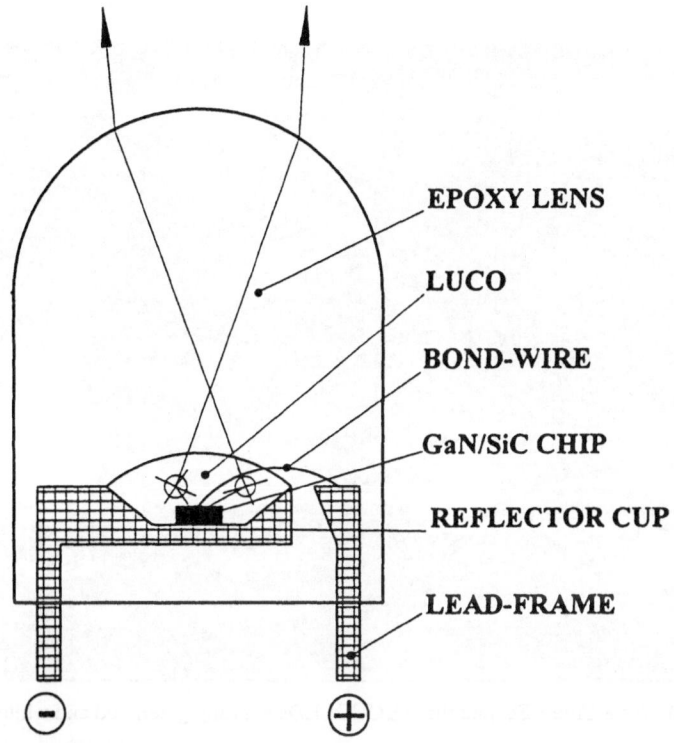

Figure 2 Schematic structure of a GaN based luminescence conversion light emitting diode (LUCOLED).

The emission spectrum of such a white LUCOLED now exhibits three spectral peaks, as seen in Fig. 4: The blue-peak (430 nm) is that of the primary blue Cree LED pump, whereas the green-peak (498 nm) and the red-peak (613 nm) arise from the two different dye species. The corresponding CIE (Commission Internationale de l'Eclairage) chromaticity coordinates [4] are close to the equal energy point $x = y = 1/3$, characteristic for white light.

Incomplete absorption of the primary light by the luminescence converter, also allows generation of mixed bluish-green (i.e. turquoise) colors by additive color mixing. Additive color mixing of the primary LED-blue with secondary photoluminescence red results in magenta, a color not accessible by a single conventional pn-junction LED.

In the same way, LUCOLEDs using inorganic phosphors as converter can be realized. For white LUCOLEDs the phosphor $Y_3Al_5O_{12}:Ce^{3+}(4f^1)$, in short: YAG:Ce, is ideally suited, since it emits yellow light under blue photo-excitation. Consequently, only one converter species is needed for white light generation, since blue and yellow are complementary colors, adding to white light after proper additive mixing. The emission spectrum of a YAG:Ce based LUCOLED is shown in Fig. 5, its visual impression

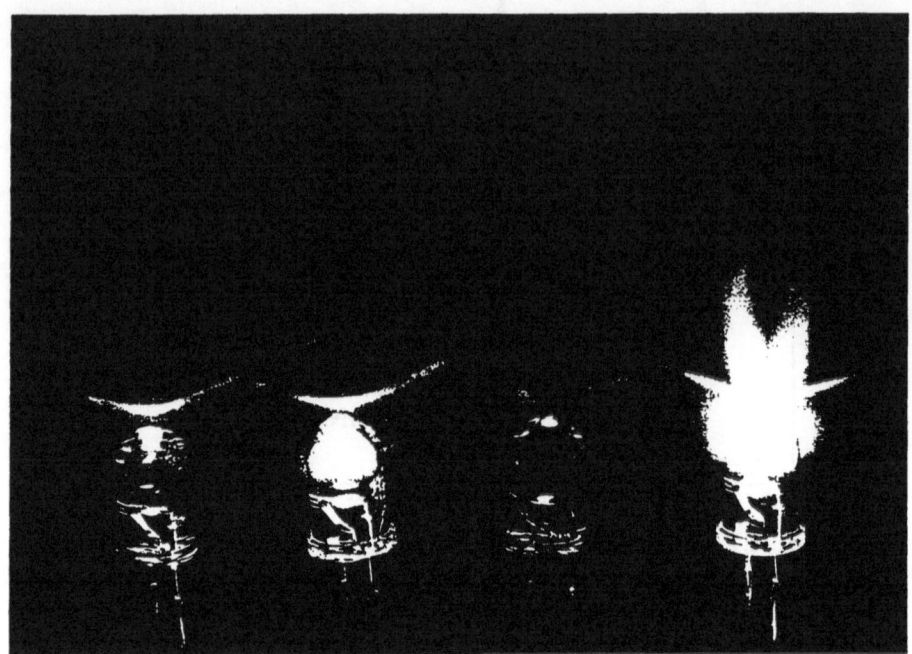

Figure 3 Blue Cree LED and three LUCOLEDs emitting green, red and white light.

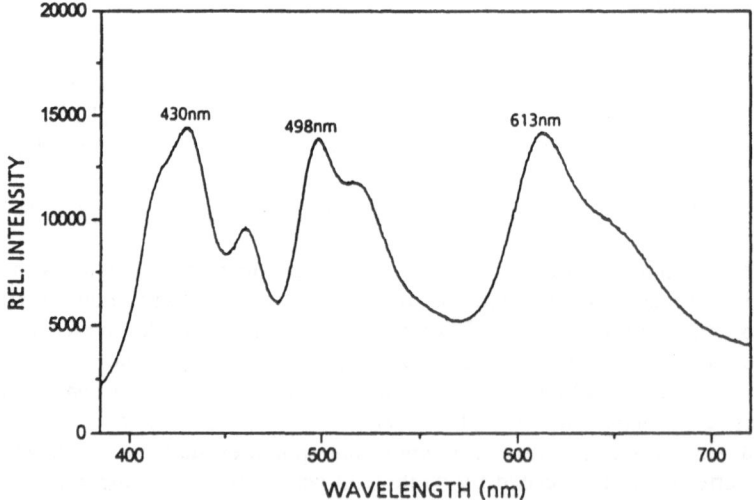

Figure 4 Emission spectrum of a three color LUCOLED.

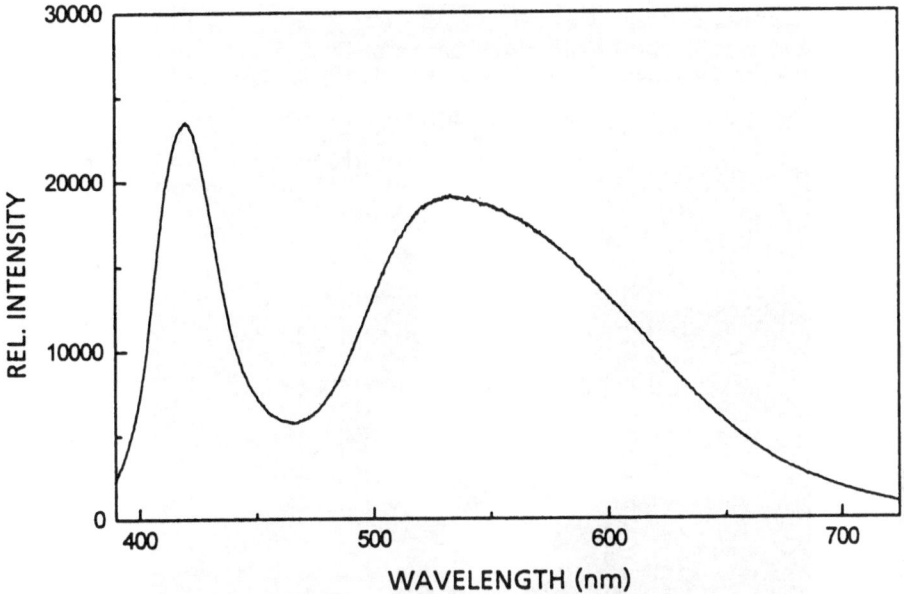

Figure 5 Emission spectrum of a white emitting LUCOLED, using a Cree LED pump and an $Y_3Al_5O_{12}:Ce^{3+}(4f^1)$ converter (Osram L175).

in Fig. 6. We mention that the above results have been shortly summarized recently elsewhere [3].

3 The YAG:Ce Converter

In view of the great importance of the YAG:Ce converter for inorganic white LU-COLEDs, a thorough understanding of the electronic structure of the cerium-correlated luminescent centers in YAG is compulsory. For this goal, very ancient and very recent experimental data on the $YAG:Ce^{3+}(4f^1)$ phosphor will now be highlighted:

Already three decades ago, Blasse and Bril [5] reported, that $Ce^{3+}(4f^1)$-activated $Y_3Al_5O_{12}$ (YAG) stuns by its intense yellow luminescence under blue photo-excitation. Such behavior is rather unique within the large family of Ce^{3+}-activated phosphors, where luminescence usually occurs in the UV/blue spectral region.

On the other hand, this unusual feature of the YAG:Ce phosphor is ideally suited for blue → yellow converters in GaN/6H-SiC (Cree)or $Ga_{1-x}In_xN/Al_2O_3$ (Nichia) pumped LUCOLEDs since, as pointed out already, additive color mixing of blue and yellow may result in white light, in its many hues of appearance. The spectral overlap of the electroluminescence spectrum of a GaN/6H-SiC and a $Ga_{1-x}In_xN/Al_2O_3$ blue LED

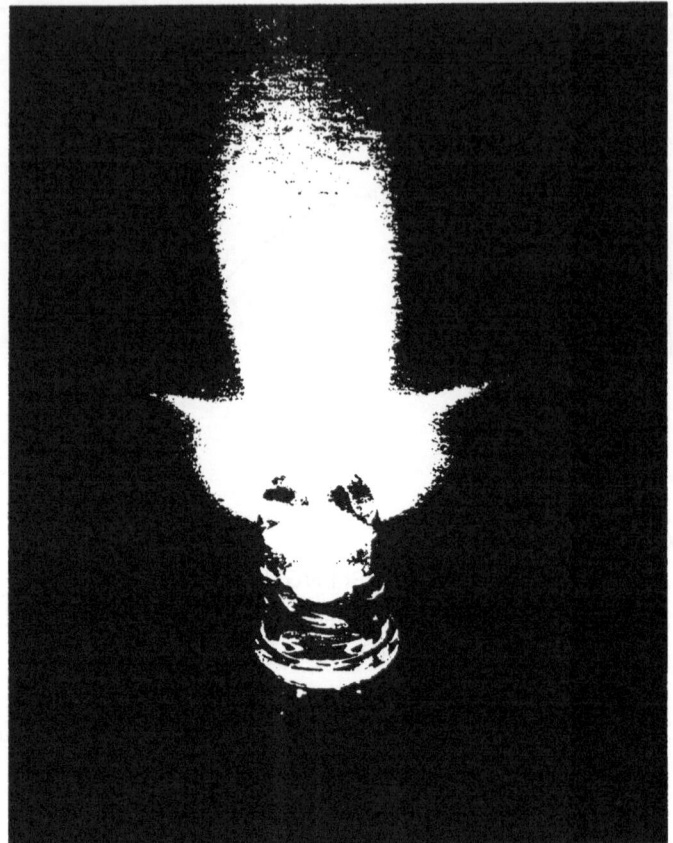

Figure 6 Color photograph of the white emitting YAG:Ce LUCOLED. Its emission spectrum is shown in Fig. 5.

with the absorption spectrum of a Czochralski-grown YAG:Ce^{3+} (1%) crystal, kindly provided by Dr. K. Petermann, is illustrated in Fig. 7.

The bandgap of $Y_3Al_5O_{12}$ is near $50\,000$ cm^{-1} ≈ 6.2 eV, as suggested by ultraviolet absorption [6]. The valence band of $Y_3Al_5O_{12}$ is formed by oxygen orbitals, $O^-(2p^5)$. However, electrons in the conduction band of YAG have multiple atomic choice: (Frenkel) localisation at either an yttrium-site, or at an aluminum site. Physical intuition favors the former. If this is accepted, the conduction band of $Y_3Al_5O_{12}$ is predominately formed by 4d-electrons of $Y^{2+}(4d^1)$, naturally subject to strong crystal-field splitting and electron-phonon interaction.

Cerium, as all other rare-earth elements, is incorporated on the eightfold-oxygen coordinated yttrium sites in the garnet lattice. Here, the point symmetry is only D_2, in the overall cubic host of space group O_h^{10}. Cerium in YAG can exist in (at least) two charge states: $Ce^{3+}(4f^1)$ and $Ce^{4+}(4f^0)$. Consequently, cerium forms a deep (0/+), i.e.··

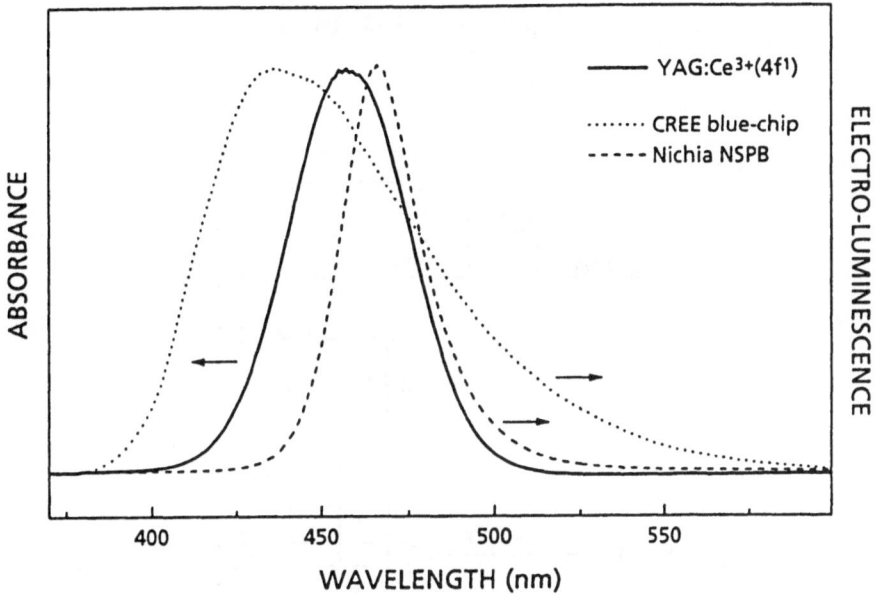

Figure 7 Absorption spectrum of an $Y_3Al_5O_{12}:Ce^{3+}(4f^1)$ single crystal, contrasted to the emission spectra a blue Cree and Nichia LEDs.

$Ce^{3+}(4f^1)/Ce^{4+}(4f^0)$, donor state. Its energy level is located at $E_V + 2.4$ eV, as evidenced by an onset of n-type photo-conductivity [7] in YAG:Ce. We should mention, that Owen et al. [8] have also discussed the possible occurrence of *di*valent cerium in YAG; if this is true, cerium in YAG would also form an $(-/0)$, i.e. $Ce^{2+}(4f^2)/Ce^{3+}(4f^1)$, acceptor state.

The free Ce^{3+}-ion has the electronic configuration $4f^1 5s^2 5p^6$. Its ^2F-state is split by spin-orbit interaction into a $^2F_{7/2}$ excited and a $^2F_{5/2}$ ground state with an energetical separation of 2253 cm^{-1}. At the orthorhombic (D_2) yttrium sites in YAG both the $J = 5/2$ ground state and the $J = 7/2$ excited state are completely split into Kramers doublets. The magnitude of this crystal field splitting is less than the spin-orbit splitting and is of the order of several 100 cm^{-1}, as typical for rare-earth ions in related oxide hosts. However, exact values have not yet been determined for YAG:$Ce^{3+}(4f^1)$, e.g. by far-infrared absorption. One should be aware that such analysis may be impeded by interference of the electronic 4f-levels with the numerous low-energy phonon modes [6] of the YAG lattice.

The first excited electronic configuration of the Ce^{3+}-ion is $5s^2 5p^6 5d^1$. Here, the single 5d-electron is no longer shielded from the environment by the complete $5s^2$ and $5p^6$ shells, as it is the case in the $4f^1 5s^2 5p^6$ ground state. Consequently, strong crystal-field splittings and vibronic coupling are characteristic for this excited ^2D-state. It is also very important for the following, that the optical absorptions $^2F \rightarrow {}^2D$ are electrically

$Y_3Al_5O_{12} : Ce^{3+}(4f^1)$

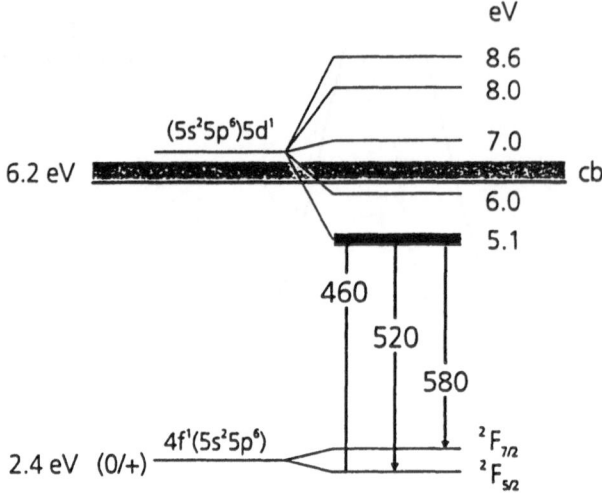

Figure 8 Energy level scheme of $Ce^{3+}(4f^1)$ in $Y_3Al_5O_{12}$ (YAG). In this host the luminescent 5d-level can be excited already by blue (460 nm) photons. Subsequent luminescence
into the $^2F_{5/2}$-ground state and the split-off $^2F_{7/2}$-state occurs under the emission of green
(520 nm) and yellow/orange (580 nm) photons, respectively.

dipole allowed. The optical absorption of YAG:Ce^{3+}, caused by these intra configura-
tional transitions, has been studied by many workers. Transitions into the five Kramers
doublets, expected for a ^2D-state in orthorhombic symmetry, have been observed [9,10].
The resulting energy level scheme of the ^2D-state is illustrated in Fig. 8.

Of particular importance for white LUCOLED applications is the fact that the lowest
Kramers doublet of the excited ^2D-state is accessible already by blue photons. This
seems to be a unique feature of the YAG:Ce^{3+} phosphor, since in almost all other oxidic
hosts this 4f $\rightarrow$ 5d transition occurs in the ultraviolet.

The lowest Kramers level of the ^2D-state is also the radiant level for the characteristic
yellow emission of YAG:Ce^{3+}. The corresponding luminescence and absorption spectra
have been studied in great detail, also at cryogenic temperatures [11]. The 5d $\rightarrow$ 4f
type transitions exhibit a pronounced Stokes-shift, caused by vibronic coupling of the
excited 5d-level. In addition, a further red-shift of the emission is caused by the spin-
orbit splitting (≈ 2200 cm^{-1}) of the ^2F ground state, see Fig. 8.

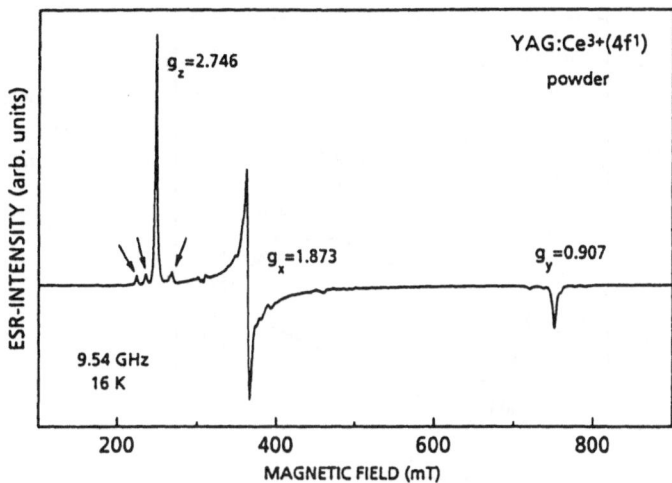

Figure 9 EPR-spectrum of an $Y_3Al_5O_{12}:Ce^{3+}(4f^1)$ powder phosphor (Osram L175). The three sharp signals arise from $Ce^{3+}(4f^1)$ on unperturbed orthorhombic yttrium sites. The weak signals marked by arrows are assigned to defect complexes involving $Ce^{3+}(4f^1)$.

We have found that Electron Paramagnetic Resonance (EPR) is a very direct and sensitive method to determine the concentration of luminescent-active Ce^{3+}-ions in YAG:Ce phosphor powders. EPR of $Ce^{3+}(4f^1)$ in YAG single crystals was observed and analysed already three decades ago [12]. The g-tensor components for the lowest Kramers doublet of the $^2F_{5/2}$-state were determined as $g_z = 2.738$, $g_x = 1.872$ and $g_y = 0.910$, at 9.4 GHz. In Fig. 9 the corresponding powder EPR-spectrum of a commercial YAG:Ce phosphor (Osram L175) is shown. The field positions of the three sharp EPR-signals are seen to be very close to those expected from the single crystal g-tensor values quoted above. In addition, weak satellite lines, marked by arrows in Fig. 9, are detected. Apparently, these arise from $Ce^{3+}(4f^1)$-ions in slightly disturbed lattice sites. The exact microscopic structure of these defects remains to be elucidated.

4 Color Engineering

The CIE chromaticity coordinates [4] of the yellow emitting YAG:Ce phosphor (Osram L175) are $x = 0.462$ and $y = 0.517$, those of the blue emitting GaN/6H-SiC Cree LED were determined as $x = 0.153$ and $y = 0.117$. The positions of the primary blue "pump" and the LUCO-phosphor in the chromaticity diagram are shown in Fig. 10. Several LUCOLEDs with different YAG:Ce content have been fabricated. Their chromaticity coordinates are close to the straight line interconnecting the points of the blue pump and the yellow phosphor, see Fig. 10. In this way, LUCOLEDs with different whitish hues can be realized.

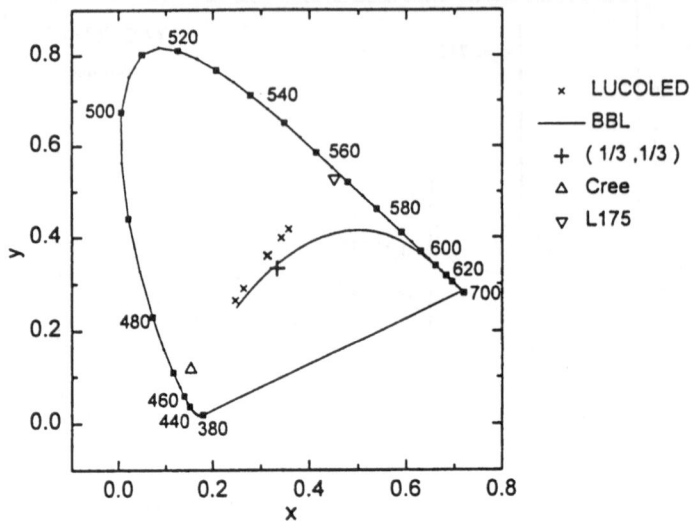

Figure 10 Position of five YAG:Ce based LUCOLEDs in the CIE (1964) chromaticity diagram. Corresponding points for the Cree "blue chip" and the yellow emitting YAG:Ce converter are shown as open triangles. The connecting straight line almost touches the point of equal energy, $x = y = 1/3$. Also plotted is the black body locus (BBL).

A wider range of colors can be realized, if various substitutions in the $Y_3Al_5O_{12}$ garnet host are made, which may change the crystal-field parameters of the Ce^{3+}-ion, thus shifting the spectral position of the yellow Ce^{3+}-emission ("color engineering"). This is most easily achieved [13,14] by replacing part of the Y^{3+}-ions in YAG by rare-earth ions with larger ionic radius, as Gd^{3+} (red-shift), or with smaller ionic radius, as Lu^{3+} (blue-shift). Alternatively, substituting gallium for aluminum in YAG was found to cause a blue-shift of the yellow Ce^{3+}-emission [14]. These old recipes have very recently been exploited by the Nichia group in their successful attempts to tune the color of $InGaN/Al_2O_3$ pumped white emitting LUCOLEDs based on the $(Y_{1-x}Gd_x)_3(Al_{1-y}Ga_y)_5O_{12}:Ce^{3+}$ garnet converter [15].

Color tuning of the YAG:Ce phosphor may also be achieved by co-doping with other fluorescent rare-earth or transition metal ions, as $Eu^{3+}(4f^6)$ and $Cr^{3+}(3d^3)$. In this case energy transfer from the excited Ce^{3+}-levels to the luminescent levels of Eu^{3+} or Cr^{3+} may occur [16], either by exciton diffusion or by photon reabsorption. The total light output of such phosphor will then contain an additional red component, thus shifting the LUCOLEDs emission into the "warm-white" spectral range. Also infrared-emitting LUCOLEDs could be realized in this way by codoping YAG:Ce with infrared-active rare-earth ions, as Nd^{3+} and Er^{3+}.

5 Outlook

The human eye can distinguish several hundred thousand different colors, including the many hues of white light. Most of these visual impressions can now be synthesized via the LUCOLED principle. In contrast, conventional pn-junction LEDs can emit light only in its quasi-monochromatic "rainbow-colors", according to $h\nu \approx E_{gap}$. Of course, multicolor display is also possible by conventional LEDs. However, in this case three different (blue, green, red) LEDs are needed, which also require different operational voltages. Obviously, multicolor display by the LUCOLED principle is more easily achieved.

For white LUCOLEDs wide fields of application can be foreseen and have been found already, e.g. in the car industry for dashboard and indoor illumination. Replacement of the incandescent light bulbs in the white car reversing lights by white LUCOLEDs would enable a full LED solution, since for the other colors required, red and orange, efficient LEDs are already on the market.

The long lifetime ($\approx 100\,000$ h) of a (LUCO)LED, combined with its small size, low electric power consumption and negligible infrared emission are important advantages. Novel concepts for the illumination industry are therefore possible. Thus almost 120 years after Edisons epoch-making invention [17], the incandescent light bulb now faces a serious challenger.

Acknowledgments

We wish to thank the Volkswagen-Stiftung for financial support of this work. Very helpful discussions with K. Petermann on various aspects of garnet luminescence are also gratefully acknowledged. We highly appreciate the important contributions of R. Schmidt and C. Hielscher to the LUCOLED design and fabrication. We wish to thank the Siemens AG for supplying the bonded Cree LED chips mounted in reflector cups.

Bibliography

[1] Nichia Chemical Industries Ltd., Anan, Tokushima 774, Japan

[2] Cree Research Inc., Durham, NC 27713, USA

[3] P. Schlotter, R. Schmidt and J. Schneider, Appl. Phys. A **64**, 417 (1997)

[4] e.g. D. E. Judd, G. Wyszecki, Color in Business, Science and Industry, 3rd ed., John Wiley&Sons,(1975), p.125

[5] G. Blasse and A. Bril, Appl. Phys. Lett. 11, **53** (1967)

[6] G. A. Slack, D. W. Oliver, R. M. Chrenko and S. Roberts, Phys. Rev. **177**, 1308 (1969)

[7]   C. Pedrini, F. Rogemond and D. S. McClure, J. Appl. Phys. **59**, 1196 (1986)

[8]   J. F. Owen, P. B. Dorain and T. Kobayasi, J. Appl. Phys. **52**, 1216 (1981)

[9]   T. Tomiki, H. Akamine, M. Gushiken, Y. Kinjoh, M. Miyazato, N. Togokawa, M. Hiraoka, N. Hirata, Y. Ganaha and T. Futemma, J. Phys. Soc. Japan **60**, 2437 (1991)

[10] S. R. Rotman, phys. stat. sol. (a) **132**, K61 (1992)

[11] D. J. Robbins, J. Electrochem. Soc. **126**, 1550 (1979)

[12] H. R. Lewis, J. Appl. Phys. **37**, 739 (1966)

[13] W. W. Holloway, Jr., and M. Kestigian, J. Opt. Soc. Am. **59**, 60 (1969)

[14] T. Y. Tien, E. F. Gibbons, R. G. DeLosh, P. J. Zacmanidis, D. E. Smith and H. L. Stadler, J. Electrochem. Soc. **120**, 278 (1973)

[15] S. Nakamura and G. Fasol, The Blue Laser Diode, GaN Based Light Emitters and Lasers, Springer-Verlag, Berlin Heidelberg New York (1997), pp. 216-221

[16] G. Blasse and A. Bril, J. Chem. Phys. **47**, 5139 (1967)

[17] Th. A. Edison (1879), US-Patent No. 223898

Acoustic Phonon Spectroscopy with Superconductor Tunnel Junctions of Low-Energy Defect Excitations in Semiconductors

K. Laßmann

Universität Stuttgart, 1. Physikalisches Institut,
Pfaffenwaldring 57, D-70550 Stuttgart, Germany

Abstract: Superconductor tunnel junctions as tunable sources of quasimonochromatic acoustic phonons and as broadband phonon detectors facilitate a straightforward and sensitive spectroscopic technique of high resolution appropriate for the investigation of defect excitations in semiconductors virtually in the entire acoustic band.

1 Introduction

The introduction of superconducting tunnel junctions as bias tunable phonon generators and as phonon detectors with a threshold energy corresponding to the superconductor gap [1] has opened up the possibility of acoustic phonon spectroscopy of defects in solids [2,3] ranging from some tenths of meV up to Debye energies with high sensitivity, good resolution, and convenient tunability [4,5]. It is the acoustic analogue to submillimeter and FIR spectroscopy in this energy range and turned out to be a good or even superior alternative to these techniques in many instances.

Transition or resonance energies in this range may belong to various types of defect states in semiconductors:

- Electronic levels in ground state multiplets of shallow acceptors or donors reflecting the (lifting of the) multiplicity of the neighbouring bands and the local symmetry of the impurity.

- Closely spaced multiplets of transition metal impurities.

- Stress-induced splittings of (nearly) degenerate states.

- (Pseudo-)Jahn-Teller effects connected with the (near) degeneracy of electronic states.

- Transition energies between bound and continuum states of carriers.

- Motional states of off-center defects with low lying tunnelling barriers or rotational potential ditches.

- Acoustic resonances of nanoscopic precipitates or superlattices of dimensions corresponding to the small wavelength of the high frequency phonons.

2   Some Basics of Phonon Spectroscopy

A small variety of superconductors such as Al, Sn, PbBi, Nb, Mo has been used so far for tunnel junctions appropriate as phonon generators and detectors. The choice is mainly determined by the possibility to produce an extremely thin non-porous isolating (oxide) layer as a tunnelling barrier between the two overlapping films evaporated onto the substrate in question. Substrate temperature during junction preparation may be room temperature or below, so, thermally induced changes of the substrate properties are negligible in general.

A (symmetric) junction at bias U may serve either as a phonon detector or as a phonon generator if the bias is such that eU is either smaller or greater than the superconductor gap 2Δ: At small bias the tunnelling current is a measure of the number of (thermally) excited quasiparticles and thus will increase when phonons of energy larger than the detector gap $2\Delta_d$ will be absorbed in the junction by Cooper-pair breaking. At large bias the tunnelling is mainly via pair breaking with the creation of quasiparticles with kinetic energies up to $eU - 2\Delta_g$. These excited carriers by relaxation and recombination emit a continuous phonon spectrum which as the most distinct bias tunable feature has a sharp step at $eU - 2\Delta_g$ as an upper limit for relaxation. By modulation techniques the emission band is 'differentiated' resulting mainly in a narrow quasi-monochromatic line corresponding to the upper limit.

A typical generator-detector characteristic obtained this way is shown in Fig. 1 for the case of phonon transmission from an Al-generator junction to a Sn-detector junction through Si:O$_i$. Any rapid signal variation such as the steep increase at the detector threshold $2\Delta_d$ or the narrow depression at 3.63 meV due to a scattering resonance of the interstitial oxygen are associated with the quasimonochromatic line $\hbar\Omega$ at $eU - 2\Delta_g$ which is therefore taken as the reference on the energy scale. The spectral resolution by this line is limited de facto by the effective gap anisotropy of the junction films. In the case of Al-junctions the ultimate resolution is as small as 10 μeV or 2.4 GHz. (Even better resolution of about 50 MHz at 250 GHz has been obtained under special conditions by taking profit of a one to one frequency conversion of Josephson photons to phonons [6].)

Another, much smaller distinct feature of the emission spectrum is the so-called thermal precursor around eU (i.e. at $\hbar\Omega + 2\Delta_g$). It results from direct recombination of an injected with a thermally excited quasiparticle as well as from the tunnelling and subsequent relaxation of quasiparticles thermally excited above the gap. In the spectra it appears as a hump $2\Delta_g$ below the detector threshold of the main line and, less distinctly,

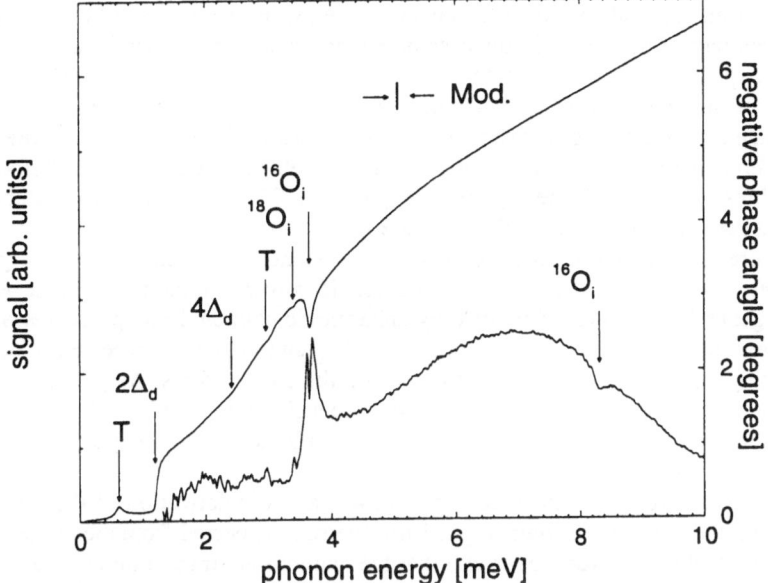

Figure 1 Phonon transmission spectrum through Si:O_i. The quasimonochromatic line at $eU - 2\Delta$ of the Al-generator is taken as a reference of the energy scale. It induces the steep signal rise at 1.2 meV, which is the Sn-detector threshold, whereas the small structure at 0.6 meV is due to a small spectral precursor emitted by the Al-generator at eU. The phonon scattering by an O_i-related resonance at 3.63 meV (see next paragraph) shows up as a dip in the component in-phase with the emission modulation (100 kHz) and the concomitant time delay as a rotation of the phase in the quadrature signal. The weak scattering by the next higher oxygen resonance at 8.3 meV is seen here only as a phase shift in the quadrature signal. The broad-band variation of the phase may result in part from isotope scattering and also from scattering at oxygen-related agglomerates since this sample has been annealed for 70 hours at 750 °C after a short 2 h/1 100 °C preanneal. (F. Zeller, Stuttgart, private communication.)

below other sharp features of the spectra. It is marked by 'T' in Fig.1. An additional continuous component of the emission spectrum below $\hbar\Omega$ may lead to a positive slope of the signal beyond the detector threshold as for the Al → Sn generator-detector set-up of Fig.1. With increasing nominal phonon energy $\hbar\Omega$ there is also an increasing signal contribution from lower energy phonons with energies above the detector threshold originating from inelastically scattered, down-converted quasimonochromatic phonons. For certain parameters of the detector junction there is an increased sensitivity for phonon energies above $4\Delta_d$ giving rise to additional quasiparticle excitation from relaxation phonons and showing up in the spectra as an increase of the slope of the signal. The effective emission and detection characteristics of the junctions are modified to some extent by the quality of the junction/substrate interface. Details are given in the review article by Eisenmenger [4], and, especially for Sn, Pb, and PbBi emitter junctions by

Berberich et al. [7], and for Al-junctions by Welte [8], Forkel [9], and Scheitler [10]. The delayed part of the signal in the case of elastic scattering will appear as a component in the quadrature signal as seen in Fig. 1 at 3.63 meV and at 8.3 meV, a higher resonance of the interstitial oxygen. (In the case of time-resolved measurements a time delay from elastic scattering will show up as a slow decay of the pulse signal.) Inelastically scattered phonons will not contribute to a time delay if down-converted below the detector threshold or the scattering frequency regime. This allows to some extent the distinction between elastic and inelastic scattering.

The theoretical upper limit for monochromatic phonon emission is of course given by the Debye frequency of the superconductor. However, the monochromatic emission is in general limited to a few meV by anharmonic phonon decay processes or the down-conversion of high frequency phonons to $2\Delta_g$ and below by (repeated) Cooper-pair breaking and quasiparticle relaxation within the emitter films. As yet only Al as a weak coupling superconductor with Debye frequencies comparable to that of silicon has been shown to be useful for spectroscopy up to at least 14 meV [11] if the films are sufficiently thin (about 20 nm).

Time-and space-resolved measurements show that the polarization of the phonons emitted from the junctions is mainly quasi-transverse as expected from the higher density of states of these slower modes and that the phonons are transmitted virtually into the complete half-space modified, however, by bunching or spreading of phonon k-vectors in certain crystal directions due to the elastic anisotropy of the crystal (so-called phonon focussing, see e.g. [12]). The precision of selecting a certain phonon k-vector and polarisation is therefore not only determined by the finite sizes of and the distance between generator and detector but also by phonon focussing.

The spectra may be taken either in transmission or reflection to observe scattering as a reduction or increase of the signal, respectively. For the analysis of the observed signal changes the specific experimental conditions have to be taken into account: By the finite sizes of the emitter and detector junctions as compared to the path length there will be a partial balance of in- and outscattering for elastic scattering events whereas the decay products of inelastic scattering may fall in energy below the detector threshold resulting in a stronger signal reduction per scattering in the case of transmission and to reduced backscattering. Comparison of transmission and reflection will thus help to identify the type of scattering. Strong elastic scattering, on the other hand, by phonon trapping may even result in a reduction instead of an increase of the backscattered signal. This is observed e.g. in the centre of the sharp O_i resonance as shown in Fig. 2. In the case of small homogeneous volume scattering and short pulse excitation the maximum of the backscattering signal comes by a factor of 1.3 to about 1.8 after the initial rise corresponding to the direct path length d between generator and detector [13,14]. Thus, for small d the main signal contribution comes from a region close to the generator-detector surface. For thin samples, a transmission-like component is superimposed on the backscattering signal from those phonons that are not transmitted through the opposite surface into the helium bath.

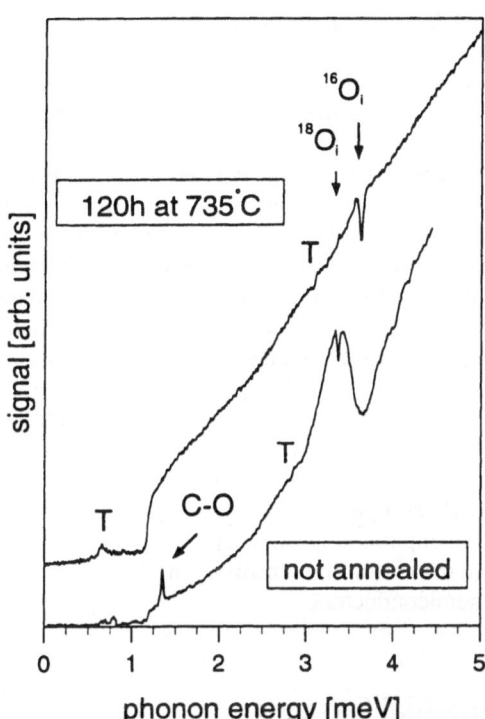

Figure 2
Backscattering spectrum of silicon with high oxygen and high carbon content before and after annealing: In the untreated sample there is a small peak at 330 GHz due to a C-O-complex of unknown microscopic nature. The intrinsic width of this line is below 2.5 GHz i.e. below that of the O_i-related lines. The overall rise of the backscattering with increasing energy beyond the detector threshold is partly due to isotope scattering partly due to the increasing scattering in the low-energy wing of the oxygen resonance. The concentration of O_i is so high that near the centre of the resonance the phonons are completely trapped in the volume without reaching the detector so that there is an inversion of the ^{16}O as well as the ^{18}O lines. After annealing the O_i-concentration and -scattering is much reduced whereas the low-frequency signal near the detector threshold is increased by elastic scattering. (G. Schrag, Diplomarbeit, Stuttgart, 1993)

For the evaluation of scattering cross-sections and linewidths from Monte-Carlo simulations [15,10] analytical approximations have been obtained for elastic scattering [10,16] in the transmission configuration.

In semiconductors phonon detection is possible also by phonon-induced conductivity (PIC) after detachment of shallow bound carriers [17]. In this case the experimental situation is similar to that of photoconductivity. In contrast to FIR-irradiation, however, the phonon density may be strongly inhomogeneous over the sample especially at higher energies because of anharmonic phonon decay and elastic isotope scattering. The position of the conductivity contacts is therefore of relevance for signal form and sensitivity [18].

As compared to the method of low-temperature thermal conductivity measurement – often applied to the investigation of phonon scattering by defects in semiconductors (see e.g. [19]) – phonon spectroscopy with superconducting tunnel junctions has, of course, much better frequency resolution, the possibility to distinguish between the various phonon modes by time-resolved ballistic propagation, and to some extent also between elastic and inelastic scattering. In special cases it may have as a drawback the restriction to relatively low working temperatures of, say, below 4.2 K as given by the superconducting materials appropriate for 'phonon junctions' with the present state of the art. Also, the magnetic field may be used as an experimental parameter only with restrictions [20,21]. An interesting new development, proposed in [4], is the use of specially designed normalconducting Al-junctions for phonon spectroscopy in high magnetic fields [22]. The resolution depends on the working temperature. It is found to be about .5 meV at 1.5 K.

Pulsed heating of a semiconductor surface layer or an evaporated metal film by e.g. current or laser pulses for the investigation of the ensuing ballistic or diffusive transport of broad-band superthermal phonons through the medium may be regarded as a technique intermediate between (cw) thermal conductivity and (cw or pulsed) phonon spectroscopy with superconducting tunnel junctions and is applied in many instances [12,23–28]. Phonon detection in such experiments has been by fast bolometers (superconductor films at the resistive transition, semiconductor layers in the hopping regime), by superconductor junctions, or by optical means. In such experiments high-resolution spectroscopy may not be of primary importance: Phonon emission from hot electrons in bulk or 2D material, phonon-exciton interaction, and also, for the investigation of single particle detection, phonon emission characteristics from impact by energetic particles in semiconductors. (See e.g. [29] for the latter topic.)

In the following, examples will be given of phonon spectroscopy with superconductor junctions demonstrating the merit of high resolution and sensitivity in a broad frequency band for the investigation of defects in semiconductors.

3   Off-Centre Impurities: Interstitial Oxygen in Silicon and Germanium

Oxygen as a point defect in Si and Ge is an electrically inactive interstitial bond-centred slightly off-axis between two nearest neighbour lattice atoms in a $\langle 111 \rangle$-direction. Information on the details of the structure has come from the investigation of the IR bands related to its motional modes and especially from the sequence of lowest-lying states in the range of FIR and acoustic phonons. These latter states depend sensitively on the microscopic parameters such as the height of the axial barrier against radial motion or the angular modulation of the potential ditch due to the two neighbouring tripods as a hindrance for the rotational motion. (See [30,31] for recent reviews.)

3.1   Silicon

In Si a sequence of lowest-lying levels between 3.63 meV and 14.93 meV above the ground state has been first found in FIR absorption [32]. It has been analysed in terms of a two-dimensional harmonic oscillator hindered by a small axial barrier [32,33], or, alternatively, in terms of a non-rigid rotator [32]. More recently, on the basis of cluster calculations [34], a quartic potential was used to calculate the dependence of

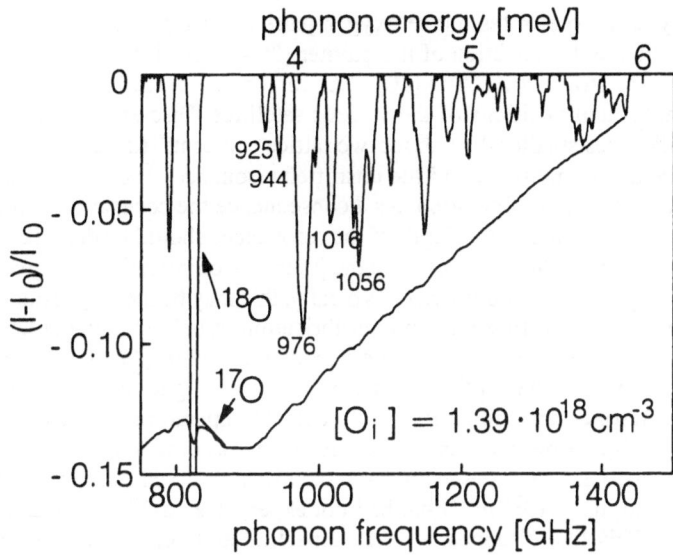

Figure 3 Transmission spectrum around 3.63 meV in silicon with higher content of interstitial oxygen. The main resonance at 3.63 meV is broadened instrumentally and partly overlapping satellites due to resonance shifts of close O_i-O_i pairs can be distinguished. The line at 790 GHz appears only in samples that have not experienced a 1 090 °C anneal after crystal growth by the supplier. It vanishes after a short anneal at 1 100 °C that does not much reduce the other satellites. The upper curve is obtained by subtraction of a base line and magnification (omitting a range around the main resonance). (From E. Dittrich, Thesis, Stuttgart, 1987)

the energy level position for varying height of a parabolic central barrier for the determination [35] especially of the oxygen isotope effect. The analysis [32] leads to a quasi-linear configuration with a distance of only 22 pm of the radial potential minimum from the axis as compared to the 235 pm for the Si-Si distance of the unperturbed lattice. (38 pm/308 pm [34], and 35 pm/311 pm [36] have been obtained from cluster calculations for the radial distance of the static total potential minimum and for the Si-Si distance with the oxygen in between.) The coupling to the lattice of this stretched configuration is weak. This is evident from the rather narrow linewidth of about 15 μeV of the lowest transition at 3.63 meV as obtained with FIR-laser spectroscopy [37], backward-wave oscillator spectroscopy [38], and with phonon spectroscopy [39,16].

Despite the weak coupling the high resolution of phonon spectroscopy allows the detection of concentrations as low as $5 \cdot 10^{19}$ m^{-3} [16]. This high sensitivity has facilitated at higher oxygen concentrations the detection of a series of satellite lines to the 3.63 meV transition as shown in Fig. 3. From the quadratic dependence of their depths on oxygen concentration it has been concluded that it is the mutual strain interaction of the statistically few, more or less close oxygen pairs that shifts the resonance depending on the location and orientation of the partner [39,40]. An isotropic continuum approximation for the strain interaction [40] gives reasonable overall agreement with the observed strengths and positions of the various satellites. Discrete lines are expected only for pair distances smaller than about two lattice constants, for larger distances the lines merge into a continuum around the unshifted resonance. The pair interaction energy depends on the pair configuration. As a consequence the occupancy of the various configurations and thus in part the depth of the resonances should be determined by the temperature range where during cool-down the jump rate of interstitial oxygen becomes too small with respect to the cooling rate. Typically, this may be the case around 700 °C. It appears to be difficult to freeze-in an equilibrization in bulk material at appreciably higher temperatures because the jump rate becomes too high [42,41]. Since the pair interaction energy within the continuum approximation is higher in general in the case of down-shifted resonances a relatively low freeze-in temperature will favour the occupation of pair configurations with up-shifted resonances as observed in the experiment. A line at 790 GHz, i.e. below the main resonance was observed only in samples with oxygen concentrations $> 10^{23}$ m^{-3} that had not experienced a 2 h 1 090 °C annealing by the supplier. This line disappeared after annealing at 1 000 °C for 30 min which would be consistent with the above reasoning. The potential of phonon spectroscopy for the investigation of the oxygen interaction and agglomeration has perhaps not yet been fully exploited as regards annealing (under stress)/quenching procedures. It has e.g. been suggested [43] that a cloud of interstitial oxygen develops during the first hours of annealing around 750 °C before collapsing. This stage should show up in the depth and the distribution of the satellites. It should be noted that the sensitivity of FIR is not sufficient to detect these satellites (effective concentrations of equivalent pairs at most $5 \cdot 10^{20}$ m^{-3}) nor have corresponding satellites been reported for IR transitions.

The high sensitivity of phonon spectroscopy allowed to locate the resonance frequency not only of ^{18}O at 822 GHz (3.40 meV) but also of ^{17}O at 848 GHz (3.51 meV). Quite close values, namely, 823 GHz for ^{18}O and 849 GHz for ^{17}O are obtained within the model of Chen and Schroder [33] with the same set of parameters that fits the series of excited states of ^{16}O. From the fit to a model potential as discussed in [35] 882 GHz and 825 GHz are obtained for ^{16}O and ^{18}O, respectively.

Transitions from the ground state directly to the higher excited states at 8.3 meV and 9.6 meV have also been observed [44]. In contrast, FIR-transitions to these states from the ground state are not allowed by the selection rules. Therefore these states originally were only observed at elevated temperatures by transitions from thermally occupied levels [32].

3.2 Germanium

For Ge:O_i a larger off-axis distance for the radial potential minimum was estimated originally from high-resolution IR-results [45] and later from cluster calculations [36]. As a consequence the lowest-lying levels should be lower in energy in Ge than in Si. This was ascertained and precised by detection with phonon spectroscopy of transitions between 0.18 meV and 4.08 meV, i.e. roughly one order of magnitude below those of interstitial oxygen in silicon [46]. The state at 0.18 meV, which is below the detector threshold of the junctions, showed up indirectly as a doublet partner 0.18 meV below the transitions from the ground state because of its thermal occupation. The sequence could be fitted quite well to the states $l = \pm 1$ to $l = \pm 5$ above the ground state $l = 0$ of a rigid rotator (RR). A small angular hindrance by a 0.28 meV modulation of the potential ditch due to the influence of the six next-nearest neighbours is manifest by a splitting of the $l = \pm 3$ state. The coupling to the lattice is also evidenced by the rather strong splitting of the $l = \pm 1$ state by $\langle 100 \rangle$-and $\langle 110 \rangle$-stress. An improved fit to the experimental data is possible in terms of a non-rigid rotator model taking into account the centrifugal distortion [47] as was similarly done in the case of Si:O_i [32]. From this fit the next rotational level $l = \pm 6$ is expected at ≈ 5.7 meV and the first excited radial level at ≈ 9.2 meV. A correspondingly large central barrier of about 236 meV is reported [47] to result from first-principles total-energy calculations.

Strong isotope scattering in natural Ge makes phonon spectroscopy beyond 4 meV rather difficult. Only in a 300 μm thin sample of natural germanium with high O_i concentration further small resonance dips could be detected at 4.57 meV and 4.74 meV, i.e. again nearly 0.18 meV apart [5]. The corresponding O_i-level at 4.74 meV is distinctly below the value estimated for the $l = \pm 6$ level. Recent experiments with isotopically enriched germanium revealed a distinct further doublet at about 5.8 meV (position depending on Ge-isotope) [48] (Fig. 4). The interpretation of these levels is not yet clear: One might think of the level $l = \pm 6$ being down-shifted and largely split by a 12-fold hindering potential from the tetrahedral neighbours but it is not straightforward to understand the down-shift and the large splitting [49]. Another possibility is given by extending to Ge:O_i the above-mentioned model calculation of Yamada-Kaneta et al. [35]. A good fit to the level sequence is possible for a relatively low central barrier of 12.1 meV. With this fit the state at 4.74 meV would correspond to the first excited radial level whereas the state at 5.6 meV would correspond to the level $l = \pm 6$. However, one would then expect the $l = \pm 1$ state of the first excited radial series near 5.1 meV. Calculation of the respective transition probabilities and experiments with (combined) optical and acoustic excitation and measuring the stress dependence of the level positions may clarify the situation.

A second important advance by the experiments with isotopically enriched germanium is the observation of a rather large Ge-isotope shift. Within the RR approximation the contribution of the Ge and O masses to the moment of inertia will depend on the position of the rotation axis. For rigid rotation e.g. of a free Ge-O-Ge quasi-molecule around its inertial axis parallel to $\langle 111 \rangle$ the momentum of inertia would be given by the reduced mass and by the distance r_0 of the oxygen from the Ge-Ge axis. From the fit to

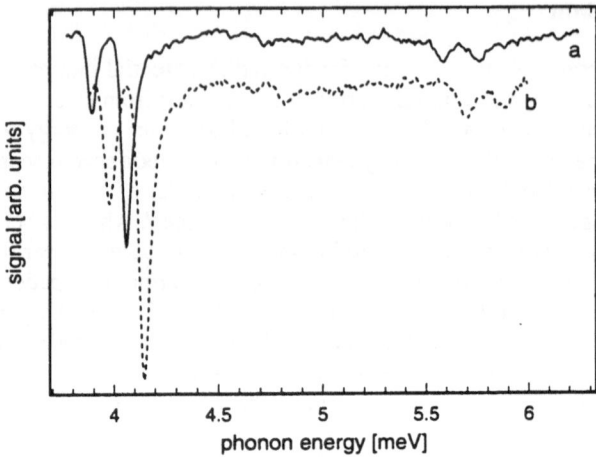

Figure 4 Transmission spectrum (baseline subtracted) at higher energies through isotopically enriched ^{73}Ge:O_i (upper curve) and ^{70}Ge:O_i. The two doublets around 4 meV belong to $l = \pm5$ [46] and the doublets around 5.8 meV possibly to $l = \pm6$ of the ground state series. The weak doublets around 4.7 meV could be due to transitions to the first excited radial state. If so, the next rotational state of this series would be expected near 5.1 meV. Its observation may be hampered by small phonon transition probabilities to these states [48].

the resonance energies this would lead to $r_0 = 93$ pm (as for the fit with the Yamada-Kaneta potential) and to small Ge-isotope shifts lying within the observed linewidth. A larger contribution of the Ge mass to the momentum of inertia would imply larger Ge-isotopic shifts as well as a smaller r_0; from the above-mentioned cluster calculations [47] a value of 58 pm is obtained. To clarify the contribution of the oxygen, its immediate germanium neighbours, and the lattice to these states it is necessary to determine the level positions in several isotopically enriched germanium crystals as well as for the oxygen isotopes. The measurements so far show already that the Ge-associated shifts are distinctly larger and that the ^{18}O-associated shift is possibly smaller than calculated with the above assumption of quasi-free rotation. The Ge-associated shifts appear to be even larger than compatible with an isotope-induced inhomogeneous linewidth of the resonances in natural germanium [48]. This could indicate some average coupling of the lattice to these low-energy excitations.

The question of phonon transition probabilities has not yet been addressed specifically. In the experiment distinct differences have been found for various transitions [11]. Their calculation may provide an additional means to distinguish between the proposed model potentials. Remarkably, the first five excited rotational levels (at least) are reached from the ground state by the phonons, whereas FIR selection rules only allow transitions between neighbouring rotational states. These restrictions together with the smallness of the level separations may be the reason why FIR investigations of these resonances have not been successful as yet. Indirectly, the positions of the first three excited rotational levels have been obtained recently from the temperature dependence of the satellites belonging to the coupled-mode IR-transitions of the 860 cm^{-1} band [50].

4   One-Phonon Ionization of Donors in Germanium

The interaction of very high frequency phonons with shallow donors and acceptors is strongly reduced due to the mismatch between the small phonon wavelength and large extent of the shallow state wave function or, alternatively, the large phonon wavevector and the confinement of the state function in k-space. This has been discussed in context with the question of short lifetimes of excess carriers at low temperatures. The restriction by k-conservation de facto prevents in general the one-step recombination with the ground state; therefore alternative routes have been proposed. For a recent review see [51]. However, in principle the shallow donors and especially Sb in Ge could be an exception: The many-valley structure of the conduction band allows the k-conservation by an effective intervalley scattering in the transition: The energy of the lowest phonon branch at the X-point in the Brillouin-zone (which corresponds to the L-L-valley scattering vector) nearly perfectly matches the binding energy of $\approx$ 10 meV of Sb in Ge [52]. The intervalley process in the transition becomes possible since at zero stress the bound states contain equal proportions of the four conduction band minima. The possibility of such a one-phonon transition has been shown [11] by irradiating Ge:Sb with phonons of appropriate energy from an Al-junction and measuring the phonon-induced change in conductivity (PIC or phonoconductivity) as shown in Fig. 5. Though

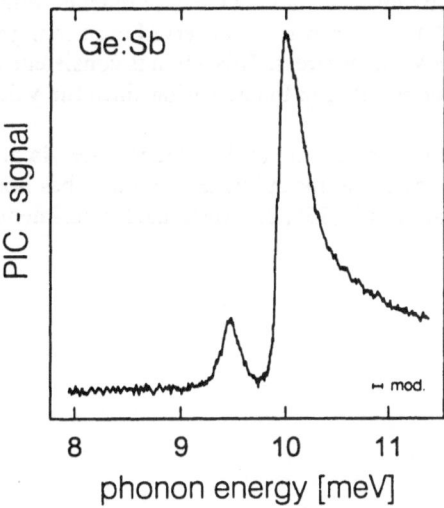

Figure 5
Phonon-induced change of (photo)conductivity (PIC) in Ge:Sb ($4 \cdot 10^{20}$ m^{-3}). The steepness of the rise at 9.95 meV is probably determined by the peak in the density of states of the TA-phonons at the X-point coinciding nearly with the Sb binding energy. (See text for details.) The peak in front of the steep rise is again due to the thermal precursor. The sample was illuminated with white light to provide a finite conductivity. (M. Gienger, Thesis, Stuttgart, 1990)

the strong isotope scattering in germanium together with anharmonic phonon decay shortens the diffusion length of very high frequency phonons to about $10\,\mu m$ less than 10^{18} m^{-3} of Sb were detectable in a volume of 10^{-11} m^{-3}. Since the TA$_X$-phonons have zero group velocity the transport into the Ge-bulk will be by LA-phonons of the same energy which by mass defect scattering are transformed into TA-phonons.

The steepness of the Sb-detection threshold shows that even at Debye energies the sharp feature in the Al-emitter spectrum is not smeared out by inelastic processes. The distance of the thermal precursor T to the main threshold is equal to the Al-gap as determined at the low gap current from the I/U characteristic, i.e. the gap is not reduced by the injection current at 10 meV and the energetic position of the threshold energy can be taken as $eU - 2\Delta_g$. The importance of the intervalley scattering process in the transition was demonstrated by the fact that the height of the threshold signal is strongly reduced under stress since also the proportion of the upshifted conduction band minima is reduced in the ground state wave function. Only if an excited bound state (associated with an upshifted band) crosses the lowest conduction band valley a bound-to-bound intervalley transition is possible with subsequent release of the bound carrier resonant with the band. For such crossings an increase of the PIC signal is observed. The above-mentioned match of energy and momentum conservation is less perfect in the case of the donors P (≈ 13 meV) and As (≈ 14 meV). Nevertheless phono-ionization was possible since the conservation conditions are somewhat relaxed for these deeper donors more confined in real space. Though these results fit well into the general picture of donors in germanium two results remain unexplained [11,53]: The binding energies of P and As as obtained from the threshold are smaller than the optical values by 0.36 meV and 0.64 meV, respectively, and there is an unexpectedly large decrease of the P- and As-binding energy for $\langle 100\rangle$-stress [54]. The Sb-threshold in contrast does not shift with stress in any of the main stress directions, probably because the threshold energy is smaller than that of the DOS-peak of the TA$_X$-phonons which thus determines the position and the steepness of the threshold. Shallow acceptors in Ge have likewise binding energies around 10 meV. A PIC threshold, however, was not observed even though the binding energy was reduced to below 8 meV under stress. This again is consistent with the fact that a large k-vector transfer is not possible in the transition since the valence band maxima are at $k = 0$.

The possible significance of a one-step recombination to the ground state via an intervalley process for the nonequilibrium carrier lifetime at low temperatures has not yet been discussed in the theoretical considerations [51,55] nor investigated experimentally e.g. by measuring its stress dependence.

5 A^+-States in Silicon and Germanium

Not depending on the band structure simultaneous momentum and energy conservation is sufficiently fulfilled for phonon interaction with the very shallow H^--like states in semiconductors metastable at low temperatures. Since the weakly bound additional carrier has to optimally share the Coulomb-attraction of the ion the wave function is by far not blown up to the same proportion as the binding energy is reduced. Correspondingly, it is found that phonon spectroscopy with detection by the phonon-induced conductivity change due to detachment of the carriers is a rather sensitive and flexible possibility for the investigation of these states [17,18,56,57]. In the case of single acceptors in silicon there is a specific interest to look for a possible fine structure of the associated A^+-states to be compared with the multiplet structures of the corresponding ground states of acceptor-bound excitons (A^0X). To understand the multiplicity and chemical shift of the latter it has been argued that the A^0X can be approximated by an A^+-state with small perturbation by the electron of the bound exciton (see references in [57]). The A^+ ground state splitting is not easily accessible. Advantage has been taken of the fact that transitions between bound states are detectable as a peak structure in the response when the upper states under uniaxial stress cross the valence band. From the analysis of the stress dependence of the PIC response an apparently good correspondence is obtained with A^0X for the shallower acceptors B, Al, and Ga. For the deeper acceptors In and Tl, however, quite a different stress dependence is found possibly related to a Jahn-Teller effect. The monotonous dependence of the A^0X properties on the A^0 binding energy is also not reproduced for the PIC threshold energies of the corresponding A^+-states: Whereas for Al^+ and Ga^+ they are even somewhat smaller than for B^+ (1.8 meV) the threshold of In^+ is at higher (5.9 meV) that of Tl^+ at intermediate (3 meV) energy. This finding seems to be at variance with the above model for the A^0X.

For the double acceptors Be and Zn in Ge the binding energies of the A^0- and the A^+-states are also anticorrelated: We have 24.8 meV and 33 meV for Be^0 and Zn^0, respectively, but 4.5 meV and 1.9 meV for Be^+ and Zn^+. These findings are in good agreement with various optical results (see references in [57]). From the stress variation of the valence band crossing peaks a zero level separation is obtained at zero stress in the case of Be whereas for Zn back-extrapolation to low stress of a crossing peak is compatible with a rather large ground state splitting of 1.4 meV.

By FIR only boron, the shallowest of the single acceptors in Si has been investigated under stress but unfortunately only for stresses much beyond the level crossing [58]. There are more FIR data for the double acceptors in Ge. From comparison with the PIC results it appears that the observation of the level crossing peaks – useful for the investigation of the fine structure of these states – is a merit of phonon transition probability. Despite a multitude of investigations there is to date no consistent understanding of these multihole states.

Only few PIC experiments have been reported for D^--states [17,11,52,59]. An interesting possibility would be the investigation of the D^--state in quantum wells (see e.g. [60]).

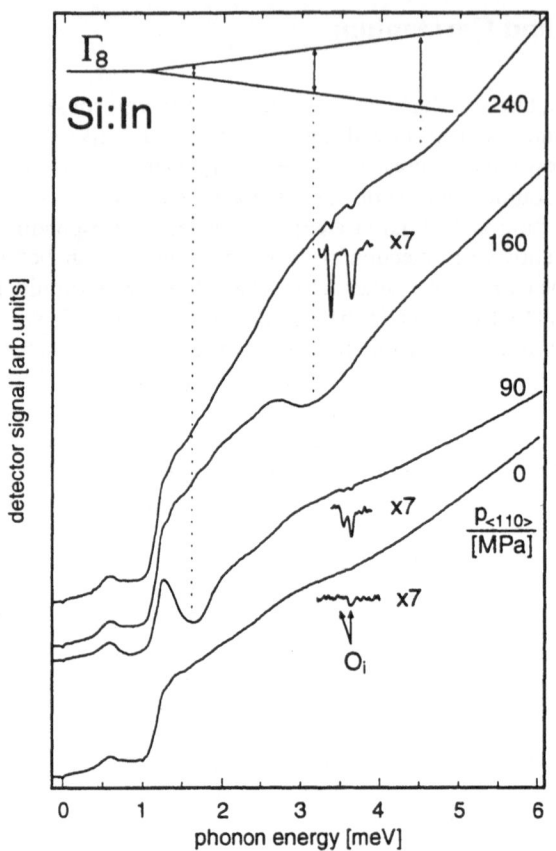

Figure 6
Phonon transmission spectrum for Si:In ([In] $= 1.2 \cdot 10^{22}$ m^{-3}). At zero stress there is a broad depression of the signal around 4 meV interpreted to be due to a dynamic Jahn-Teller effect of the In ground state. With increasing stress the strong resonant scattering by the split ground state evolves and shifts beyond the detector gap. At the same time scattering by the stress split O$_i$ resonance ([O] $<$ $1 \cdot 10^{22}$ m^{-3}) becomes more and more visible at ≈ 3.6 meV showing that the inelastic JT-scattering is reduced under stress. (F. Maier, private communication, Stuttgart 1997)

6   Ground State Multiplets of Neutral Acceptors in Silicon and Germanium

Under moderate uniaxial stress the fourfold degenerate Γ_8-state of single acceptors may be separated into two Kramers doublets with energies in the meV-range allowing to determine e.g. the deformation potential constants by phonon spectroscopy unambiguously and with high accuracy from which the expected phonon-acceptor interaction can be derived. The values obtained this way are in good agreement with those from piezo-optic spectroscopy [61]. Corresponding to the large strain interaction the lifetime broadening of these transitions is large, in the range of a tenth of a meV at about 2 meV. Acceptor concentrations as low as 10^{19} m^{-3} are readily observed. The scattering by the stress-split ground state of the acceptor boron in silicon has been used to evaluate the emitter spectrum of Sn, Pb, PbBi [7], and of Al [10] junctions.

In principle a Jahn-Teller instability should be connected with the Γ_8-state. It should be more important for the deeper acceptors with smaller extent of the wave function. Especially for Si:In various types of experiments such as ultrasonic attenuation [62], thermal conductivity [63], bound exciton luminescence [64] indicate a broad phonon scattering resonance around 4 meV. In phonon spectroscopy it is seen as a broad depression of the signal. For large stress splitting the JTE should vanish. This is demonstrated in Fig. 6 where the scattering at the stress-split oxygen resonance becomes visible if the stress splitting of the ground state becomes comparable with the JT-resonance.

For the two-hole state of double acceptors the situation is more complicated by the concurrence between hole-hole and hole-lattice interaction with additional central cell effects. This is similar to the situation of A^+-states. Despite a series of (mostly optical) investigations there is as yet no clear picture as to the relevance of the various parameters [65–67].

Phonon spectroscopy is promising in this context since the observed level separations are in the range of meV and thus directly accessible whereas with optical measurements in the IR differences from transitions to excited states have to be taken. In Ge only the double acceptors Be and Zn have been investigated [68]. The results essentially confirm those obtained by optical means: For the shallower acceptor Be a near zero ground state splitting is obtained from the stress dependence whereas a large zero stress level separation of 2.4 meV is found. In Si the situation is additionally complicated by the fact that doping with group II elements does not or not only result in substitutional double acceptors but by pairing or complexing to a series of other electrically active or inactive defects that may or may not be detected by the various experimental methods such as DLTS, ESR, or IR. By phonon spectroscopy as yet only Si:Zn has been investigated with some systematics [69] at relatively low doping concentrations. As with other experimental techniques not only the substitutional Zn is found but a number of other defects the microscopic nature of which is more ore less subject to speculation as yet.

There have been only a few frequency-resolved experiments concerning the phonon scattering off ground state multiplets of donors in Ge: In an early piezospectroscopic investigation Dynes et al. [70] determined the coupling of LA and TA phonons to the Sb ground state and by that the recombination spectrum of a tin junction driven at the gap. Later Gienger et al. [11] reported the resonant enhancement of the P^- and As^- phonoconductivity signal by the trapping of phonons due to elastic scattering within the stress-split ground state quartet.

7   Geometrical Resonances: Superlattices and Precipitates

The wavelength of, say, 3 meV TA phonons in silicon is about 6 nm, i.e. geometric resonances of nanoscopic entities should be observable. A first example has been given in semiconductors for GaAs/AlGaAs-superlattices [71] and later it was shown that sharp phonon stop-bands can be obtained with amorphous Si/SiO_2 superlattices on silicon [73], see Fig. 7. Further investigations were concerned with the problem of phonon

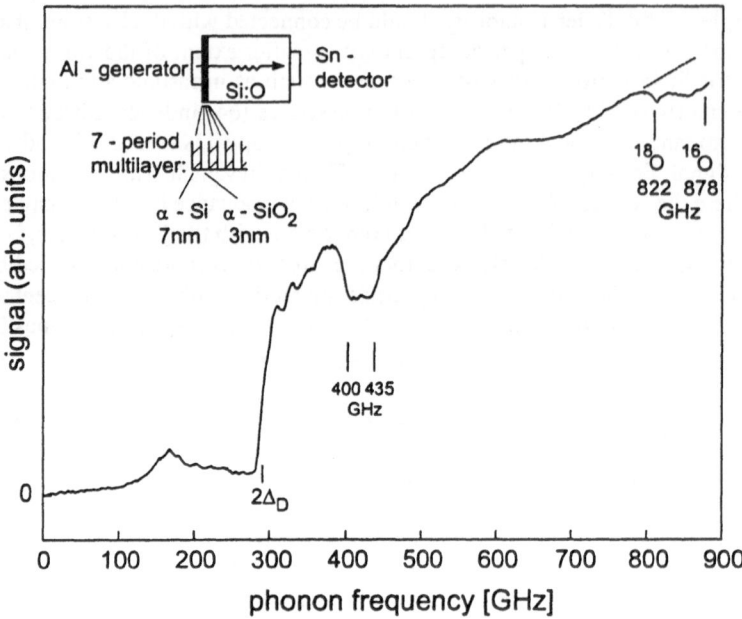

Figure 7 Al-generator/Sn-detector transmission spectrum through an α-Si/α-SiO₂ super-lattice. The broad depression around 420 GHz is the second phonon stop-band according to the superlattice parameters. It is preceded by a first stop-band at 210 GHz as measured by a Sn-generator/Al-detector set-up. The narrower wiggles preceding the 420 GHz band can be associated with side-lobes of the multilayer transmission function. The oxygen resonance of the substrate shows that high-frequency phonon transmission is not damped by the layers of the lattice. (After [73])

bands in superlattices [74]. (A multiplicity of single layer resonances within the junction superconductor films is often observed in the low-energy region of the transmission spectra [2,9].)

Phonon spectroscopy of superlattices may be also regarded as a model for the investigation of nanoscopic precipitates. A first example was given for the case of Ca clusters in irradiated CaF₂ [75]. In silicon the precipitation of oxygen is of scientific interest and technological relevance. Phonon spectroscopy can provide information on size distributions of clusters and, by the difference of elastic and anelastic scattering, to some extent on the inner structure. A systematic study has been started to investigate the precipitation in carbon-lean and carbon-rich material depending on the variation of annealing steps for nucleation and precipitation. The results so far show that there is a broad-band elastic scattering after long-term nucleation annealing around 750 °C (that may be preceded by a short-term step at 1 100 °C) and narrow scattering bands developing after further short-term annealings above 1 000 °C. Differences are observed for material with high and low carbon content [76,77]. Fig. 2 demonstrates the effect of annealing on the backscattering spectra.

8 Summary

The foregoing examples show that phonon spectroscopy with superconductor tunnel junctions provides high sensitivity, good resolution and wide-range tunability for the investigation of various types of defects in Si and Ge. The technique is not restricted to these materials; it has been applied also to III-V compounds which show quite a number of different phonon resonances [78]. There are up to now only a few investigations of implanted or epitaxial layers [56,48]. For the experiments reported here only Sn and Al tunnel junctions have been used. Depending on the specific experimental demand junctions made of superconductors like Nb, α-Mo, or PbBi are possible and may be advantageous. However, as regards upper frequency limit, high resolution, and ease of fabrication Al-junctions as phonon generators are unchallenged as yet.

Acknowledgment

Most of the work reviewed in the foregoing is an outcome of the common effort of many colleages of the 1. Physikalisches Institut at Stuttgart University among which I should like to name M. Welte, W. Forkel, E. Dittrich, W. Scheitler, M. Gienger, W. Burger, P. Groß, M. Glaser, C. Wurster, F. Maier, F. Zeller, and, with his continuous creative and experienced involvement, W. Eisenmenger. Many of the investigated Ge- and Si-samples were kindly supplied by E. E. Haller, Berkeley, and W. Zulehner, Burghausen, respectively. We profited from helpful discussions with H. Yamada-Kaneta, Kawasaki, and E. Artacho, Madrid concerning the O_i-related levels in Ge. Most of the projects have been financially supported by the Deutsche Forschungsgemeinschaft, which is gratefully acknowledged.

Bibliography

[1] W. Eisenmenger, A. Dayem, Phys. Rev. Lett. **18** 125 (1967)

[2] H. Kinder, Phys. Rev. Lett. **28** 1564 (1972)

[3] W. Forkel, M. Welte, W. Eisenmenger, Phys. Rev. Lett. **31** 215 (1973)

[4] W. Eisenmenger in *Physical Acoustics* **12**, ed. by W.P. Mason and R.N. Thurston, (Academic Press 1976), p. 79

[5] K. Laßmann, Mat. Sci. Forum **196-201** 1563 (1995), Proc. of the 18th Int. Conf. on Defects in Semiconductors, Sendai 1995, ed. by M. Suezawa and H. Katayama-Yoshida

[6]   P. Berberich, H. Kinder in *Phonon Scattering in Condensed Matter*, ed. by W. Eisenmenger, K. Laßmann, and S. Döttinger, Springer Series in Solid-State Sciences **51** 19 (1984) *and* A. Schick, P. Berberich, W. Dietsche, H. Kinder *l.c.*, p. 40

[7]   P. Berberich, M. Schwarte, Z. Phys. **B64** 1 (1986)

[8]   M. Welte, Thesis, Universität Stuttgart (1976)

[9]   W. Forkel, Thesis, Universität Stuttgart (1977)

[10] Scheitler, W., Thesis, Universität Stuttgart (1989)

[11] M. Gienger, P. Groß, K. Laßmann, Phys. Rev. Lett. **64** 1138 (1990)

[12] J.P. Wolfe in *Advances in Solid State Physics* **29**, ed. by U. Rössler, Vieweg 1989, p. 75.

[13] C. Schoell, Diplomarbeit, Universität Stuttgart (1986)

[14] A. Mrzyglod, O. Weis, Z. Phys. B **97** 103 (1995)

[15] M. Schwarte, P. Berberich, J. Phys. C **15** 3225 (1985)

[16] C. Wurster, E. Dittrich, W. Scheitler, K. Laßmann, W. Eisenmenger, W. Zulehner, Physica B **219&220** 763 (1996)

[17] W. Burger, K. Laßmann, Phys. Rev. Lett. **53** 2035 (1984)

[18] W. Burger, K. Laßmann, Phys. Rev. B **33** 5868 (1986)

[19] N. Butler, L.J. Challis, M. Sahraoui-Tahar, B. Salce, W. Ulrici, J. Phys.: Condens. Matter **1** 1191 (1989)

[20] R.C. Dynes, V. Narayanamurti, Phys. Rev. B **6** 5143 (1972)

[21] J. Lang, W. Eisenmenger, Phys. Rev. Lett. **77**, 2546 (1996)

[22] S. Roshko and W. Dietsche, Sol. St. Comm. **98** 453 (1996)

[23] R.J. von Gutfeld in *Physical Acoustics* **V**, ed. by W.P. Mason, (Academic Press 1965), p. 233

[24] W. Reupert, K. Laßmann, P. de Groot, in *Phonon Scattering in Solids*, ed. by L.J. Challis, V.W. Rampton, and A.F.G. Wyatt, Plenum Press, New York 1976, p.315

[25] A.Yu. Blank, N.N. Zinov'ev, D.I. Kovalev, L.P. Ivanov, I.D. Yaroshetskii, Pis'ma Zh. Eksp. Teor. Fiz. **52** 1225 (1990) [JETP Lett. **52** 644 (1990)]

[26] A.V. Akimov, A.A. Kaplyanskiĭ, E.S. Moskalenko J. Luminesc. **45** 135 (1990)

[27] B. Danilchenko, S. Roshko, Fiz. Tverd. Tela **32** 984 (1990) [Sov. Phys. Solid State **32** 579 (1990)]

[28] B. Danilchenko, S. Roshko, M. Asche, R. Hey, M. Hörike, H. Kostial, J. Phys.: Condens. Matter **9** 3169 (1993)

[29] B. Cabrera, P.L. Brink, B. Chugg, B.L. Dougherty, K.D. Irwin, S.W. Nam, A.T. Lee, J.G. Pronko, S. Tamura, B.A. Young, Physica B **219&220** 744 (1996)

[30] B. Pajot in *Semiconductors and Semimetals* **42** 191 (1994), ed. by R.K. Willardson, A.C. Beer, and E.R. Weber

[31] P. Clauws, Mat. Sci. & Eng. **B36** 213 (1996)

[32] D. R. Bosomworth, W. Hayes, A. R. L. Spray and G. D. Watkins, Proc. Roy. Soc. Lond. A **317** 133 (1970)

[33] C.S. Chen and D.K. Schroder, Appl. Phys. A **42** 257 (1987)

[34] C. Kaneta, H. Yamada-Kaneta, A. Ohsawa, Mat. Sci. Forum **38-41** 323 (1989)

[35] H. Yamada-Kaneta, C. Kaneta, T. Ogawa, Phys. Rev. **B42** 9650 (1990)

[36] A. Lizón-Nordström, F. Ynduráin, Sol. St. Comm. **89** 819 (1994)

[37] U. Werling, K.F. Renk, Phys. Rev. **B42** 1286 (1990)

[38] A.A. Volkov, Yu.G. Goncharov, V.P. Kalinushkin, G.V. Kozlov, A. M. Prokhorov, Fiz. Tverd. Tela **31** 262 (1989) [Sov. Phys. Solid State **31** 1249 (1989)]

[39] E. Dittrich, W. Scheitler, W. Eisenmenger, Jap. Journ. Appl. Phys. **39**, Supplement **26-3** 873 (1987)

[40] E. Dittrich, Thesis, Universität Stuttgart (1989)

[41] J.W. Corbett, R.S. McDonald, G.D. Watkins, J.Phys. Chem. Solids **25** 873 (1964)

[42] M. Stavola, J.R. Patel, L.C. Kimerling, P.E. Freeland, Appl. Phys. Lett. **42** 73 (1986)

[43] M. Reiche, in *Defect Control in Semiconductors*, ed. by K. Sumino, North-Holland 1990, p. 239

[44] F. Zeller, Stuttgart, 1997, private communication

[45] B. Pajot, P. Clauws, Proc. of the 18th Int. Conf. on the Physics of Semiconductors, Stockholm 1986, ed. by O. Engström, World Scientific, Singapore, Vol. 2, 911 (1987)

[46] K. Laßmann, M. Gienger, M. Glaser, Sol. St. Comm. **86** 285 (1993)

[47] E. Artacho, F. Ynduráin, Mat. Sci. Forum **196-201** 103 (1995), Proc. of the 18th Int. Conf. on Defects in Semiconductors, Sendai 1995, ed. by M. Suezawa and H. Katayama-Yoshida

[48] N. Aichele, U. Gommel, K. Laßmann, F. Maier, F. Zeller, E.E. Haller, K.M. Itoh, L.I. Khirunenko, V. Shakhovtsov, B. Pajot, E. Fogarassy, H. Müssig, accepted for contribution at the ICDS-19, Aveiro, Portugal, 1997

[49] E. Artacho, private communication, 1997

[50] A.J. Mayur, M. Dean Sciacca, M.K. Rudo, A.K. Ramdas, K. Itoh, J. Wolk, E.E. Haller, Phys. Rev. **B49** 16293 (1994)

[51] V.N. Abakumov, V.I. Perel, I.N. Yassievich: *Nonradiative Recombination in Semiconductors*, North Holland, 1991

[52] K. Laßmann, M. Gienger, P. Groß, Physica B**169** 377 (1991)

[53] M. Gienger, Thesis, Universität Stuttgart 1990

[54] M. Gienger, P. Groß, K. Laßmann, Inst. of Physics Conf. Proc. **95** 173 (1988), Proc. 3rd Int. Conf. on Shallow Impurities in Semiconductors, ed. by B. Monemar

[55] R.A. Brown, S. Rodriguez, Phys. Rev. **153** 890 (1967)

[56] K. Laßmann, W. Burger, in *Phonon Scattering in Condensed Matter* V, ed. by A.C. Anderson and J.P. Wolfe, Springer Series in Solid-State Sciences **68** 116 (1986)

[57] P. Groß, K. Laßmann, Ann. Phys. **4** 503 (1995)

[58] N. Sugimoto, S. Narita, M. Taniguchi, M. Kobayashi, Solid State Commun. **30** 395 (1979)

[59] B. Danilchenko, S.I. Komirenko, in *Phonon Scattering in Condensed Matter* VII, ed. by M. Meissner and R.O. Pohl, Springer Series in Solid-State Sciences, **112** 137 (1993)

[60] S. Huant, A. Mandray, G. Martinez, B. Etienne, in *Proc. 23rd Int. Conf. on the Physics of Semiconductors*, Berlin 1996, ed. by M. Scheffler and R. Zimmermann, p. 2789, World Scientific, 1996

[61] H.R. Chandrasekhar, A.K. Ramdas, S. Rodriguez, Phys. Rev. **B12** 5780 (1975)

[62] Hp. Schad, K. Laßmann, Phys. Lett. **56A** 409 (1976)

[63] A. Ambrosy, K. Laßmann, A.M. de Goër, B. Salce, H. Zeile, in *Phonon Scattering in Condensed Matter* ed. by W. Eisenmenger, K. Laßmann, and S. Döttinger, Springer Series in Solid-State Sciences **51** 361 (1984)

[64] R. Sauer, W. Schmid, J. Weber, Sol. St. Comm. **27** 705 (1978)

[65] F. Ham, C.-H. Leung, Phys. Rev. Lett. **71** 3186 (1993)

[66] F. Ham, C.-H. Leung, Sol. St. Comm. **93** 375 (1995)

[67] M.L.W. Thewalt, D. Labrie, I.J. Booth, B.P. Clayman, E.C. Lightowlers, E.E. Haller, Physica **146B** 47 (1987)

[68] B. Altreuther, Diplomarbeit, Universität Stuttgart (1989)

[69] J. Staiger, P. Groß, K. Laßmann, H. Bracht, N.A. Stolwijk, Mat. Sci. Forum **143-147** 675 (1994), Proc. of the 17th Int. Conf. on Defects in Semiconductors, Gmunden 1993, ed. by H. Heinrich and W. Jantsch

[70] R.C. Dynes, V. Narayanamurti, M. Chin, Phys. Rev. Lett. **26** 181 (1971)

[71] V. Narayanamurti, H.L. Störmer, M.A. Chin, A.C. Gossard, W. Weigmann, Phys. Rev. Lett. **43** 2012 (1979)

[72] M. Glaser, Diplomarbeit, Universität Stuttgart (1992)

[73] O. Koblinger, J. Mebert, E. Dittrich, S. Döttinger, W. Eisenmenger in *Phonon Scattering in Condensed Matter V*, ed. by A.C. Anderson and J.P. Wolfe, Springer Series in Solid-State Sciences **68** 156 (1986)

[74] P.V. Santos, L. Ley, J. Mebert, O. Koblinger, Phys. Rev. **B36** 1306 (1987)

[75] C. Wurster, K. Laßmann, W. Eisenmenger, Phys. Rev. Lett. **70** 3451 (1993)

[76] G. Schrag, M. Rebmann, C. Wurster, F. Zeller, K. Laßmann, W. Eisenmenger, submitted for publication in phys. stat. sol. (a)

[77] F. Zeller, C. Wurster, K. Laßmann, W. Eisenmenger, accepted for contribution at the ICDS-19, Aveiro, Portugal, 1997

[78] F. Maier, R. Eilenberger, W. Beck, K. Laßmann, Mat. Sci. Forum **196-201** 219 (1995), Proc. of the 18th Int. Conf. on Defects in Semiconductors, Sendai 1995, ed. by M. Suezawa and H. Katayama-Yoshida

Ab Initio Lattice Dynamics: Methods, Results, and Applications

D. Strauch, P. Pavone, A. P. Mayer, K. Karch, H. Sterner,
A. Schmid, Th. Pletl, R. Bauer, and M. Schmitt

Theoretische Physik, Universität Regensburg, D-93040 Regensburg

Abstract: The density-functional perturbation theory (DFPT) introduced by Baroni and co-workers, allows for the *ab initio* (parameter-free) calculation of lattice-dynamical properties. The method is sketched and set into relation to other approaches. Biased by personal view, we present current applications of the DFPT to the calculation of the harmonic lattice dynamics, i.e., phonon dispersion curves and eigenvectors for various systems ranging from insulators to metals. Since the phonon frequencies and eigenvectors are in extremely good agreement with the available experimental data the calculations have a very reliable predictive power for a number of crystals which have not yet been experimentally investigated. Within and beyond the harmonic approximation several thermal quantities can be calculated such as heat capacity, Debye-Waller factors, cumulants (for EXAFS experiments), and others, often with a precision exceeding the experimental one. Last but not least, the combination of the results of the DFPT with frozen-phonon techniques allows a straightforward calculation of nonlinear coefficients of the lattice potential such as anharmonic coupling coefficients, nonlinear dipole-moment coefficients, and Raman coupling constants. Within this context, numerical results are presented for Grüneisen constants, thermal expansion coefficients, various pressure and temperature effects, and Raman and infrared two-phonon spectra.

1 Introduction

Most of the temperature effects in solid-state physics have their origin in the anharmonicity of the lattice potential energy. When attempting to describe those effects quantitatively with the help of lattice-dynamical models one is usually confronted with the following fact: With increasing order of the anharmonic interactions the number of model parameters is increasing more rapidly than are the available experimental data, rendering the theoretical treatment extremely difficult. Morever, different models often have led to different conclusions.

With the implementation of methods based on the density-functional theory (DFT) [1] in efficient numerical codes and with the help of ever faster and bigger computers the electronic structure of (most) materials has become tractable. In the case of crystals, group-theoretical methods are extremely useful in reducing the storage and speed requirements for computers.

The extension of *ab initio* methods to lattice dynamics has followed various paths, the most successful being the density-functional perturbation theory (DFPT) due to Baroni *et al.* [2] and Zein [3]. The DFPT method has been put to work [4] for the calculation of the linear coupling parameters between external forces or electric fields and resulting displacements and electrical polarization, allowing the calculation of the harmonic force constants, effective charges, and dielectric susceptibilities, see Section 3.1. Variants to Ref. [4] have been developed in the way the perturbation is treated [5–8].

Apart from DFPT, successful applications of *ab initio* methods [9,10] to lattice dynamics have been achieved by the dielectric screening method [10,12–19], the frozen-phonon technique (also called direct method) [20–28], force methods [29–37], partly using full potentials (e.g., in the LAPW scheme [35,38,39] or LMTO scheme [8,25,27,30,34]), and *ab initio* molecular dynamics [40]. A recent review of the response method with special emphasis on the LMTO method is given by Savrasov [41].

Bridging the gap between lattice-dynamical models like the shell model or the bond-charge model on the one hand and *ab initio* techniques on the other, semi-empirical methods have been developed and applied to lattice dynamics (e.g., the partial-density method by Falter *et al.* [42], the embedded-atoms method [43], or the semi-empirical tight-binding approach [44–47]). They have the advantage of being more easily handable from a computational point of view. However, either they still need empirical input or they are based on restrictive assumptions the validity of which has to be tested in the specific case of application. In order to release the dependence upon the experimental input attempts have been made to combine lattice-dynamical models and *ab initio* frozen-phonon results to extend or even replace the set of experimental data to which the model parameters are fitted [33,48,49].

It would be beyond the scope of this review to discuss in detail all those different methods and their applications to lattice dynamics. Instead, we focus on the DFPT method as it is the most suited for the kind of problems we have in mind.

Activities in the application of the DFPT to lattice dynamics have initially been concentrated on testing the usual approximations, i.e., the local-density approximation (LDA) [1,50] and others [1] made in the technical implementation of the scheme. These tests have been very successful for the elemental and III-V semiconductors as well as for a number of other materials, and they are ongoing for more complex crystals.

The availability of realistic phonon eigensolutions for the first time has initiated applications to various experimental situations. The extension of the method to the calculation of nonlinear coupling constants has barely begun.

An account of the application of the method to the statics and dynamics of surfaces has been given recently in this series [51], and we refer the reader to that article for details and references. A similar procedure is used for the description of adsorbate vibrations. In particular, small adsorbate masses cause high-frequency localized vibrations with large amplitudes bringing along the necessity of treating anharmonic corrections [52].

This review is organized as follows: In Section 2 we list the ingredients of the DFT and DFPT schemes. Finally, in Sections 3–5 we review a number of results that have already come out of this new development.

2 Methods

2.1 Density-Functional Theory and Approximations

The currently most successful methods of computing electronic properties of solids are based on the density-functional theory [1]. In application to the electronic ground state the theorem of Hohenberg and Kohn (HK) [53] states that the ground-state energy E of the system under consideration is a functional of the single-particle electron density $\rho(r)$ and that there is a one-to-one correspondence between the density and the electron-ion interaction potential

$$v_{el-ion}(r) = \sum_i \frac{Z_i}{|r - R_i|} . \tag{2.1}$$

Furthermore, the HK theorem states that the ground-state energy functional has a global minimum at the exact ground-state electron density. The ground-state energy functional $E[\rho]$ can be written in terms of different energy contributions

$$E[\rho] = T[\rho] + E_H[\rho] + E_{el-ion}[\rho] + E_{xc}[\rho] . \tag{2.2}$$

While the Hartree energy

$$E_H[\rho] = \frac{1}{2} \int \frac{\rho(r)\rho(r')}{|r - r'|} d^3r \, d^3r' \tag{2.3}$$

and the electron-ion interaction energy

$$E_{el-ion}[\rho] = \int \rho(r) \, v_{el-ion}(r) \, d^3r \tag{2.4}$$

are obvious functionals of the density, the kinetic, $T[\rho]$, and exchange-correlation (xc), $E_{xc}[\rho]$, energy functionals are unknown. The difficulty with the unknown kinetic-energy functional can be circumvented by using the Kohn-Sham (KS) scheme [54], in which the minimization of the ground-state energy functional is performed in terms of *single-particle* wavefunctions. The ground-state electron density is obtained from the solution of a set of self-consistent coupled equations which are known as Kohn-Sham (KS) equations (atomic units are used):

$$\hat{H}_{KS}\,\psi_n(r) \equiv \left(-\nabla^2 + v_{KS}[\rho(r)]\right)\psi_n(r) = \varepsilon_n\,\psi_n(r) \tag{2.5}$$

$$v_{KS}[\rho(r)] = v_{el-ion}(r) + v_H[\rho(r)] + v_{xc}[\rho(r)] \tag{2.6}$$

$$v_H[\rho(r)] = \frac{\delta E_H[\rho(r)]}{\delta\rho(r)}\,; \qquad v_{xc}[\rho(r)] = \frac{\delta E_{xc}[\rho(r)]}{\delta\rho(r)} \tag{2.7}$$

$$\rho(r) = \sum_{\text{occ.}} |\psi_n(r)|^2\,. \tag{2.8}$$

The final expression for the total energy is then

$$E[\rho(r)] = \sum_{\text{occ.}} \varepsilon_n + E_I[\rho(r)] - \int \rho(r)\frac{\delta E_I[\rho(r)]}{\delta\rho(r)}\,\mathrm{d}^3 r\,. \tag{2.9}$$

where the *interaction* energy $E_I = E_H + E_{xc}$ has been introduced. However, the calculation of expression (2.9) requires an explicit form for the unknown XC energy functional $E_{xc}[\rho(r)]$. In this context, the most-commonly used approximation is the so-called local-density approximation (LDA)

$$E_{xc}[\rho] \approx E_{xc}^{LDA}[\rho] = \int \rho(r)\,\varepsilon_{xc}^{hom}(\rho(r))\,\mathrm{d}^3 r\,, \tag{2.10}$$

where $\varepsilon_{xc}^{hom}(x)$ is the exchange-correlation energy per particle of the homogeneous electron gas with density x. Variants to the LDA can be found in Ref. [1].

2.2   Plane-Wave Expansion and ab initio Pseudopotentials

The numerical implementation of the minimization of the ground-state energy requires the representation on a *finite* basis of the Kohn-Sham equations. The optimal choice of the basis set is problem dependent: For systems with different physical properties (in particular with different degree of localization of the single-particle wavefunction) different basis sets have to be chosen. For molecular systems the most natural choice is to expand the wavefunctions in atomic orbitals localized on the different atoms. On the other hand, the periodic character of the wavefunctions in a solid suggests in this case a plane-wave (PW) basis set. The latter is the basis set used in most solid-state calculations, nevertheless, in some special cases, mixed sets of plane waves and localized functions, such as Gaussians or atomic orbitals, are used to improve the description of partially localized states (as those coming from atomic d states). In recent years, basis sets including the equivalent of PW's in *adaptive* coordinates have been successfully introduced for describing quasi-localized systems [55].

For the actual periodic potential in a solid two different types of single-particle eigenstates can be found: core states which are mainly atomic-like localized states and valence states at higher energies which extend to the whole crystal. The description of the core states in terms of a plane-wave expansion requires very large basis sets. However, the core states are supposed to have only a negligible influence on the chemical and physical properties of a solid. In fact, in this case the actual potential can be replaced by a *pseudo*-potential which does not possess core states but reproduces exactly the valence (extended) states of the original potential.

Very accurate results are obtained by using atomic so-called *norm-conserving* pseudopotentials [56]. Various construction schemes for the norm-conserving pseudopotentials have been designed [57–63]. The ingredients are typically the following: (a) The valence energies coincide with the exact energies (calculated with DFT and LDA); (b) the pseudopotential is nonlocal, often given in the semilocal form (in angular coordinates),

$$v_{ps}(\boldsymbol{r},\boldsymbol{r}') = \delta(\boldsymbol{r}-\boldsymbol{r}')\, v_{loc}(\boldsymbol{r}) + \delta(\boldsymbol{r}-\boldsymbol{r}')\sum_l |l\rangle\, v_l(\boldsymbol{r})\, \langle l| \; ; \qquad (2.11)$$

(c) the pseudopotential (for a given angular momentum l) is identical to the Coulomb potential for $r > r_l$, i.e., outside an assumed cut-off radius r_l (typically somewhat larger than the outer wavefunction maximum); the pseudo-wavefunctions are thus identical to the true wavefunctions for $r > r_l$; (d) the eigenfunctions are nodeless for $r < r_l$; (e) the eigenfunctions are normalized; (f) the eigenfunctions and various derivatives of them are continuous at the cut-off radius r_l. *Ab initio* pseudopotentials are computed completely without experimental input. Therefore, one and the same pseudopotential is used for the atom, the crystal and the compounds of the atom under consideration.

The use of pseudopotentials is essential for numerical implementations based on the plane-wave expansion. However, it can be avoided when completely different basis sets are introduced, e.g., in the case of the full-potential methods [8,38,39,64–67].

2.3 Hellmann-Feynman Forces

The minimum energy determines the structure parameters of the crystal lattice. Because of anharmonic effects these are actually not the equilibrium values (which are given by the minimum of the free energy rather than of the energy). Indeed, neglecting the anharmonic effects causes in the structure parameters relative deviations of the order of some fraction of a percent.

If an atom is displaced from the potential minimum it exerts a direct force on the other atoms from the Coulomb interaction and an indirect force mediated by the electrons, the latter being given by the Hellmann-Feynman theorem

$$\boldsymbol{F}_{el} = -\left\langle \Psi \left| \frac{\partial \hat{\mathcal{H}}}{\partial \boldsymbol{u}} \right| \Psi \right\rangle = -\int \rho(\boldsymbol{r})\, v_{\boldsymbol{u}}^{el-ion}(\boldsymbol{r})\, d^3 r \;, \qquad (2.12)$$

where Ψ $(\hat{\mathcal{H}})$ is the total manybody electron wavefunction (Hamiltonian) and $v_u^{\text{el}-\text{ion}} \equiv \partial v_{\text{el}-\text{ion}}/\partial u$ (a local pseudopotential is assumed here for the sake of simplicity).

Equation (2.12) is the starting point of the molecular dynamics simulations [40]. This method aims directly at correlation functions and less at phonon dispersion curves. Equation (2.12) can also be used to compute forces as a function of displaced-atom positions u. The part linear (nonlinear) in u gives the harmonic (anharmonic) force constants. The drawback of this method is the loss of translational symmetry due to the atomic displacements from the regular lattice positions. The general method is thus well applicable to molecules.

The more common so-called *frozen-phonon* procedure [20–26,28] for crystals consists of the computation of the total energy as a function of atomic displacements. To retain partial symmetry one uses displacements corresponding to superstructures which restricts the practical application of the method to phonons with high-symmetry wavevectors.

2.4   Response Theory

Like the forces so have the force constants K direct (Coulombic) and indirect (electronic) contributions. Taking another derivative of Eq. (2.12) one obtains the electron-mediated contribution to the (harmonic) force constants,

$$K_{\text{el}} = \int \rho(r)\, v_{uu'}^{\text{el}-\text{ion}}(r)\, \mathrm{d}^3r + \int \rho_{u'}(r)\, v_u^{\text{el}-\text{ion}}(r)\, \mathrm{d}^3r. \tag{2.13}$$

(The first term is diagonal with respect to the atom positions and accounts for the acoustic sum rule.) With

$$\rho_u(r) = \int \chi(r,r')\, v_u^{\text{el}-\text{ion}}(r')\, \mathrm{d}^3r' \tag{2.14}$$

(neglecting retardation) the result can be expressed in terms of the density response function $\chi(r,r')$ [10–19]. The drawback of this formulation is the necessity of inverting the huge dielectric matrix and the explicit occurrence of all the excited-state energies and wavefunctions in the expression for χ, while the standard DFT is exact for the ground-state properties only.

2.5   Density-Functional Perturbation Theory

First-order perturbation theory of the Kohn-Sham equations (2.5) yields

$$\left(\hat{H}_{\text{KS}} - \varepsilon\right) \psi_u(r) = - \left(\hat{H}_u^{\text{KS}} - \varepsilon_u\right) \psi(r), \tag{2.15}$$

with the index u referring to the derivative with respect to u taken at the potential minimum. This equation can be viewed as an inhomogeneous differential equation for the first-order wavefunction. This kind of formulation has the huge advantage of the crystal symmetry being retained [2,3]. Equation (2.15) can be rephrased so as to involve the electronic valence states only [68],

$$P_c \psi_u^v(r) = - \left(\hat{H}_{KS} - \varepsilon_v - aP_v \right)^{-1} P_c v_u^{KS} \psi_v(r), \qquad (2.16)$$

with the projector P_v on the valence states, $P_c = 1 - P_v$, and a a number sufficiently large so as to avoid zeroes in the denominator. The linear density variation can then be obtained as

$$\rho_u(r) = \sum_{occ.} \psi_n(r)^* \psi_u^n(r) + c.c. \qquad (2.17)$$

The computations are usually performed in reciprocal space, in this case the spectral representation of the operators in Eq. (2.16) involves summations over intermediate (valence) states, and the corresponding integration over the Brillouin zone is done by summing over special-point sets [69]. The direct application of Eqs. (2.13), (2.16) and (2.17) yields the transverse part (at vanishing macroscopic electric field) of the (electronic) dynamical matrix.

As a straightforward extension of the previous procedure, the effective-charge and high-frequency dielectric-constant tensors can be obtained as the displacement and polarization response, respectively, to the macroscopic electric field. In this way the complete dynamical matrix (with the inclusion of macroscopic electric field effects) can be calculated. Diagonalization of the dynamical matrices finally yields the phonon frequencies and eigenvectors, including the LO–TO splitting.

As a last remark, most of the actual implementations of DFPT rely on the use of pseudopotentials, but the use of all-electron ("full") potentials has begun [8,38,39,64–67].

3   Results

3.1   Phonon Dispersion Curves in Various Systems

Figure 1 demonstrates the excellent agreement of the theoretical phonon dispersion curves [4] with the experimental data points [70] of GaAs. The dispersion curves for GaAs are typical for most of the III-V compounds and elemental semiconductors, see Fig. 8 (upper panel) below for the case of Si, except those involving elements from the first row of the periodic table. Notably, the transverse-acoustic branches are very flat near the Brillouin-zone boundary indicating long-range interatomic forces. As long as they are sophisticated enough, i.e., as long as they contain a sufficiently large number

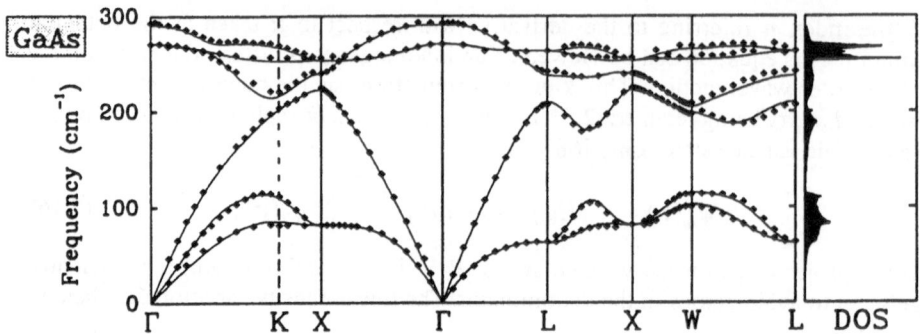

Figure 1 Phonon dispersion curves (left panel) and phonon density of states (DOS) (right panel) for GaAs. The experimental data points are taken at low temperatures with error bars typically of the size of the dots (or mostly less) [70]. The lines are the harmonic result of a DFPT calculation [4]. (From Ref. [4].)

of parameters models are able to describe the branches along the main-symmetry directions [001], [110], and [111] similarly well. These same models fail, however, in fitting or predicting the data for wavevectors along the off-symmetry directions [70]. Even more discrepancies occur for the eigenvectors, as can be seen in Section 3.2. The (harmonic) calculations [4] using the DFPT are in perfect agreement with experiment. This type of agreement is found for all of the semiconductors for which experimental data are available, and the method thus can be considered to have a reliable predictive power for those compounds for which experimental data are nonexistent or inaccessible. The phonon frequencies of Si have become the test grounds for nearly all of the various methods of computation.

The dispersion curves of semiconductors involving first-row elements are distinctly different. In particular, the topmost (longitudinal-optic) dispersion sheet in diamond has a minimum (and not a maximum, in contrast to most cases) at the zone center [71], see Fig. 2. This anomaly had been suspected [72–74], but none of the various models constructed in the past had been able to produce it. This anomaly leads to a sharp peak in the phonon density of states (DOS), and this peak is of relevance to the Raman data, see Section 5.4. The original neutron data [75] did not give any indication, and recent attempts [76] to verify the anomaly by inelastic synchrotron scattering have been unsuccessful due to large scatter of the data, but more recent inelastic neutron scattering experiments [77] have verified the theoretical predictions. In this context the extremely precise theoretical results have triggered an essential improvement of the data-analysis procedure, and it has been shown [77] that the instrumental resolution function not only broadens the phonon line but also leads to a shift of the peak frequency from the actual theoretical frequency. There may be a small absolute frequency shift due to instrument-alignment inaccuracies. In addition, the anharmonic correction has been shown to enhance the optical frequency at the Γ point by about 2% [78].

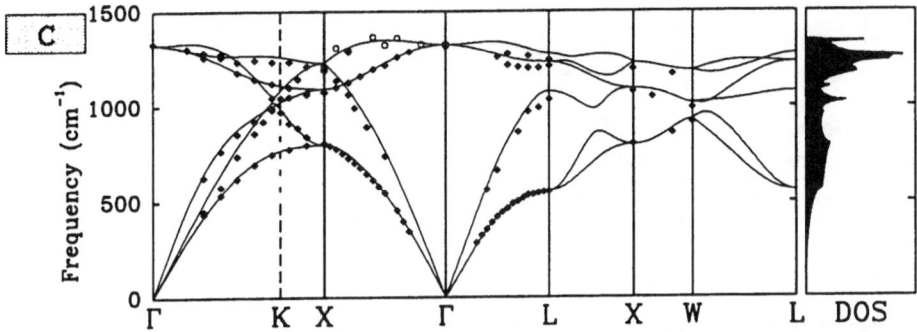

Figure 2 Phonon dispersion curves (left panel) and phonon density of states (DOS) (right panel) for diamond. The experimental data points are taken with inelastic neutron (full symbols [75]) and synchrotron (open symbols [76]) spectroscopy. The lines are the harmonic results of a DFPT calculation [71]. (From Ref. [71].)

Other systems have been investigated by DFPT (to give only a few references): SiC [79,80], III-V compounds [4], in particular the nitrides [81,82], II-VI compounds [83], alkaline earth oxides [84], solid Cl [85], copper halides [39,46], quartz [7], graphite [86], ferroelectrics [87], CsI [88], LiH [89,90], NaH [89], and LiD [90], solid C_{60} and K_3C_{60} [91], various metals [6,8,37,64–67,92,93], icosahedral α-boron [94], superlattices [95], mixed crystals [96], and others. The precision of the method can be seen from Fig. 3 where not only the theoretical phonon dispersion curves [97] are in overall good agreement with experiment [98] but also the slight oscillations of the group velocity.

The method is, however, not fool-proof. Some systems require improved approximations. In particular, the van der Waals-bound crystals (noble-gas and molecular crystals) require approximations better than the LDA at the DFT stage of the calculation. Surprisingly, the inter-layer force constants in graphite seem to be well described within the LDA [86]. Finally, the heavy atoms require a (pseudo)relativistic version of the DFT, and magnetic systems a spin-density version.

3.2 Phonon Eigenvectors

The mere fact that the eigenvalues of a matrix (here the dynamical matrix) do not determine all of the matrix elements has the consequence that there is an infinitely large set of matrices with the same eigenvalues but different eigenvectors [99] (which are related to each other by unitary transformations). This leads, in fact, to quite different eigenvectors from different models, even though the eigenvalues are (practically) the same (in the main symmetry directions). Since essentially all physical properties (except the specific heat) involve the eigenvectors in addition to the eigenfrequencies, the eigenvectors must be precise, if the quantity under consideration is to come out precise.

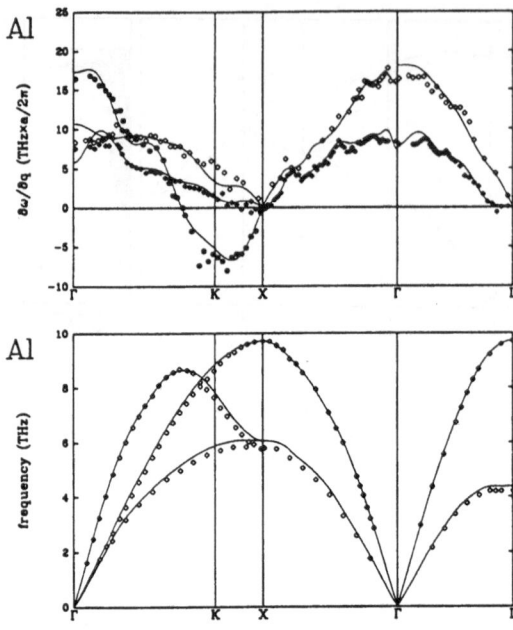

Figure 3
Phonon dispersion curves (lower panel) and group velocities (upper panel) in Al (with the theoretical lattice constant $a = 7.63\ a_B$). The lines are the results of a DFPT calculation [97]. The data points are obtained from inelastic neutron scattering results [98].

In crystals with one atom per unit cell the eigenvectors of modes with wavevectors along the main-symmetry directions are completely determined by symmetry. The simplest non-trivial case is that of the diamond structure where the displacements for the two (identical) atoms have the same amplitudes but a relative phase difference $\phi_j(q)$; along the Δ and Λ directions the eigenvectors are polarized either perpendicular or parallel to the wavevector direction. The eigenvectors are thus

$$e(\kappa|q,j) = \frac{e_j}{\sqrt{2}} \times \begin{cases} 1 & \text{for } \kappa = 1 \\ \pm e^{i\phi_j(q)} & \text{for } \kappa = 2. \end{cases} \tag{3.18}$$

The two different signs refer to acoustic and optic modes, respectively, and e_j is the polarization unit vector. The calculated [71,100] and currently available measured [101–104] eigenvector phases are shown in Fig. 4 for longitudinal modes with wavevector along the [111] direction. While different models give very different eigenvector phases in Si [101], the results of the DFPT are in excellent agreement with experiment. The eigenvector phase of Si is typical for most of the semiconductors. For diamond, the theoretical prediction [71] of an anomalous limit of the phase angle at the Brillouin-zone boundary has been confirmed by experiment [104]. The phase for SiC is predicted [79] to be intermediate between Si and diamond.

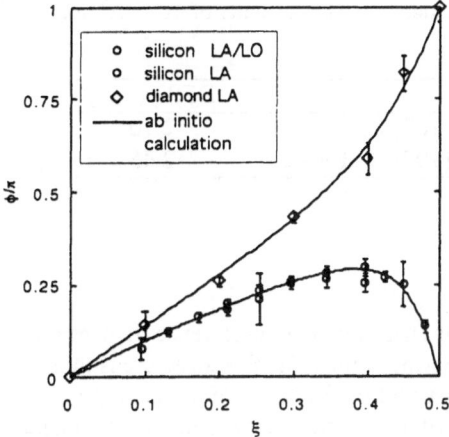

Figure 4
Phonon eigenvector phase in Si (lower set of data) and diamond (upper set of data) for longitudinal modes with wavevector along the [111] direction. Data points are determined from inelastic neutron scattering intensities in different Brillouin zones [101,102,104]. The lines are the results of DFPT calculations [71,100]. (From Ref. [104].)

4 Applications

4.1 Debye-Waller Exponents

For cubic systems the Debye-Waller exponent W_κ can be expressed in terms of displacement correlations $\langle u^2(\kappa) \rangle$ or temperature factors B_κ as

$$W_\kappa = \frac{1}{6} Q^2 \langle u^2(\kappa) \rangle = \frac{1}{2} B_\kappa \frac{\sin^2\theta}{\lambda^2} \tag{4.19}$$

$$= \frac{1}{6} Q^2 \frac{1}{NM_\kappa} \sum_{q,j} \frac{\hbar}{2\omega(q,j)} |e(\kappa|q,j)|^2 (1 + 2n_B(q,j)) . \tag{4.20}$$

Here λ is the wavelength of the radiation, θ is the scattering angle, Q is the momentum transfer, n_B is the Bose factor, and $e(\kappa|q,j)$ is the eigenvector of atom κ in the mode with phonon wavevector q and branch index j.

As a typical and simple example of calculation of the Debye-Waller factor we consider GaAs. One might think that differences between values of W_κ as an integral quantity calculated from different models might average out; quite the contrary is true. Furthermore, even though GaAs is one of the most thoroughly investigated semiconductors the experimental Debye-Waller exponents exhibit large scatter. In Table 1 the different experimental [105–107] and model [70] results of the Debye-Waller exponents for GaAs are compared with DFPT values [108], the differences of the model results originating from different phonon frequencies and in particular quite different eigenvectors for wavevectors away from the high-symmetry directions.

Table 1 Thermal displacement fluctuations $\langle u^2(\kappa) \rangle$ in units of 10^{-2} Å^2 for GaAs at 300 K.

κ	experiments	models [70]	DFPT [108]
Ga	2.4 [105], 2.3 [106], 2.8 [107]	1.5 – 2.8	2.10
As	1.7 [105], 1.9 [106], 2.3 [107]	1.9 – 3.0	2.45

4.2 EXAFS Cumulants

In EXAFS experiments the fine structure is caused by the interference of electron emission with and without scattering at neighboring atoms. The fine structure is thus a measure of the thermal average of the pair distribution function (of atoms at positions R_i and R_j). In the harmonic case the pair distribution function is a Gaussian. In the general case the exponential of the Gaussian is replaced by a superposition of so-called cumulants C_n^{ij}; the one with $n = 2$ is related to a relative-displacement correlation function,

$$C_2^{ij} \sim \langle (u_i - u_j)^2 \rangle . \tag{4.21}$$

While the Debye-Waller exponent describes the displacement self-correlation of an atom, the cumulants as obtained in EXAFS experiments are the displacement correlations of different atoms [110]. The theoretical expression has a close resemblance to Eq. (4.20). Figure 5 shows the results for C_2^{ij} in Ge with the pairs i and j referring to first, second, and third-nearest neighbor shells.

For the data evaluation one assumes that the harmonic and anharmonic contributions are given by the Gaussian and non-Gaussian part, respectively, of the pair distribution function [111]. Figure 5 shows the experimental [111] anharmonic and harmonic cumulants together with DFPT results [108]. The conclusion from the excellent agreement between the "experimental" and theoretical harmonic data is that the phenomenological procedure of separating the harmonic from the anharmonic contributions to the experimental data is a valid one. Similar calculations have been performed for Ge and diamond [108] and quartz [109].

4.3 Electron-Phonon Interaction and Transport Properties

The interaction of the electrons with phonons leads to a frequency-dependent modulation of the electrical current [112]. This effect is present in both the superconducting and the normal state of metals. Within this context, physical properties such as the electrical and thermal resistivity can be obtained from the knowledge of the transport spectral function $\alpha_{tr}^2 F(\omega)$ [113]. Measurements of the spectral function for superconducting metals can be provided by tunneling spectroscopy [112]. On the other hand, the electron-phonon coupling for non-superconductors can be measured by point-contact spectroscopy, as explained in Ref. [114]

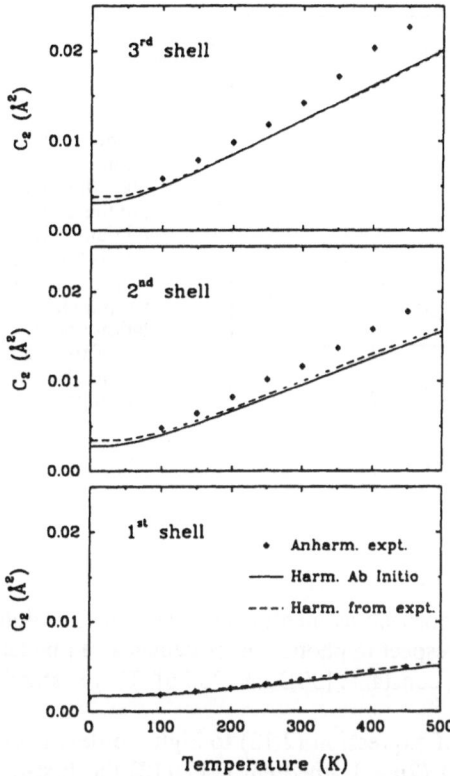

Figure 5
Temperature dependence of the harmonic and anharmonic EX-AFS cumulants for first, second, and third-nearest-neighbor shells (from bottom to top) in Si. The symbols are the experimental data [111], and the broken lines are the harmonic parts of the experimental data. The full lines are the harmonic results obtained using phonon data from a DFPT calculation [108]. (From Ref. [108].)

From the theoretical point of view, the tunneling spectra of superconductor or point contacts can be calculated from the electron band structure, the deformation potentials, and lattice dynamics without any further information [23,67,93,97,115]. An example of the results of a DFPT calculation [97] of the point-contact spectral function $\alpha_{pc}^2 F(\omega)$ of Na in comparison with experimental data [116] is shown in Fig. 6. Apart from the experimental broadening the agreement is very good in this case.

The spectral functions $\alpha_{tr}^2 F(\omega)$ and $\alpha_{pc}^2 F(\omega)$ are related to the electronic contribution to the phonon damping function. It turns out, however, that this contribution to the phonon line width [97] is negligible in comparison with the anharmonic effects.

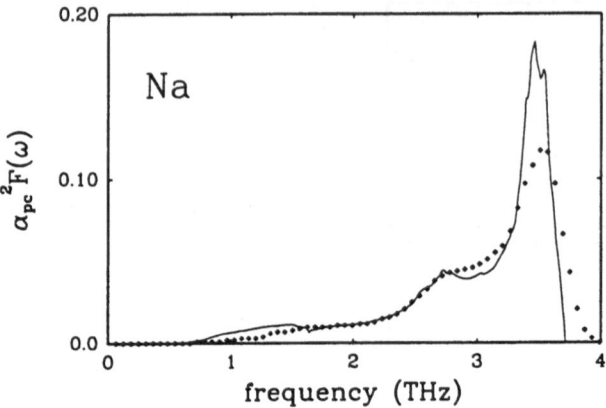

Figure 6
Point-contact tunneling spectrum of Na. The symbols are the experimental data [116]. The line is the result of using electron data from a DFT calculation and phonon data from a subsequent DFPT calculation [97].

5 Nonlinear Extensions

5.1 Nonlinear Coupling Constants

In the framework of the frozen-phonon approach as mentioned in Section 2.3 higher-order derivatives of the total energy with respect to phonon displacements can be taken in order to compute anharmonic coupling constants [30,35,34,33,36]. This is standard for molecular vibrations.

In the DFPT approach, the extension of expression (2.13) to higher orders is quite involved [14]. According to the so-called $(2n + 1)$-theorem [117,118] the first-order wavefunction is sufficient to calculate the third-order energy, i.e., the third-order anharmonic coupling constants. Results for the third-order DFPT coupling constants and application to the temperature dependence of the Raman oscillator line width for the elemental semiconductors have been obtained first by Debernardi et al. [119] in very good agreement with experiment. The extension to higher orders has not yet been done.

A method combining the frozen-phonon and DFPT approaches is based on the DFPT calculation of the linear coupling constants (force constants, effective charges, and dielectric susceptibility) for various *frozen* phonon displacement patterns with subsequent numerical differentiation. Expanding in terms of phonon field operators $A(\lambda)$ one obtains, e.g., the third-order anharmonic coupling coefficients

$$V_3(\lambda_1,\lambda_2,\lambda_3) = \frac{1}{\hbar} \frac{\partial^3 E}{\partial A(\lambda_1)\,\partial A(\lambda_2)\,\partial A(\lambda_3)} \qquad (5.22)$$

(with $\lambda_i = (q_i,j_i)$) and likewise the higher-order terms [120]. In particular, the coupling constants between the $q = 0$ phonons and one or more arbitrary phonons can be computed this way. Results for the parameters coupling the Raman mode with displacements along the (111) direction to other phonons with (q,j) are shown in Fig. 7. Even though this approach is reasonably straightforward and suffices for the calculation of the

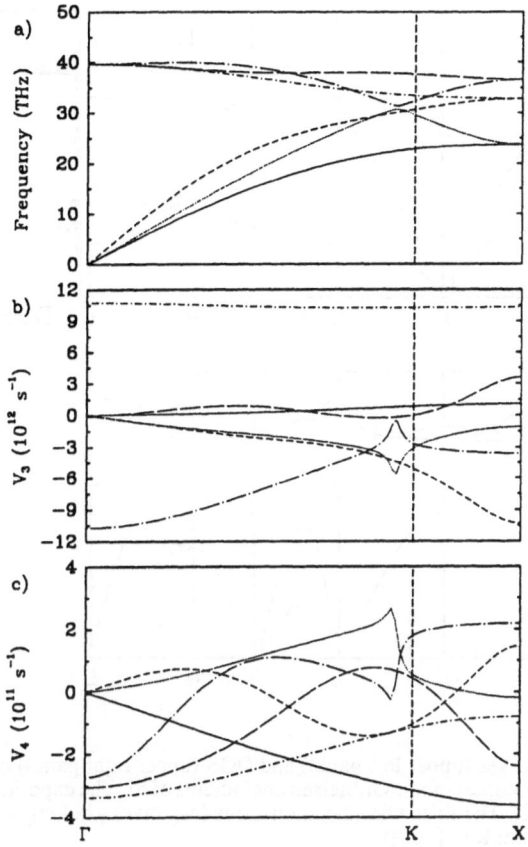

Figure 7
Phonon dispersion curves $\omega(q,j)$ (upper panel) and third and fourth-order coupling-constant dispersion curves, $V_3(0,0;q,j;-q,j)$ and $V_4(0,0;0,0;q,j;-q,j)$ (middle and lower panels, respectively) in diamond. The wavevector q is along the [110] direction, and the atomic displacements for the Raman mode (0,O) is chosen along [111]. Corresponding branches have the same line types. Note the anomaly of the anharmonic coefficients around the near-degeneracy close to the K point. (Partially from Ref. [78]b.)

anharmonic properties of the Raman oscillator, it has the drawbacks mentioned in Section 2.3 and cannot be easily extended to the anharmonicity of modes with wavevectors other than those with very high symmetry.

5.2 Anharmonicities: Volume, Pressure, and Temperature Dependence

In most cases, the volume, pressure, or temperature dependence of various quantities has its origin in the anharmonicity. These effects can often be discussed qualitatively and quantitatively in the framework of the quasiharmonic approximation (QHA). Within the QHA the form of the harmonic free energy is retained allowing the volume dependence of the phonon frequencies. In particular, the mode-Grüneisen parameters defined as

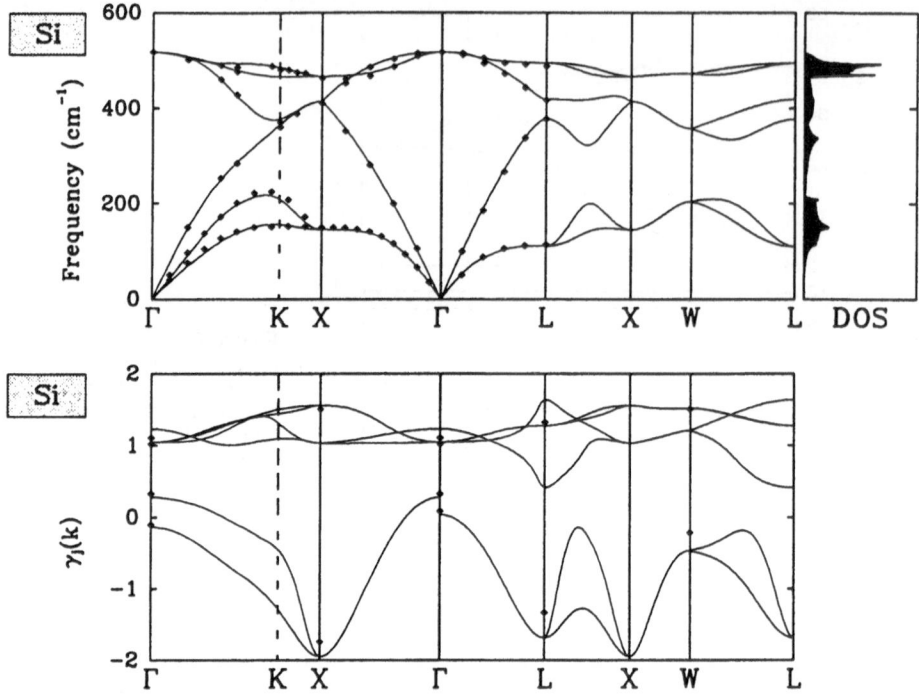

Figure 8 Phonon dispersion curves (upper left panel) and DOS (upper right panel) of Si (from Ref. [4]). Lower panel: (Volume) mode-Grüneisen coefficients of Si. The experimental data are from Refs. [121,122]. The lines with the anomalous (negative) γ belong to the transverse acoustic branches (from Ref. [100]).

$$\gamma(q,j) = -\frac{d\ln\omega(q,j)}{d\ln V} \qquad (5.23)$$

are easily obtained by computing the phonon frequencies $\omega(q,j)$ for different volumes V and taking the derivative. The results obtained for silicon are shown in Fig. 8. The largely negative values of the Grüneisen constants of the transverse acoustic branches are typical of most of the tetrahedrally coordinated semiconductors (except those with elements from the first row of the periodic table).

The elements of the mode-Grüneisen tensor

$$\gamma_{\alpha\beta}(q,j) = -\frac{d\ln\omega(q,j)}{d\eta_{\alpha\beta}} \qquad (5.24)$$

(with the strain $\eta_{\alpha\beta}$) is less straightforward, but can be expressed in terms of the anharmonic coupling parameters, which in turn can be obtained by the non-linear extension of the DFPT. Results of such a calculation for silicon [123] are shown in Fig. 9. The

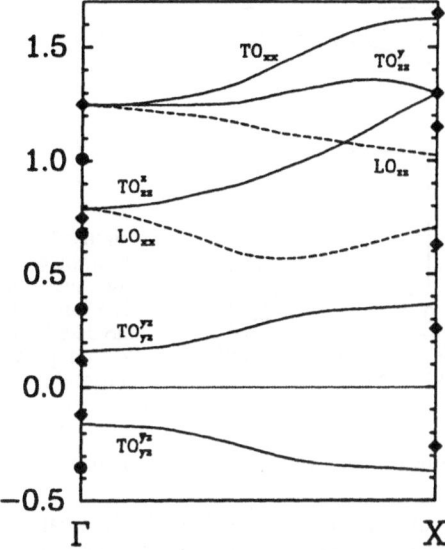

Figure 9
Dispersion of the mode Grüneisen tensor elements for longitudinal (dashed lines) and transverse (solid lines) optical phonons in silicon along the Δ direction. The sub-(super-)scripts denote the Cartesian (phonon-polarization) indices of the tensor elements. Filled circles are taken from Refs. [125,126]. The filled diamonds denote the results from a bond-charge-shell model [124].

mode-Grüneisen coefficients can be expressed by the tensor elements and compare very well with the experimental data, see Fig. 8, thus giving confidence in the theoretical predictions.

Once the mode-Grüneisen constants $\gamma(q,j)$ are known, the linear thermal expansion coefficient α can be obtained in the quasiharmonic approximation as

$$\alpha(T) = \frac{1}{3B}\sum_{q,j}c(q,j,T)\,\gamma(q,j)\,. \qquad (5.25)$$

Here B is the bulk modulus and $c(q,j,T)$ is the contribution of the mode (q,j) to the heat capacity. A result for Si is shown in Fig. 10. At very low temperatures ($T < 15$ K) only the lowest-frequency modes with mostly positive Grüneisen constants γ are excited, and the thermal expansion coefficient is positive. At intermediate temperatures (15 K $< T <$ 100 K) modes with mostly negative γ contribute to α causing it to be negative. At high temperatures ($T > 100$ K) finally the contribution of the modes with positive Grüneisen constants outweigh those with negative ones.

The dependence of various quantities upon pressure is relatively easy to obtain by performing the calculation at various volumes; by using the theoretical bulk modulus and its pressure derivative the pressure dependence can be obtained from the volume dependence. It is surprising that many quantities (including the bulk modulus) turn out to be nonlinear functions of the volume but approximately linear functions of the pressure. On the other hand, the dielectric constant and the Born effective charge behave quite nonlinearly [80,127,128]. As a further application, the calculated dependence upon pressure of the phonon frequencies of the so-called "soft" modes can provide

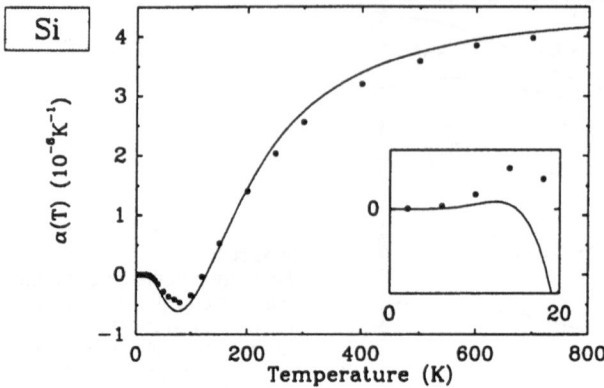

Figure 10
Thermal expansion coefficient of Si. The insert shows the details at very low temperatures. (From Ref. [100].)

information about phonon-induced structural phase transitions in tetrahedral semiconductors [129].

Explicit temperature dependence of many physical properties of solids can be accounted for using the equation of state, which is obtained from the free energy within the QHA by differentiation with respect to the temperature. The free energy is obtained by adding the phonon contribution to the static total energy, the latter being supplied by the DFT. Among others, this leads to a temperature dependence of the lattice constant [130], of the bulk modulus and the dielectric susceptibility [78], and of the effective charges. The anharmonic zero-point corrections of the bulk modulus, of the (high-frequency) dielectric constant, and of the Raman frequency of diamond amount to more than 2% [78]. This field is under active research, partly with regard to isotope effects.

Further developments towards the understanding of temperature-induced phase transitions can be achieved by comparing the free energy of different structures as in the prototype case of the α to β phase transition in Sn [131]. Morover, this approach can be used to estimate the temperature dependence of the transition pressure in simple semiconductors [132].

5.3 Nonlinear Dipole Moment and Infrared Absorption

For a quantitative calculation of the infrared absorption spectra of elemental semiconductors, the second-order dipole moments have to be known in addition to the phonon frequencies and eigenvectors over the whole Brillouin zone. These dipole coefficients are the second-order expansion coefficients of the dipole moment M with respect to the atomic displacement component $u_\alpha(R)$ of the atom at position R [120]

$$M_\alpha = M_\alpha^{(0)} + \sum_{R,\beta} M_{\alpha\beta}^{(1)}(R)\, u_\beta(R) + \frac{1}{2} \sum_{R,R',\beta,\gamma} M_{\alpha\beta\gamma}^{(2)}(R,R')\, u_\beta(R)\, u_\gamma(R') + \cdots . \quad (5.26)$$

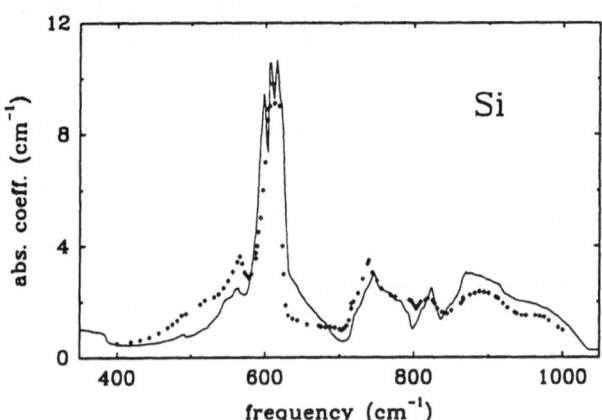

Figure 11
Infrared spectrum of Si at 293 K in the upper two-phonon frequency region. Diamonds, experimental spectrum [134]. Full line, theoretical spectrum [133] with the inclusion of the effects of multiple reflections and reflection losses at the surface of a film of 1 cm thickness (corresponding to the experimental situation).

The first-order expansion coefficients are the elements of the effective-charge tensor. The nonlinear dipole moment can also be viewed as the derivative of the effective charge tensor with respect to phonon displacements. We have thus calculated the effective-charge tensors for various displacements and taken the derivative. In this way, we have determined all the real-space nonvanishing elements of $M^{(2)}_{\alpha\beta\gamma}(\boldsymbol{R}, \boldsymbol{R}')$ up to the fourth shell of neighbors and partially those up to the eighth shell. The result of such calculation [133] is shown in in Fig. 11. Note that there is no fitting parameter in the theoretical result; instead, both relative and absolute intensities in the spectrum are calculated completely from first principles and compare well with the experimental data [134].

It should be mentioned that the positions of the critical points in the spectrum are related to those of the density of states (DOS). Thus, the analysis of the critical points of the spectrum provides information about phonon frequencies throughout the entire Brillouin zone. In addition, the relative intensities within the spectrum and the overall intensity depend strongly on the phonon eigenvectors on the one hand and on the nonlinear dipole coefficients on the other.

5.4 Raman Coupling Coefficients and Off-Resonance Raman Scattering

A similar procedure can be carried out for the derivatives of the high-frequency dielectric susceptibility to obtain the first- and second-order (off-resonance) Raman coupling constants. This has been carried out for diamond [135] and other substances [136], and in this way, a complete first- and second-order (off-resonance) Raman spectrum has been calculated fully *ab initio* on an absolute scale. It turns out that the real-space coupling constants of the atomic displacements to the photons are very short-ranged, extending not farther than to second neighbors. The absolute scattering intensities as well as the spectral forms are in very good agreement with experiment.

The Γ_1 (A_{1g}) two-phonon Raman spectrum is largely (but not exclusively, neglecting eigenvector influences) determined by the phonon overtone spectrum and thus resembles the one-phonon DOS (with the frequency scaled by a factor of 2). As shown in the right panel of Fig. 2 the DOS of diamond exhibits a sharp peak at the one-phonon cut-off frequency. This peak shows up in the experimental Γ_1 two-phonon Raman spectrum and is reproduced by the calculations using the *ab initio* Raman coupling coefficients and phonon frequencies and eigenvectors. This *ab initio* result has finally ended a long-standing controversy in the literature about the origin of the "anomalous" peak in the two-phonon Raman spectrum of diamond [135]. Neither experimentally nor theoretically is this peak found in the isostructural Si or Ge [136].

6   Conclusions

For a quantitative understanding of a variety of effects in condensed-matter physics, precise knowledge of phonon frequencies and eigenvectors as well as coupling coefficients for the electron-phonon interaction, anharmonic phonon-phonon interaction, and the coupling of the lattice to the electromagnetic field is paramount. The application of *ab initio* techniques to lattice dynamics, especially the DFPT reviewed in this article, opens up the possibility of making predictions for these quantities that do not rely on model assumptions.

The potential of *ab initio* lattice dynamics reaches far beyond the examples described here. They include complex crystal structures, structural phase transitions, currently actively studied surface statics and dynamics, electrical and thermal transport properties, defect vibrations, etc. With ever increasing computer power, studies of the lattice-dynamical properties of artificially manufactured nanostructures are coming into reach, which are relevant for a number of spectroscopic data on such systems as for disordered and composite materials with the goal of predicting reliably material properties and their dependence on temperature and pressure.

Recently, anharmonic localized vibrations in crystal lattices have received a considerable amount of interest. They are mostly investigated on the basis of simple models. Here, DFPT can provide a firm basis for such investigations.

In view of these potentials and of the progress made in combining DFT with molecular dynamics, one may rightly say that the field of *ab initio* lattice dynamics has only passed its initial stage.

Bibliography

[1] There are severals review books and articles on the density-functional theory: (a) *Theory of the Inhomogeneous Electron Gas*, S. Lundqvist and N. H. March (eds.) (Plenum, New York, 1983). (b) R. M. Dreizler and J. da Providencia, (eds.) *Density Functional Methods in Physics*, NATO ASI Series B, Vol. 123 (Plenum, New York, 1985); (c) R. O. Jones and O. Gunnarson, *Rev. Mod. Phys.* **61**, 689 (1989); (d) R. M. Dreizler and E. K. U. Gross, *Density-Functional Theory* (Springer, Berlin etc., 1990); (e) E. K. U Gross and R. M. Dreizler (eds.), *Density Functional Theory*, NATO ASI Series B, Vol. 337 (Plenum, New York, 1995); (f) H. Eschrig, *The Fundamentals of Density Functional Theory* (Teubner, Stuttgart, 1996).

[2] S. Baroni, P. Giannozzi, and A. Testa, *Phys. Rev. Lett.* **58**, 1861 (1987).

[3] N. E. Zein, *Fiz. Tverd. Tela* **26**, 3028 (1984) [*Sov. Phys.—Solid State* **26**, 1825 (1984)].

[4] P. Giannozzi, S. de Gironcoli, P. Pavone, and S. Baroni, *Phys. Rev. B* **43**, 7231 (1991).

[5] R. D. King-Smith and R. J. Needs, *J. Phys. Condens. Matter* **2**, 3431 (1990).

[6] A. A. Quong and B. M. Klein, *Phys. Rev. B* **46**, 10 734 (1992); A. A. Quong, *Phys. Rev. B* **49**, 3226 (1994).

[7] X. Gonze, D. C. Allan, and M. P. Teter, *Phys. Rev. Lett.* **68**, 3603 (1992); X. Gonze, J. C. Charlier, D. C. Allan, and M. P. Teter, *Phys. Rev. B* **50**, 13 035 (1994).

[8] S. Y. Savrasov, *Phys. Rev. Lett.* **69**, 2819 (1992); S. Y. Savrasov, *Phys. Rev. B* **54**, 16 470 (1996).

[9] For a series of articles on methods prior to the DFPT see *Electronic Structure, Dynamics, and Quantum Structural Properties of Condensed Matter*, J. T. Devreese and P. E. van Camp (eds.), NATO ASI series, Vol. 121 (Plenum, New York, 1985).

[10] For a review of the state-of-the-art in the 80's, see *Ab Initio Calculations of Phonon Spectra*, J. T. Devreese, V. E. van Doren, and P. E. van Camp (eds.) (Plenum, New York, 1983).

[11] The application of the tight-binding method ot transition metals has been reviewed by S. K. Sinha, in *Dynamical Properties of Solids*, G. K. Horton and A. A. Maradudin (eds.), Vol. 3 (North-Holland, Amsterdam, 1980), p. 1.

[12] L. J. Sham, *Phys. Rev.* **188**, 1431 (1969).

[13] R. M. Pick, M. H. Cohen, and R. M. Martin, *Phys. Rev. B* **1**, 910 (1970).

[14] F. A. Johnson, *Proc. Roy. Soc. London* A **310**, 79, 89, and 101 (1969).

[15] D. C. Wallace, *Thermodynamics of Crystals* (Wiley, New York, 1972).

[16] E. G. Brovman and Yu. M. Kagan, *Usp. Fiz. Nauk* **112**, 369 (1974) [*Sov. Phys.—Usp.* **17**, 125 (1974)]; E. G. Brovman and Yu. M. Kagan, in *Dynamical Properties of Solids*, G. K. Horton and A. A. Maradudin (eds.), Vol. 1 (North-Holland, Amsterdam, 1974), p. 191.

[17] J. T. Devreese, P. E. van Camp, and V. E. van Doren, in Ref. [10], p. 157, and references therein.

[18] A. R. Williams and U. v. Barth, in Ref. [1](a), p. 189.

[19] M. S. Hybertson and S. G. Louie, *Phys. Rev. B* **35**, 5585, 5606 (1987).

[20] R. M. Martin and co-workers in Ref. [9], p. 175, and references therein.

[21] K. Kunc and co-workers in Ref. [9], p. 227, and references therein; K. Kunc and E. Tosatti, *Phys. Rev.* B **29**, 7045 (1984).

[22] J. R. Chelikowsky and S. G. Louie, *Phys. Rev.* B **29**, 3470 (1984).

[23] M. T. Yin and M. L. Cohen, *Phys. Rev. Lett.* **45**, 1004 (1980); M. T. Yin and M. L. Cohen, *Phys. Rev.* B **25**, 4317 (1982); P. K. Lam and M. L. Cohen, *Phys. Rev.* B **25**, 6139 (1982); P. K. Lam, M. M. Dacarogna, and M. L. Cohen, *Phys. Rev.* B **34**, 5065 (1986).

[24] B. N. Harmon, W. Weber, and D. R. Hamann, *Phys. Rev.* B **25**, 1109 (1982); K. M. Ho, C. L. Fu, and B. N. Harmon, *Phys. Rev.* B **28**, 6687 (1983); *Phys. Rev.* B **29**, 1575 (1984).

[25] M. Methfessel, C. O. Rodriguez, and O. K. Andersen, *Phys. Rev.* B **40**, 2009 (1989).

[26] S. Wei and M. Y. Chou, *Phys. Rev. Lett.* **69**, 2799 (1992).

[27] V. P. Andropov, O. Gunnarson, and A. I. Liechtenstein, *Phys. Rev.* B **48**, 7651 (1993).

[28] A. Garcia and D. Vanderbilt, *Phys. Rev.* B **54**, 3817 (1996) and references therein for perovskite-structure ferroelectrics.

[29] J. Ihm, A. Zunger, and M. L. Cohen, *J. Phys.: Solid State Phys.* **12**, 4409 (1979).

[30] M. T. Yin and M. L. Cohen, *Phys. Rev.* B **26**, 2359 (1982).

[31] J. P. Vigneron, in *Festkörperprobleme—Advances in Solid State Physics*, Vol. 25, P. Grosse (ed.) (Vieweg, Braunschweig, 1985), p. 195.

[32] K. Kunc and P. Gomez Dacosta, *Phys. Rev.* B **32**, 2010 (1985); G. P. Srivastava and K. Kunc, *J. Phys.: Solid State Phys.* **21**, 5087 (1988).

[33] D. Vanderbilt, S. G. Louie, and M. L. Cohen, *Phys. Rev.* B **33**, 8740 (1986); D. Vanderbilt, S. H. Taole, and S. Narasimhan, *Phys. Rev.* B **40**, 5657 (1989).

[34] C. O. Rodriguez, A. I. Liechtenstein, I. I. Mazin, O. Jepsen, O. K. Andersen, and M. Methfessel, *Phys. Rev.* B **42**, 2692 (1990).

[35] R. Yu, D. Singh, and H. Krakauer, *Phys. Rev.* B **43**, 6411 (1991).

[36] S. Yu. Savrasov and D. Yu. Savrasov, *Phys. Rev.* B **46**, 12 181 (1992).

[37] W. Frank, C. Elsässer, and M. Fähnle, *Phys. Rev. Lett.* **74**, 1791 (1994), and references to earlier work on metals therein.

[38] K. H. Weyrich, *Phys. Rev.* B **37**, 10 269 (1988).

[39] R. Yu, and H. Krakauer, *Phys. Rev.* B **49**, 4467 (1994).

[40] R. Car and M. Parrinello, *Phys. Rev. Lett.* **55**, 2471 (1985).

[41] S. Y. Savrasov and E. G. Maksimov, *Usp. Fiz. Nauk* **165**, 773 (1995) [*Sov. Phys.—Usp.* **38**, 737 (1995)].

[42] C. Falter, M. Selmke, W. Ludwig, and W. Zierau, *J. Phys. C: Solid State Phys.* **17**, 21 (1976); C. Falter, M. Selmke, W. Ludwig, and K. Kunc, *Phys. Rev.* B **32**, 6518 (1985).

[43] M. S. Daw and M. I. Baskers, *Phys. Rev.* B **29**, 6443 (1984).

[44] D. J. Chadi and R. M. Martin, *Solid State Commun.* **19**, 643 (1976); D. J. Chadi, *Phys. Rev. Lett.* **41**, 1062 (1978).

[45] A. Mazur and J. Pollmann, *Phys. Rev.* B **39**, 5261 (1989);

[46] C. Z. Wang, C. T. Chan, and K. M. Ho, *Phys. Rev.* B **39**, 8586 (1989); **42**, 11 276 (1990).

[47] C. Falter, G. A. Hoffmann, and M. Klenner, *Phys. Rev.* B **53**, 14 917 (1996) and references therein.

[48] U. Scherz, S. Biernacki, J. Schöpp, and D. Weider, *Z. Phys. B: Condens. Matter* **85**, 69 (1991).

[49] I. I. Mazin, S. N. Rashkeev, V. P. Andropov, O. Jepsen, I. A. Liechtenstein, and O. K. Andersen, *Phys. Rev.* B **45**, 5114 (1992).

[50] D. M. Ceperley and B. J. Alder, *Phys. Rev. Lett.* **45**, 566 (1980); J. P. Perdew and A. Zunger, *Phys. Rev.* B **23**, 5049 (1981).

[51] J. Fritsch, C. Eckl, P. Pavone, and U. Schröder, in *Festkörperprobleme—Advances in Solid State Physics*, Vol. 36, R. Helbig (ed.) (Vieweg, Braunschweig, 1997), p. 135.

[52] R. Honke, A. P. Mayer, and U. Schröder, unpublished.

[53] P. Hohenberg and W. Kohn, *Phys. Rev.* **136**, B864 (1964).

[54] W. Kohn and L. J. Sham, *Phys. Rev.* **140**, A1133 (1965).

[55] F. Gygi, *Phys. Rev.* B **48**, 11692 (1993); D. R. Hamann, *Phys. Rev.* B **51**, 7337 (1995).

[56] D. R. Hamann, M. Schlüter, and C. Chiang, *Phys. Rev. Lett.* **43**, 1494 (1979); G. B. Bachelet, D. R. Hamann, and M. Schlüter, *Phys. Rev.* B **26**, 4199 (1982).

[57] L. Kleinman and D. M. Bylander, *Phys. Rev. Lett.* **48**, 1425 (1982).

[58] G. P. Kerker, *J. Phys. C: Solid State Phys.* **13**, L189 (1980).

[59] E. Shirley, D. Allan, R. Martin, and J. Joannopoulos, *Phys. Rev.* B **40**, 3652 (1989).

[60] D. Vanderbilt, *Phys. Rev.* B **32**, 8412 (1985); *Phys. Rev.* B **41**, 7892 (1990).

[61] N. Troullier and J. L. Martins, *Phys. Rev.* B **43**, 1993 (1991).

[62] M. Teter, *Phys. Rev.* B **48**, 5031 (1993).

[63] A. Filippetti, D. Vanderbilt, W. Zhang, Y. Cai, and G. B. Bachelet, *Phys. Rev.* B **52**, 11 793 (1995).

[64] S. Y. Savrasov, *Solid State Commun.* **74**, 69 (1990).

[65] N. E. Zein, *Phys. Lett.* A **161**, 526 (1992).

[66] C. Z. Wang, R. Yu, and H. Krakauer, *Phys. Rev. Lett.* **72**, 368 (1994).

[67] S. Y. Savrasov, D. Y. Savrasov, and O. K. Anderson, *Phys. Rev. Lett.* **72**, 372 (1994); S. Y. Savrasov and D. Y. Savrasov, *Phys. Rev.* B **54**, 16 487 (1996).

[68] R. M. Sternheimer, *Phys. Rev.* **96**, 951 (1954); *ibid.* **107**, 1565 (1957); *ibid.* **115**, 1198 (1959).

[69] D. J. Chadi and M. L. Cohen, *Phys. Rev.* B **8**, 5747 (1973); H. J. Monkhorst and J. D. Pack, *Phys. Rev.* B **13**, 5188 (1976).

[70] D. Strauch and B. Dorner, *J. Phys.: Condens. Matter* **2**, 1457 (1990).

[71] P. Pavone, K. Karch, O. Schütt, W. Windl, D. Strauch, P. Giannozzi, and S. Baroni, *Phys. Rev.* B **48**, 3156 (1993).

[72] M. J. P. Musgrave and J. A. Pople, *Proc. Roy. Soc. London* A **268**, 474 (1962).

[73] K. Uchinokura, T. Sekine, and E. Matsuura, *J. Phys. Chem. Solids* **35**, 171 (1974).

[74] R. Tubino and J. L. Birman, *Phys. Rev.* B **15**, 5843 (1977).

[75] J. L. Warren, J. L. Yarnell, G. Dolling, and R. A. Cowley, *Phys. Rev.* **158**, 805 (1967).

[76] E. Burkel, *Inelastic Scattering of X Rays with Very High Energy Resolution*, Springer Tracts in Modern Physics, Vol. 125 (Springer, Berlin, 1991).

[77] J. Kulda, B. Dorner, B. Roessli, H. Sterner, R. Bauer, Th. May, K. Karch, P. Pavone, and D. Strauch, *Solid State Commun.* **99**, 799 (1996).

[78] K. Karch, T. Dietrich, W. Windl, P. Pavone, A. P. Mayer, and D. Strauch, *Phys. Rev.* B **53**, 7259 (1996). K. Karch, A. P. Mayer, T. Dietrich, G. Lang, W. Windl, P. Pavone, D. Strauch, and F. Bechstedt, in *The Physics of Semiconductors*, M. Scheffler and R. Zimmermann (eds.), (World Scientific, Singapore, 1996) p. 301.

[79] A rather extensive overview can be found in K. Karch, P. Pavone, W. Windl, O. Schütt, and D. Strauch, *Phys. Rev.* B **50**, 17 054 (1994): K. Karch, P. Pavone, W. Windl, D. Strauch, and F. Bechstedt, *Intl. J. Quantum Chem.* **56**, 801 (1995).

[80] C. Z. Wang, R. Yu, and H. Krakauer, *Phys. Rev.* B **53**, 5430 (1996).

[81] K. Karch, G. Portisch, F. Bechstedt, P. Pavone, and D. Strauch, in *Silicon Carbide and Related Materials* (Institute of Physics, Bristol, 1996), p. 967; K. Karch, F. Bechstedt, P. Pavone, and D. Strauch, *Physica* B **219 & 220**, 445 (1996).

[82] K. Karch, F. Bechstedt, and T. Pletl, *Phys. Rev.* B **56**, 3560 (1997).

[83] A. Dal Corso, S. Baroni, and R. Resta, *Phys. Rev.* B **47**, 3588 (1993).

[84] O. Schütt, P. Pavone, W. Windl, K. Karch, and D. Strauch, *Phys. Rev.* B **50**, 3746 (1994).

[85] R. Bauer, O. Schütt, P. Pavone, W. Windl, and D. Strauch, *Phys. Rev.* B **51**, 210 (1995).

[86] P. Pavone, R. Bauer, K. Karch, O. Schütt, S. Vent, W. Windl, D. Strauch, S. Baroni, and S. de Gironcoli, *Physica* B **219 & 220**, 439 (1996).

[87] R. Yu and H. Krakauer, *Phys. Rev. Lett.* **74**, 4067 (1995); R. Yu, C. Z. Wang, and H. Krakauer, *Ferroelectrics* **164**, 161 (1995).

[88] M. Buongiorno Nardelli, S. Baroni, and P. Giannozzi, *Phys. Rev. Lett.* **69**, 1069 (1992).

[89] D. K. Blat, N. E. Zein, and V. I. Zinenko, *J. Phys.: Condens. Matter* **3**, 5515 (1991).

[90] G. Roma, C. M. Bertoni, and S. Baroni, *Solid State Commun.* **98**, 203 (1996).

[91] K. P. Bohnen, R. Heid, K. M. Ho, and C. T. Chan, *Phys. Rev.* B **51**, 5805 (1995).

[92] S. de Gironcoli, *Phys. Rev.* B **51**, 6773 (1995).

[93] A. Y. Liu and A. A. Quong, *Phys. Rev.* B **53**, R7575 (1996).

[94] N. Vast, S. Baroni, G. Zerah, J. M. Besson, A. Polian, M. Grimsditch, and J. C. Chervin, *Phys. Rev. Lett.* **78**, 693 (1997).

[95] S. Baroni, P. Giannozzi, and E. Molinari, *Phys. Rev.* B **41**, 3870 (1990).

[96] S. Baroni, S. de Gironcoli, and P. Giannozzi, *Phys. Rev. Lett.* **65**, 84 (1990).

[97] R. Bauer, A. Schmid, P. Pavone, and D. Strauch, to be published.

[98] J. W. Wittmouth and R. Stedman, *Phys. Rev.* B **2**, 4743 (1970).

[99] R. S. Leigh, B. Szigeti, and V. K. Tewary, *Proc. Roy. Soc. London* A **320**, 505 (1971).

[100] P. Pavone, PhD thesis, Scuola Internazionale Superiore di Studi Avanzati (SISSA), Trieste, Italy, 1991, http://www.sissa.it/cm/CMsector/PHD.html; first published in Ref. [71].

[101] D. Strauch, A. P. Mayer, and B. Dorner, *Z. Phys. B: Condens. Matter* **78**, 405 (1990).

[102] J. Kulda, D. Strauch, P. Pavone, and Y. Ishii, *Phys. Rev.* B **50**, 13 347 (1994).

[103] J. Kulda, Y. Ishii, S. Katano, *Physica* B **213 & 214**, 427 (1995).

[104] J. Kulda, R. Bauer, H. Sterner, and D. Strauch, *Physica* B **234–236**, 124 (1997).

[105] R. I. Cottam and G. A. Saunders, *J. Phys. C: Solid State Phys.* **6**, 2105 (1973).

[106] M. Gomm, PhD thesis, Erlangen 1977, unpublished.

[107] U. Pietsch, *Phys. Stat. Sol.* (b) **103**, 93 (1981).

[108] D. Strauch, P. Pavone, N. Nerb, K. Karch, W. Windl, G. Dalba, and P. Fornasini, *Physica* B **219 & 220**, 436 (1996).

[109] C. Lee and X. Gonze, *Phys. Rev.* B **51**, 8610 (1995).

[110] For a review, see E. D. Crozier, J. J. Rehr, and R. Ingalls, in *X-Ray Absorption: Principles, Applications, and Techniques*, D. C. Kroningsberger and R. Prins (eds.) (Wiley, New York, 1991), p. 373.

[111] G. Dalba, D. Diop, P. Fornasini, and F. Rocca, *J. Phys.: Condens. Matter* **6**, 7321 (1994).

[112] G. Grimvall, *The Electron-Phonon Interaction in Metals* (North-Holland, Amsterdam 1981); E. L. Wolf, *Principles of Electron Tunneling Spectroscopy* (Oxford University Press, New York, 1985).

[113] P. B. Allen, *Phys. Rev.* B **31**, 305 (1971); P. B. Allen, *Phys. Rev.* B **17**, 3725 (1978).

[114] A. G. M. Jansen, A. P. van Gelder, and P. Wyder, *J. Phys.* C **13**, 6073 (1980).

[115] O. K. Andersen, I. A. Liechtenstein, O. Rodriguez, I. I. Mazin, O. Jepsen, V. P. Andropov, O. Gunnarsson, and S. Gopalan, *Physica* C **185–189**, 147 (1991).

[116] Y. G. Naidyuk, I. K. Yanson, A. A. Lysykh, and O. I. Shklarevkii, *Fiz. Tverd. Tela* **22**, 3665 (1980) [*Sov. Phys.–Solid State* **22**, 2145 (1980)].

[117] X. Gonze and J. P. Vigneron, *Phys. Rev.* B **39**, 13 120 (1989); **44**, 3494 (E) (1991); X. Gonze, *Phys. Rev.* A **52**, 1096 (1995).

[118] Ph. M. Morse and H. Feshbach, *Methods of Theoretical Physics*, Vol. II (McGraw-Hill, New York, 1953) (p. 1120); J. O. Hirschfelder, W. Byers Brown, and S. T. Epstein, in *Advances in Quantum Chemistry* (Academic, New York 1964), pp. 267, 288, and 300.

[119] A. Debernardi and S. Baroni, *Solid State Commun.* **91**, 813 (1994); A. Debernardi, S. Baroni, and E. Molinari, *Phys. Rev. Lett.* **75**, 1819 (1995).

[120] H. Bilz, D. Strauch, and R. K. Wehner, *Vibrational Infrared and Raman Spectra of Non-metals*, Encyclopedia of Physics—Handbuch der Physik, Vol. 25/2d, S. Flügge and L. Genzel (eds.) (Springer, Berlin, 1984), in particular Section G.

[121] B. A. Weinstein, and G. J. Piermarini, *Phys. Rev.* B **12**, 1172 (1975).

[122] C. J. Buchenauer, F. Cerdeira, and M. Cardona, in: *Light Scattering in Solids*, M. Balkanski (ed.) (Flammarion, Paris 1971), p. 280.

[123] G. Birner, Diplom-Arbeit, Regensburg 1996, unpublished; M. Schmitt, Diplom-Arbeit, Regensburg 1996, unpublished.

[124] M. T. Labrot, A. P. Mayer, and R. K. Wehner, in: *Phonons 89*, S. Hunklinger, W. Ludwig, and G. Weiss (eds.) (World Scientific, Singapore, 1990), p. 181.

[125] S. Ganesan, A. A. Maradudin, and J. Oitmaa, *Ann. Phys.* **56**, 556 (1970).

[126] E. Anastassakis, A. Pinczuk, E. Burstein, F. H. Pollak, and M. Cardona, *Solid State Commun.* **8**, 133 (1970).

[127] K. Karch, F. Bechstedt, P. Pavone, and D. Strauch, *Phys. Rev.* B **53**, 13400 (1996); *J. Phys.: Condens. Matter* **8**, 2945 (1996).

[128] G. Wellenhofer, K. Karch, P. Pavone, U. Rössler, and D. Strauch, *Phys. Rev.* B **53**, 6071 (1996).

[129] K. Karch, P. Pavone, and D. Strauch, unpublished; S. Klotz, J. M. Messon, M. Braden, K. Karch, F. Bechstedt, D. Strauch, and P. Pavone, *Phys. Stat. Sol.* (b) **198**, 105 (1996).

[130] P. Pavone and S. Baroni, *Solid State Commun.* **90**, 295 (1994).

[131] P. Pavone, S. Baroni, and S. de Gironcoli, to be published.

[132] A. Bauer, M. Schmitt, K. Karch, P. Pavone, and D. Strauch, unpublished.

[133] H. Sterner, unpublished.

[134] M. Ikezawa and M. Ishigame, *J. Phys. Soc. Jpn.* **50**, 3734 (1981).

[135] W. Windl, K. Karch, P. Pavone, O. Schütt, and D. Strauch, *Intl. J. Quantum Chem.* **56**, 787 (1995).

[136] W. Windl, K. Karch, P. Pavone, O. Schütt, D. Strauch, W. H. Weber, K. C. Hass, and L. Rimai, *Phys. Rev.* B **49**, 8764 (1994).

Resonant Optical Spectroscopy of Semiconductor Microstructures

E. L. Ivchenko

A. F. Ioffe Physico-Technical Institute, 194021 St. Petersburg, Russia

Abstract: Optical spectroscopy based on light reflection and transmission measurements, Raman scattering, polarized photoluminescence and four-wave mixing is a powerful tool to investigate the fine structure of exciton levels and excitonic kinetics in semiconductor microstructures. The efficiency of spectroscopic methods is demonstrated for excitons in various kinds of nanostructures: exciton polaritons in long-period multiple quantum wells, in particular in resonant Bragg and anti-Bragg structures, localized excitons in type II GaAs/AlAs superlattices, excitons localized on anisotropic islands in type I quantum wells and confined in semiconductor nanocrystals, and exciton polaritons in microcavities with embedded quantum wells or gratings of quantum wires. The exchange and Zeeman splittings are shown to be extremely sensitive to the structure geometry and the shape of the exciton envelope function.

1 Introduction

If allowance is made for free-carrier spin degeneracy, then the exciton energy levels are degenerate, even in the case of simple bands. The ground-state degeneracy is given by the product of the conduction- and valence-band degeneracies at the extremum point. The short-range electron-hole exchange interaction partially removes this degeneracy, the long-range exchange interaction and the coupling with photons (or the polariton effect) give rise to additional splittings between the branches of the free-exciton dispersion curve or between sublevels of localized excitons. Due to the Zeeman effect, an external magnetic field can further modify the fine structure of excitonic states. The aim of the present paper is to display the rich possibilities of various spectroscopic methods to study optical properties of semiconductor nanostructures with emphasis put on the exciton fine structure and the polariton effect. The next Section deals with normal light waves in long-period multiple-quantum-well structures. Sec. 3 is devoted to the fine structure of excitonic levels for quasi-zero-dimensional excitons. The effect of the exchange interaction on the optical orientation and alignment of excitons in type II GaAs/AlAs superlattices is discussed in Sec. 4. The Zeeman effect in biased heterostructures is considered in Sec. 5. The results of the theoretical study of degenerate four-wave mixing in quantum microcavities are presented in Sec. 6.

2   Exciton Polaritons in Finite Multiple Quantum Well Structures

In bulk crystals, the exciton-photon coupling renormalizes the dispersion of bare photons and mechanical excitons leading to the formation of mixed modes – exciton-polaritons (see e.g. the review [1] and references therein). If exciton scattering by phonons and static defects is neglected then the exciton-polariton modes propagate inside the crystal without any decay. In an infinite periodic quantum-well (QW) structure or superlattice, the details of exciton-polariton formation can be different but in this case, as well as in a perfect crystal, there is no need in the terms 'radiative damping' or 'radiative lifetime'. For a single-QW structure, the notion of exciton-polaritons has undergone a substantial transformation (see [2,3]). Really, an exciton excited in an ideal QW into the state with the two-dimensional (2D) wavevector $k = (k_x, k_y)$ can annihilate and emit a photon into the barrier if $k = |k| < (\omega_0/c)n_b$, where ω_0 is the exciton resonant frequency, c is the light velocity in vacuum and n_b is the refractive index of the barrier material. The renormalization of the resonant frequency of the quasi-2D exciton is insignificant and, thus, the coupling with photons results in an appearance of exciton radiative damping. It should be noted that, in contract to excited isolated atoms which emit light inside a wide solid angle $\sim 4\pi$, the exciton radiation consists only of the two light waves with the wavevectors $q = (k_x, k_y, \pm k_z)$, where $k_z = [(\omega/c)^2 n_b^2 - k_x^2 - k_y^2]^{1/2}$ and $\hbar\omega$ is the exciton energy. Excitons with $k > (\omega_0/c)n_b$ induce an electric field which decays exponentially in the barriers. The self-consistent backward influence of the field on the exciton renormalizes the exciton resonance frequency but gives no rise to a new recombination channel.

New intriguing properties of exciton-polaritons have been revealed in long-period multiple-QW structures composed of a finite number of wells. Similarly to a single QW, in this case the exciton recombination is followed by emission of the photons $(k_x, k_y, \pm k_z)$. However, in the multilayered structure, the radiative recombination rate and the exciton-polariton dispersion $\omega(k_x, k_y)$ are effected by coherent multireflection and interference of the light. In terms of quantum electrodynamics, the excitons in different QW's are coupled via the electromagnetic field. In the same way as for N coupled oscillators, the system of N QW's for each in-plane wavevector is characterized by N eigen frequencies, ω_j $(j = 1 \ldots N)$, of exciton-polaritons. If $k < (\omega_0/c)n_b$ the frequencies ω_j are complex even if the nonradiative exciton damping is excluded. In the following the formulae describing the reflection and transmission of a finite system of QW's are presented and discussed, and a special attention is paid to the resonant Bragg QW structures.

We suppose the light frequency ω to lie close to the resonance frequency, ω_0, of the 1s exciton level and the contribution to the dielectric response from other excitonic states and electron-hole continuum is described by a background refractive index n_a which in general differs from the barrier refractive index n_b. Then the reflection and transmission coefficients for a single QW sandwiched between the semiinfinite barriers can be written as [4]

$$r_1 = \frac{E_r}{E_0} = r^{(0)} + r_{exc}, t_1 = \frac{E_t}{E_0} = t^{(0)} + r_{exc}. \tag{2.1}$$

Here $r^{(0)}$ and $t^{(0)}$ are the reflection and transmission coefficients calculated neglecting the exciton contribution:

$$r^{(0)} = e^{-ik_BL} r_{ba} \frac{1 - e^{2ik_aL}}{1 - r_{ba}^2 e^{2ik_aL}}, t^{(0)} = e^{ik_aL}(e^{-ik_bL} + r_{ab} r^{(0)}), \tag{2.2}$$

$k_{a,b} = (\omega/c)n_{a,b}$, $r_{ba} = -r_{ab} = (n_b - n_a)/(n_b + n_a)$, L is the QW width. For a symmetrical QW, the 1s-exciton contribution to r_1, t_1 has the form

$$r_{exc} = t^{(0)} \frac{i\bar{\Gamma}_0}{\omega_0 - \omega - i(\Gamma + \bar{\Gamma}_0)}, \bar{\Gamma}_0 = \frac{1 + r_{ab}e^{ik_aL}}{1 - r_{ab}e^{ik_aL}} \Gamma_0, \tag{2.3}$$

where Γ_0, Γ are respectively the radiative and nonradiative exciton damping rates in a single QW.

For a regular system of N QW's, the reflection and transmission coefficients (referred to the planes shifted along the growth direction by $\mp d/2$ with respect to the first and last QW's) are given by [4]

$$\bar{r}_N = \frac{\bar{r}_1 \sin NQd}{\sin NQd - \bar{t}_1 \sin(N-1)Qd}, \bar{t}_N = \frac{\bar{t}_1 \sin Qd}{\sin NQd - \bar{t}_1 \sin(N-1)Qd}, \tag{2.4}$$

where $\bar{r}_1 = e^{ik_bd}r_1, \bar{t}_1 = e^{ik_bd}t_1$ and the product Qd satisfies the equation

$$\cos Qd = \frac{\bar{t}_1^2 - \bar{r}_1^2 + 1}{2\bar{t}_1}. \tag{2.5}$$

Note that, for an infinite periodic heterostructure, Q means the light wavevector component along the growth direction and Eq. (2.5) is nothing more than the dispersion equation for exciton-polaritons [3].

Recently the resonant Bragg heterostructures has been attracting much attention [5–12]. The infinite *resonant Bragg multiple-QW structure* is defined as one where the normal light wave at $\omega = \omega_0$ is a standing wave with the electric field, $E(z)$, being odd with respect to the center of any QW. If the dielectric constant mismatch is neglected, i.e. $n_a = n_b$, then the above condition is equivalent to

$$\frac{\omega_0}{c} n_b d = \pi. \tag{2.6}$$

For a finite resonant Bragg structure with $n_a = n_b$, Eq. (2.4) reduces to [5]

$$\bar{r}_N = \frac{-iN\Gamma_0}{\omega_0 - \omega - i(\Gamma + N\Gamma_0)}, \bar{t}_N = (-1)^N \frac{\omega_0 - \omega - i\Gamma}{\omega_0 - \omega - i(\Gamma + N\Gamma_0)}. \tag{2.7}$$

Poles of the functions $\bar{r}_N(\omega), \bar{t}_N(\omega)$ determine complex frequencies of the eigenmodes calculated taking into account exciton-photon coupling. It follows then from Eq. (2.7) that in the N-well structure with $(\omega_0 n_b/c)d$ approaching π, $N-1$ eigenmodes do not interact with the radiation and only one mode is optically active with the oscillator strength being N times greater than that for an exciton in a single QW.

The multiple-QW structure has a mirror plane of symmetry so that every exciton-polariton solution is characterized by a particular parity with respect to the reflection in this plane. Let $\omega_i^{(+)}$ be the complex frequencies for the even solutions, where $i = 1 \ldots N/2$ if N is even and $i = 1 \ldots (N+1)/2$ if N is odd, and $\omega_j^{(-)}$ represent the frequencies for the odd exciton-polaritons, where $j = 1 \ldots N/2$ for N even and $j = 1 \ldots (N-1)/2$ for N odd. One can show that, for the structures with the same value of N and with the periods d_1, d_2 satisfying the condition

$$\frac{\omega_0}{c} n_b (d_1 + d_2) = \pi, \tag{2.8}$$

the complex eigen frequencies $\omega_j^{\pm}(d_1)$ and $\omega_j^{\pm}(d_2)$ are related by

$$\omega_j^{(\pm)}(d_2) = \begin{cases} 2\omega_0 - \omega_j^{(\pm)*}(d_1) & \text{, if } N \text{ is odd,} \\ 2\omega_0 - \omega_j^{(\mp)*}(d_1) & \text{, if } N \text{ even.} \end{cases} \tag{2.9}$$

In a particular case of the *anti-Bragg structure* with $(\omega_0/c)n_b d = \pi/2$ we obtain from Eq. (2.9) that, for N even, the frequencies $\omega_j^{(-)}$ are obtained from the set $\omega_j^{(+)}$ by the mirror reflection in the vertical line $\omega = \omega_0$ of the complex plane ω and, for N odd, the set $\omega_j^{(+)}$ or the set $\omega_j^{(-)}$ is distributed symmetrically with respect to this vertical axis.

In the general case of $n_a \neq n_b$ the resonant Bragg condition reads

$$r^{(0)} - t^{(0)} = \exp\left(-i \frac{\omega_0}{c} n_b d\right). \tag{2.10}$$

If the period of the structure deviates from that given by Eq. (2.10) or the frequency ω is shifted from ω_0 the value of

$$\rho = e^{ik_b d}(r^{(0)} - t^{(0)}) - 1 \tag{2.11}$$

becomes nonzero. For small ρ, Eq. (2.4) is reduced to

$$\bar{r}_N = \frac{\bar{r}_1}{1 + \bar{t}_1 \frac{\sin(N-1)w}{\sin Nw}}, \bar{t}_N = \frac{\bar{t}_1}{\frac{\sin Nw}{\sin w} + \bar{t}_1 \frac{\sin(N-1)w}{\sin w}}, \tag{2.12}$$

where

$$w = \left[2\rho\left(\frac{1}{2}\rho - \bar{r}_1\right)\frac{1}{\bar{t}_1}\right]^{1/2}. \tag{2.13}$$

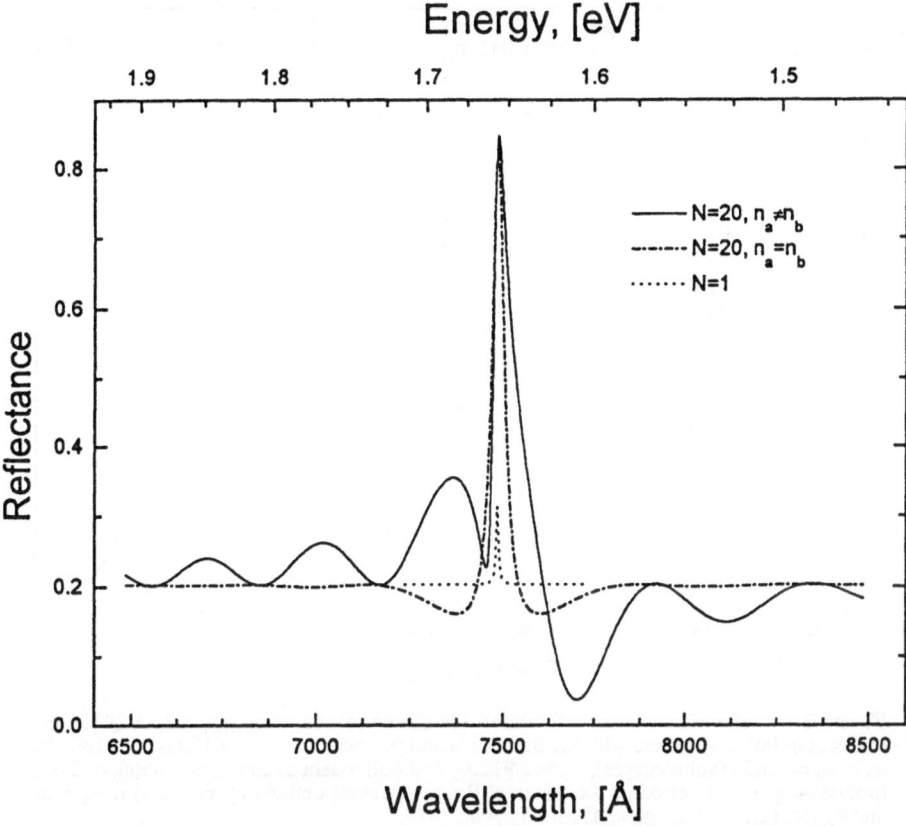

Figure 1 Reflection spectra at normal incidence calculated for a single QW structure (dotted curve) and for the resonant Bragg multiple QW's with $N = 20$ for $n_a = n_b = 2.653$ (dashed-dotted) and $n_a = 2.912$, $n_b = 2.638$ (solid). The distance between the vacuum-barrier interface and the center of the first QW was taken to equal a half of the structure period d. The other parameters are $\hbar\Gamma = 0.7$ meV, $\hbar\Gamma_0 = 0.12$ meV, $\hbar\omega_0 = 1.656$ eV. [4]

Figs. 1, 2 demonstrate effects of the dielectric constant mismatch. Curves in Fig. 1 are calculated for the normal incidence of the light from vacuum on the structure with a cap layer of the thickness $d/2$, where d satisfies the condition (2.10). The results presented in Fig. 2 describe the reflectance under normal incidence from the barrier material in order to ignore implications arising due to the reflection from the vacuum-barrier boundary. One can see that the reflection spectra calculated for the quasi-Bragg structures are very sensitive both to the detuning from the exact condition (2.10) and to the difference $n_a - n_b$.

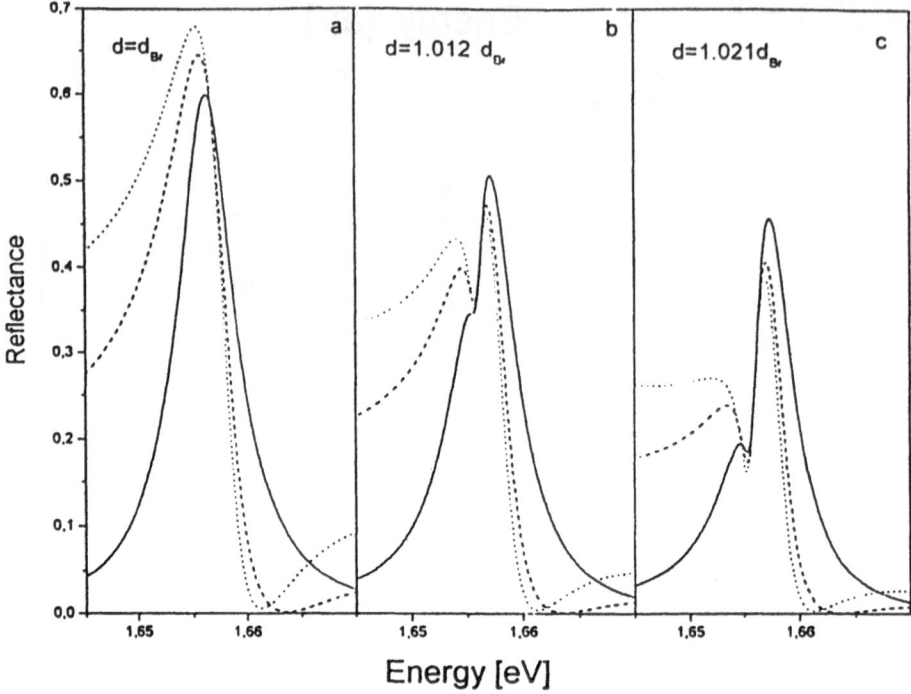

Figure 2   Reflection spectra calculated for structures containing multiple-QW's between semiinfinite barriers with the three different periods $d = d_{Br}; 1.012d_{Br}; 1.021d_{Br}$ for $n_a = n_b = 2.638$ (solid curves), $n_a = 2.912, n_b = 2.638$ (dashed) and $n_a = 3.05, n_b = 2.638$ (dotted). d_{Br} is the period of the resonant Bragg structure defined by Eq. (2.6) if $n_a = n_b$ and by Eq. (2.10) in the general case $n_a \neq n_b$. [4]

3   Effects of Electron-Hole Long-Range Exchange Interaction for Zero-Dimensional Excitons

In semiconductors the exchange interaction between an electron and a hole consists of the short-range and long-range terms [13]. The latter can be as well considered as arising due to the macroscopic electric field induced by the pair excitation. In bulk crystals the short-range contribution leads to the splitting of the mechanical-exciton states consistent with the crystal point symmetry while the long-range part is responsible for the longitudinal-transverse splitting with respect to the exciton wavevector direction. In this section we discuss the effect of long-range exchange interaction on the fine structure of quasi-zero-dimensional (0D) excitons. We first concentrate on the excitons localized on laterally-anisotropic islands in QW's and then briefly discuss the exchange splittings for excitons confined in nanocrystals.

Let us consider the heavy-hole excitons in a GaAs/Al$_x$Ga$_{1-x}$As(001) QW structure. The short-range exchange interaction splits the fourfold degenerate level e1-hh1(1s) into a radiative doublet with the angular momentum component $M = \pm 1$ and two closely-lying nonradiative singlets. The envelope function of an exciton localized into an island of well width fluctuation is approximated by

$$\Psi(r_e, r_h) = \psi(\rho_e, \rho_h)\, \varphi_{e1}(z_e)\, \varphi_{hh1}(z_h)\,, \tag{3.14}$$

where $\varphi_{e1}(z_e)$ and $\varphi_{hh1}(z_h)$ describe the electron and heavy-hole size-quantization, $\rho_{e,h}$ is the electron or hole in-plane position. If the function $\psi(\rho,\rho)$ is anisotropic the radiative doublet is split by the long-range exchange interaction into two sublevels polarized in orthogonal directions determined by the shape of $\psi(\rho,\rho)$. If $\psi(\rho,\rho)$ is invariant under the two-dimensional mirror reflection $(x,y) \to (-x,y)$, the sublevel polarizations coincide with the axes $\alpha = x,y$ and the 0D-exciton resonance-frequency renormalization, $\delta\omega_0^{(\alpha)}$, and radiative lifetime, τ_α, are given by [14]

$$\delta\omega_0^{(\alpha)} - i\,(2\tau_\alpha)^{-1} = \frac{4\pi e^2 p_{cv}^2}{S\hbar\omega_0^2 m_0^2 \varepsilon_b} \sum_q T_\alpha(q) F^2(q)\,, \tag{3.15}$$

$$T_\alpha(q) = \frac{i}{2}\frac{q_\alpha^2 - k^2}{k_z} \int\int dz dz'\, \Phi(z)\Phi(z')\, e^{ik_z|z - z'|}\,, \tag{3.16}$$

where q is the two-dimensional vector (q_x, q_y), $k_z = \sqrt{k^2 - q^2}$, $k = (\omega_0/c)n_b$, $\Phi(z) = \varphi_{e1}(z)\varphi_{hh1}(z)$, $F(q) = \int d\rho\, \exp(-iq\rho)\,\psi(\rho,\rho)$, p_{cv} is the interband matrix element $i\langle S|\hat{p}_z|Z\rangle$, $\varepsilon_b = n_b^2$, m_0 is the free electron mass, S is the sample area. Note that Eqs. (3.15, 3.16) also describe the longitudinal-transverse splitting of free heavy-hole excitons in QWs [2,3]: for a free exciton with the in-plane wave-vector K, $\psi(\rho_e, \rho_h) = S^{-1/2} \times \exp(iKR)f(\rho_e - \rho_h)$ and $F(q)^2 = Sf^2(0)\delta_{q,K}$.

Fig. 3 shows the exchange energy $E_\alpha = \hbar\delta\omega_0^{(\alpha)}$ for the exciton polarized in the $\alpha = x,y$ direction as a function of the exciton localization length along one of the in-plane directions. We assume the exciton to be confined in a 80 Å-thick quantum well, localized at the anisotropic island of interface fluctuations and described by the in-plane center-of-mass envelope function

$$\psi(\rho,\rho) = \frac{2}{\pi}\frac{1}{\bar{a}\sqrt{L_x L_y}}\exp\left[-\left(\frac{\rho_x}{L_x}\right)^2 - \left(\frac{\rho_y}{L_y}\right)^2\right]\,, \tag{3.17}$$

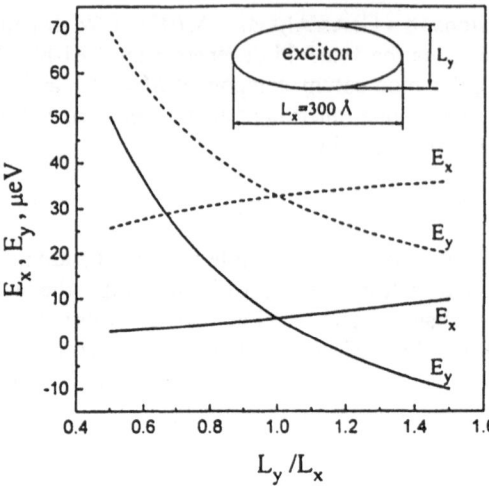

Figure 3
The long-range exchange energy as a function of the ratio L_y/L_x for the e1-hh1(1s) localized exciton polarized in the x or y direction and described by the center-of-mass envelope function (3.17). The value of L_x is fixed. The solid and dashed curves are calculated respectively taking into account and neglecting the retardation effect. [14]

where $\bar{a}$ is the effective two-dimensional Bohr radius and $L_x, L_y \gg \bar{a}$. The parameters used are as follows: $L_x = 300$ Å, the longitudinal-transverse splitting for the bulk exciton $\hbar\omega_{LT} = 80\,\mu eV$, the effective masses $m_e = 0.067\,m_0$, $m_{hh}^z = 0.5\,m_0$, $m_{hh}^{xy} = 0.15\,m_0$, and the band offsets are $V_e = 248$ meV, $V_h = 160$ meV. The solid curves are calculated by using Eqs.(3.15,3.16) while dashed curves are obtained neglecting the retardation, i.e. replacing $q_\alpha^2 - k^2$ and k_z by q_α^2 and iq respectively. Comparing the dashed and solid curves one can conclude that the retardation effect plays an important role in the exciton energy renormalization. The exchange-induced splitting of the radiative doublet has an order of tens μeV if the exciton localization lengths in the x and y directions differ by a factor of $1.5 \div 2$.

Recently Gammon et al. [15] have studied the photoluminescence (PL) of GaAs/AlGaAs QWs in the optical near-field regime and measured the PL spectrum of a single quantum dot formed by a large monolayer high island. They report a fine structure splitting of 20–50 μeV and linear polarization of the split sublevels for both the ground and excited states of the localized exciton. The splitting of radiative excitonic states observed in GaAs/AlGaAs QWs by Blackwood et al. [16] may also be related to the lateral anisotropy of exciton localization. Fig. 4 gives the energy and the long-range exchange splitting of the ground and excited states for the exciton localized by a rectangular monolayer fluctuation of the QW width [17]. While calculating the curves we used the approximation of the factorized envelope function [18]

$$\psi(\rho,\rho) = \alpha(\rho_x)\,\beta(\rho_y), \tag{3.18}$$

where the single-variable functions $\alpha(\rho_x)$ and $\beta(\rho_y)$ are found self-consistently. In Fig. 4 the value $V - \varepsilon$ is plotted where V is the energy difference between the 1s-exciton levels in the two perfect QW's differing in the width by one monomolecular layer, ε is the exciton localization energy referred to the free-exciton energy in the narrower QW

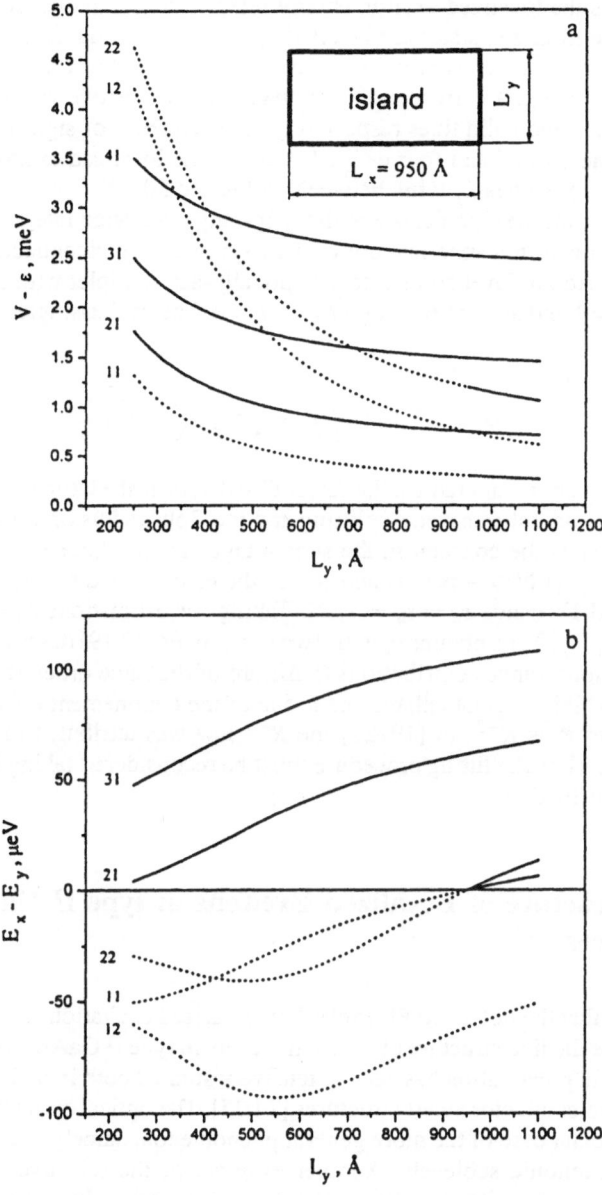

Figure 4 The energy of the localized exciton (a) and the splitting of the radiative doublet (b) as a function of L_y for a fixed value of $L_x = 950$ Å, where L_x, L_y determine the extention of the rectangular monolayer island along the x and y axes for the standard QW width of 28 Å. [17]

so that $V - \varepsilon$ is the localized-exciton energy referred to the free-exciton energy in the wider QW. The localized states are labelled by two integer quantum numbers, e.g. 11, 21, 22 etc., describing the exciton in-plane confinement in the x and y directions. The parts of curves in Fig. 4 corresponding to negative and positive values of $E_x - E_y$ are shown by dashed and solid lines respectively. The sequence of signs of this difference observed for the ground and four excited states of the localized exciton [15] is reproduced in Fig. 4 for values L_y lying between 450 Å and 500 Å.

Next we illustrate the significance of the long-range exchange interaction for the $\Gamma_6 \times \Gamma_7$ exciton confined in a spherical dot of a T_d-symmetry semiconductor. The fourfold degenerate 1s-exciton level splits into an optically-active triplet with the total angular momentum $J = 1$ and an inactive singlet with $J = 0$. The triplet-singlet splitting is given by [14]

$$\Delta E = C \left[\varepsilon_0 \left(\frac{a_0}{R} \right)^3 + \frac{\pi}{3} \hbar \omega_{LT} \left(\frac{a_B}{R} \right)^3 \right], \qquad (3.19)$$

where R is the quantum dot radius, the factor $C \approx 0.7$, a_0 is the lattice constant, $\hbar \omega_{LT}$ and a_B are the longitudinal-transverse splitting and the Bohr radius of a free exciton in the bulk material, ε_0 is the constant in the short-range electron-hole exchange interaction $\hat{V}_{short} = \varepsilon_0 a_0^3 (s_e s_h) \delta(r_e - r_h)$, s_e and s_h are the electron and the hole spin operators (for the Γ_6 and Γ_7 bands $s_e = s_h = 1/2$). Taking for estimations $\varepsilon_0 \sim 1$ eV, $\hbar \omega_{LT} \sim 10^{-3}$ eV, $a_B/a_0 \sim 10$ one obtains that the two terms in Eq. (3.19) describing respectively the short- and long-range contributions to ΔE are of the same order of magnitude. The splitting ΔE exhibits a giant enhancement due to the confinement effect and increases with decreasing R as R^{-3}. In [19–21] the R^{-3} law was attributed to the short-range exchange only. Thus the fitting procedure must be reconsidered taking into account the long-range contribution.

4 Fine Structure of Localized Excitons in type II GaAs/AlAs Superlattices

Here we show that the polarized PL excited by polarized excitation serves as a powerful tool to examine the fine structure of excitonic levels in type II GaAs/AlAs superlattices. The exciton spin polarization has been extensively studied both in bulk semiconductors [22,23] and semiconductor heterostructures [24,25]. The optical orientation of excitonic spins is a particular case of the more general phenomenon, namely, the selective optical excitation of excitonic sublevels. Another example of the selective excitation is the optical alignment of excitons by linearly polarized radiation. In contrast to the optical orientation which means the photoinduced inequality in the populations of the states $|m\rangle$ with the exciton spin m, say with $m = 1$ and $m = -1$, the linearly-polarized light can in the resonant conditions excite the excitonic states with a definite direction of oscillating electric-dipole moment. Such a state can be also described as a coherent superposition, $(|1\rangle + e^{i\Phi}| - 1\rangle)/\sqrt{2}$, of the states $|\pm 1\rangle$, where the phase Φ is determined by the orientation of the light polarization plane.

It often happens that the symmetry of the system is reduced because of microscopic or local anisotropy in the plane normal to the light propagation direction as well as due to application of an external field (magnetic field, uniaxial strain). As a result, the optical orientation and optical alignment are interconnected and one needs to consider not only circular-circular and linear-linear but also linear-circular and circular-linear configurations of the polarizer and analyzer.

It was demonstrated experimentally [26–32] and explained theoretically [33,34] that for localized heavy-hole excitons e1-hh1(1s) in type II GaAs/AlAs superlattices the degeneracy of the radiative doublet $|\pm 1\rangle$ is lifted and the two split sublevels, $E_{[110]}$ and $E_{[1\bar{1}0]}$, are dipole-active respectively along the [110] and [1$\bar{1}$0] directions in the interface plane. Moreover, in a single superlattice with fixed layer thicknesses, there exist two classes of localized-exciton states characterized by the same absolute value but opposite signs of the anisotropic exchange splitting $\hbar\omega_2 \equiv E_{[110]} - E_{[1\bar{1}0]}$. The explanation is based on the fact that excitons contributing to the low-temperature PL of the undoped type II superlattices are bound electron-hole pairs localized by the structure imperfections in the plane of interfaces with an X-electron and a Γ-hole confined inside two neighboring AlAs and GaAs layers. Thus, in fact the local symmetry of the system is C_{2v} rather than D_{2d}. According to the theory [33,34] the two classes of excitonic states with opposite signs of ω_2 are excitons localized at the AlAs-on-GaAs and GaAs-on-AlAs interfaces, i.e. localized excitons with the electron confined in the right either in the left AlAs layer relative to a GaAs layer where the hole is photoexcited. The theory predicts an appearance of remarkable linear polarization $P_l = (I_{[110]} - I_{[1\bar{1}0]})/(I_{[110]} + I_{[1\bar{1}0]})$ (I_α is the intensity of the PL component polarized along the axis α) induced by the magnetic field $B \parallel z$ under excitation with the circularly polarized light [24].

In order to make transparent the physics of the phenomena under consideration we give here a simplified description of the optical orientation and alignment and the orientation-to-alignment conversion induced by a longitudinal magnetic field in the type II superlattices. For resonant excitation conditions and in the absence of spin relaxation, the optically-inactive sublevels remain unpopulated, the only nonzero components of the exciton spin-density matrix $\rho_{mm'}$ are those with $m, m' = \pm 1$ and the 1s-hh1(1s) exciton acts as a two-level system. Recall that any two levels can be considered as two states of an effective three-dimensional pseudospin with $S = 1/2$. The 2×2 spin-density matrix $\rho_{mm'}(m, m' = \pm 1)$ is expressed in terms of the average pseudospin S as

$$\hat{\rho} = N \left(\frac{1}{2} + S\sigma \right), \tag{4.20}$$

where $\sigma_\alpha(\alpha = x,y,z)$ are the Pauli matrices and N is the steady-state exciton concentration. The pure exciton states $|1\rangle$ and $|-1\rangle$ are equivalent to the pseudospin polarized parallel or antiparallel to the z axis, respectively. The exciton states $|X\rangle = (|1\rangle + |-1\rangle)/\sqrt{2}$, $|Y\rangle = -i(|1\rangle - |-1\rangle)/\sqrt{2}$ dipole-active along the $[1\bar{1}0]$ or $[110]$ axis are described by a pseudospin with $S_x = 1/2$ or $S_x = -1/2$ respectively. Finally, the states $|X'\rangle = (|X\rangle + |Y\rangle)/\sqrt{2}$ or $|Y'\rangle = (-|X\rangle + |Y\rangle)/\sqrt{2}$ polarized in the $[100]$ and $[010]$ directions correspond to a pseudospin with nonzero component $S_y = 1/2$ or $S_y = -1/2$. The degree of the radiation circular polarization, P_c, and the degrees of linear polarization, P_l and $P_{l'}$, referred to the two pairs of rectangular axes $[110]$, $[1\bar{1}0]$ and $[100]$, $[010]$ are related to $\mathbf{S}$ as

$$P_c = 2S_z \, , P_l = -2S_x \, , P_{l'} = 2S_y \, . \qquad (4.21)$$

The pseudospin Hamiltonian is a sum of the exchange and Zeeman terms

$$\mathcal{H} = \frac{\hbar}{2}\left(\omega_2\sigma_x + \omega_\parallel\sigma_z\right) , \qquad (4.22)$$

where $\hbar\omega_2$ is the above-mentioned zero-field splitting of the radiative doublet, $\hbar\omega_\parallel = (g_h^\parallel - g_e^\parallel)\mu_0 B_z$, μ_0 is the Bohr magneton and $g_e^\parallel, g_h^\parallel$ are the electron and heavy-hole longitudinal g-factors. The pseudospin components S_x^0, S_y^0, S_z^0 describing the initial polarization of photogenerated excitons are related to the Stokes parameters $P_l^0, P_{l'}^0, P_c^0$ of the incident light similarly to Eq. (4.21). According to Eq. (4.22) the pseudospin rotates around the vector $\omega = (\omega_2, 0, \omega_\parallel)$ with the effective Larmour frequency $\omega = \sqrt{\omega_\parallel^2 + \omega_2^2}$. If the exciton lifetime τ is long enough so that $\omega_2\tau \gg 1$ (and this is the case for localized excitons in type II GaAs/AlAs superlattices) then the pseudospin Larmor precession around ω leads to a depolarization of the initial spin component perpendicular to ω while the component parallel to ω remains unchanged. As a result the steady-state pseudospin orientation is obtained by projecting S^0 onto the ω direction

$$\mathbf{S} = \frac{(S^0\omega)\omega}{\omega^2} \, . \qquad (4.23)$$

This simple model naturally explains the interconnection between the orientation and alignment in longitudinal magnetic fields. Let us consider particular cases. If excitons are excited by the light polarized along the $[1\bar{1}0]$ or $[110]$ axis then the pseudospin is initially directed parallel or antiparallel to the x axis. The vector ω lies in the (x,z) plane and makes the angle $\varphi = \arctan \omega_\parallel/\omega_2$ with the x axis. According to Eq. (4.23) the components of the average pseudospin are given by

$$S_x = S_x^0\cos^2\varphi \, , S_y = 0 \, , S_z = S_x^0\cos\varphi\sin\varphi \, . \qquad (4.24)$$

Taking into account the relation between P_l, P_c and $\mathbf{S}$ we obtain

$$P_l = P_l^0 \frac{\omega_2^2}{\omega_2^2 + \omega_\parallel^2} \ , P_c = -P_l^0 \frac{\omega_2 \omega_\parallel}{\omega_2^2 + \omega_\parallel^2} \ . \tag{4.25}$$

It follows then that, under linearly polarized excitation along $[1\bar{1}0]$ or $[110]$, the longitudinal magnetic field gives rise to the two effects: (a) suppression of the alignment and (b) polarization conversion with the appearance of circular polarization in the PL. For circularly polarized excitation the initial pseudospin S^0 is directed along z. At zero magnetic field the vectors S^0 and ω are perpendicular and, for $\omega_2 \tau \gg 1$, the exciton PL is unpolarized. The longitudinal magnetic field restores the PL circular polarization and induces the linear polarization P_l. According to Eq. (4.23), both effects are described by

$$P_c = P_c^0 \frac{\omega_\parallel^2}{\omega_2^2 + \omega_\parallel^2} \ , P_l = -P_c^0 \frac{\omega_2 \omega_\parallel}{\omega_2^2 + \omega_\parallel^2} \ . \tag{4.26}$$

Under excitation by light polarized in the $[100]$ or $[010]$ direction, the vector S^0 is orthogonal to the (x,z) plane and the spin precession suppresses the PL polarization. Since in type II GaAs/AlAs superlattices there are two kinds of localized excitons with ω_2 differing in sign, the conversion terms in Eqs.(4.25,4.26) have to be multiplied by the imbalance factor

$$f = \frac{N^{(+)} - N^{(-)}}{N^{(+)} + N^{(-)}} \ ,$$

where $N^{(\pm)}$ is the concentration of excitons localized at the AlAs-on-GaAs and GaAs-on-AlAs interfaces.

With allowance for the spin relaxation one has to include the inactive states $|\pm 2\rangle$ into consideration and analyze a four-level system. As a result Eq. (4.26) undergoes a transformation to [35]

$$P_c = b \frac{\omega_\parallel^2 P_c^0 - f\omega_2 \omega_\parallel P_l^0}{a\,\omega_\parallel^2 + \omega_2^2} \ , P_l = b \frac{-f\omega_2 \omega_\parallel P_c^0 + \omega_2^2 P_l^0}{a\,\omega_\parallel^2 + \omega_2^2} \ , P_{l'} = 0 \ , \tag{4.27}$$

where a and b are dimensionless parameters dependent on the exciton lifetimes and spin relaxation times. It is worth mentioning that equations similar to Eq. (4.27) are derived from the two-level model by introducing an anisotropic effective-pseudospin lifetime, $\tau_{s,\parallel}$ for S_z and $\tau_{s,\perp}$ for S_x, S_y. In this case the parameters a and b are given respectively by $T_\perp / T_\parallel$ and $T_\perp / \tau$, where the effective lifetimes of optical orientation and alignment are defined as $T_\parallel^{-1} = \tau^{-1} + \tau_{s,\parallel}^{-1}$ and $T_\perp^{-1} = \tau^{-1} + \tau_{s,\perp}^{-1}$.

In the following the abbreviation I_β^α is used to designate the intensity of secondary emission in the configuration (α, β) of the polarizer and analyzer where α, β are linear polarizations along the axes $[100], [010], [110], [1\bar{1}0]$ or circular polarizations σ_+, σ_-. In [35] instead of measuring the polarization degrees $P_c^\alpha = (I_{\sigma_+}^\alpha - I_{\sigma_-}^\alpha)/(I_{\sigma_+}^\alpha + I_{\sigma_-}^\alpha)$, P_l^α and $P_{l'}^\alpha$ the modulation technique is applied where the analyzer is in a fixed position and the sample is pumped by the incident light changing its polarization from circular or linear to orthogonal at a particular frequency (26.61 kHz). Measured values are the effective polarization degrees

$$\rho_\alpha^c = \frac{I_\alpha^{\sigma+} - I_\alpha^{\sigma-}}{I_\alpha^{\sigma+} + I_\alpha^{\sigma-}}, \rho_\alpha^l = \frac{I_\alpha^{110} - I_\alpha^{1\bar{1}0}}{I_\alpha^{110} + I_\alpha^{1\bar{1}0}}, \rho_\alpha^{l'} = \frac{I_\alpha^{100} - I_\alpha^{010}}{I_\alpha^{100} + I_\alpha^{010}}. \tag{4.28}$$

Fig. 5 displays the dependencies $\rho_\alpha^c(B_\parallel)$ and $\rho_\alpha^l(B_\parallel)$ measured on one of the type II GaAs/AlAs superlattice samples (called NMSL-7) at the PL spectral maximum. The polarization properties of the phonon replicas are similar to those of the no-phonon line. It is seen from Fig. 5a that $\rho_{\sigma_+}^c(B_\parallel)$ rapidly increases and saturates from 2.5% to 5% in weak magnetic fields $B_\parallel \approx 20$ G and then gradually increases up to the level of 20% at $B_\parallel = 2.5$ kG. Fig. 5b clearly demonstrates the field-induced orientation-to-alignment conversion: $\rho_{110}^c(B_\parallel)$ reaches a maximum value of 5% at $B_\parallel \approx 0.7$ kG and reverses its sign under the field inversion. Moreover, $\rho_{110}^c(B_\parallel)$ differs in sign from the measured dependence $\rho_{1\bar{1}0}^c(B_\parallel)$. The effect of longitudinal magnetic field upon the optical alignment is illustrated in Fig. 5c. Note that the main variation of ρ_{110}^l takes place at the same magnetic fields $B_\parallel \approx 0.7$ kG as for the function $\rho_{\sigma_+}^c(B_\parallel)$ in Fig. 5a. Fig. 5d shows that the orientation-to-alignment effect is reversible: the experimental dependencies $\rho_{110}^c(B_\parallel)$ and $\rho_{\sigma_+}^l(B_\parallel)$ are close to each other.

The solid curves in Fig. 5 are calculated by using Eq. (4.27). We attribute the fast low-field increase of $\rho_{\sigma_+}^c$ in Fig. 5a to a spatially separated electron-hole pairs characterized by small values of exchange splittings. In the analysis this contribution is taken into account by adding a constant value of 5% to the theoretical curve $\rho_{\sigma_+}^c(B_\parallel)$. Except for this narrow region the data obtained on the sample NMSL-7 can be described by using the one-step model and neglecting any influence of the field on the exciton polarization at the intermediate stage before the exciton is trapped into a localized state. The products bP_c^0 and bP_l^0 in Eq. (4.27) for the excitation respectively by σ_+- and [110]-linearly-polarized light are considered as fitting parameters because of possible losses in the orientation and alignment during the photoexcitation process. Note that the data obtained from another sample [35] can be understood only in terms of a cascade two-step model which allows one to deduce the exciton lifetime at the intermediate stage along with the localized-exciton parameters. In addition, the magnetic-field-induced conversion between the circular and linear polarizations suggests an effective method to measure an important structural parameter, namely, the imbalance factor f. The proposed approach has very promising perspective to be applied for other low-dimensional systems, in particular in quantum wires and quantum dots.

The anisotropic exchange splitting, $\hbar\omega_2$, of excitonic levels in type II GaAs/AlAs(001) cannot be explained in terms of the long-range exchange interaction in excitons localized by laterally-anisotropic islands because the exciton oscillator strength in the type II heterostructures is too small. The theory [33,34] is based on the interface-induced mixing between the heavy- and light-hole states under the hole normal incidence. This mixing allowed by the point symmetry C_{2v} of an ideal GaAs/AlAs(001) interface was postulated in [33] by including additional terms in the boundary conditions for the hole envelope function at interfaces

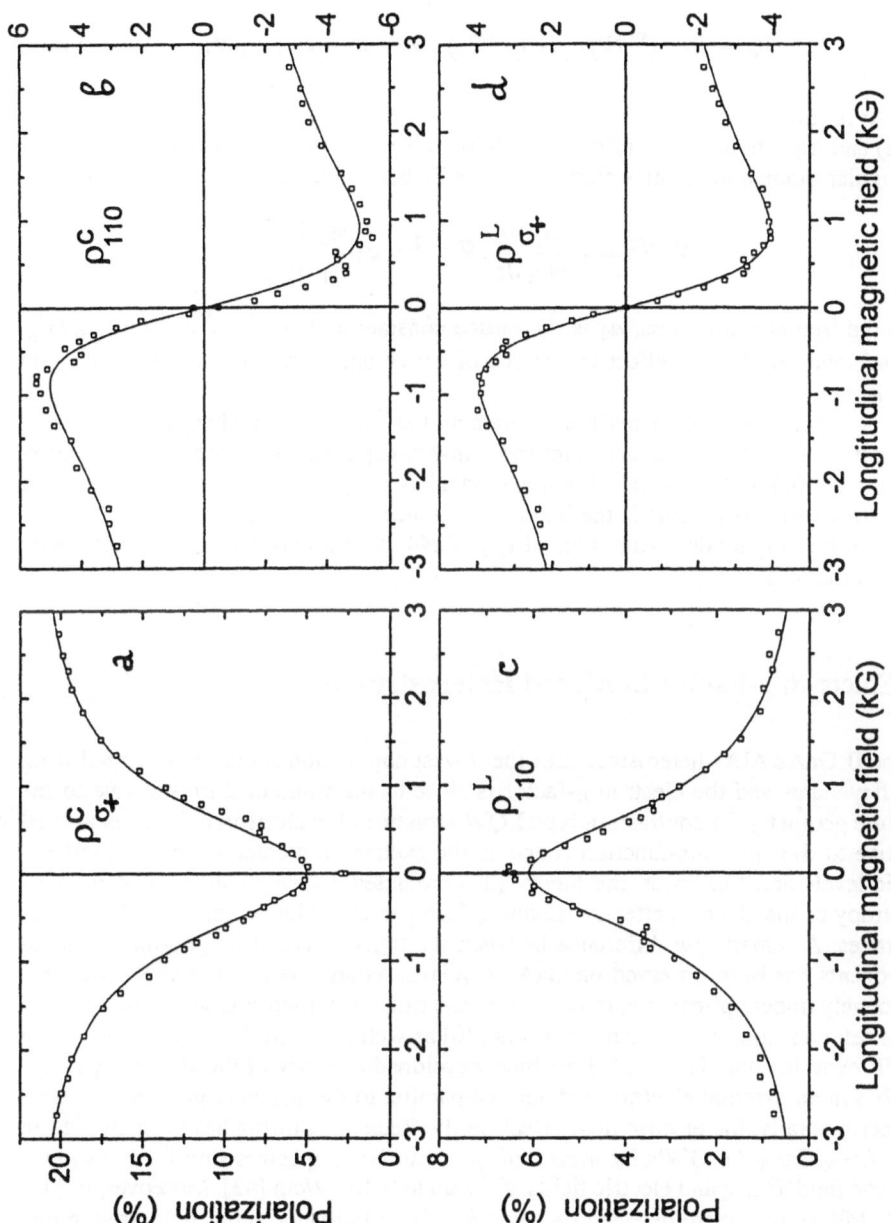

Figure 5 Effect of the longitudinal magnetic field on optical orientation and alignment of localized excitons in the 23Å/16Å GaAs/AlAs superlattice: (a) $\rho^c_{\sigma_+}(B_z)$, (b) $\rho^c_{110}(B_z)$, (c) $\rho^l_{110}(B_z)$, (d) $\rho^l_{\sigma_+}(B_z)$. Experimental data (T = 4.2 K) are shown by points. Solid curves are theoretical fits. [35]

$$\Phi_A = \Phi_B, \left(\nabla^j \Phi_j\right)_A = \left(\nabla^j \Phi_j\right)_B + \frac{2}{\sqrt{3}} t_{l\text{-}h} \{J_x J_y\}_{jj'} \Phi_{j'}. \tag{4.29}$$

Here Φ is the four-component column of the Γ_8 envelopes Φ_j for heavy-hole ($j = \pm 3/2$) and light-hole ($j = \pm 1/2$) states, J_α ($\alpha = x,y,z$) is the 4×4 matrix consisting of the angular momentum matrix elements in the Γ_8 basis, $\{J_x J_y\} = (J_x J_y + J_y J_x)/2$,

$$\nabla^{\pm 3/2} = a_0 \frac{m_0}{m_{hh}} \frac{\partial}{\partial z}, \nabla^{\pm 1/2} = a_0 \frac{m_0}{m_{lh}} \frac{\partial}{\partial z},$$

m_0 is the free electron mass, a_0 is the lattice constant and we use the notations m_{hh}^A, m_{hh}^B, m_{lh}^A and m_{lh}^B for the effective masses of heavy and light holes in the GaAs and AlAs layers.

An agreement between experimental data on the anisotropic exchange splitting and the theory that also takes into account the spin-orbit split valence band has been achieved for $t_{l\text{-}h} = 0.5$. In [34], a tight-binding model has been used to relate the microscopic parameters with coefficients in the boundary conditions for the hole envelope function. The tight-binding model estimation of $t_{l\text{-}h} = 0.44$ is in a reasonable agreement with other estimations of $t_{l\text{-}h}$.

5 Electron g-Factor in Biased Heterostructures

In type II GaAs/AlAs heterostructures the lowest conduction subband is formed from X_z-valley states and the electron g-factor is close to the value of 2 irrespective to the structure geometry. In contrast, in type I QW structures the electron g-factor is as well sensitive to the envelope-function shape as the exchange interaction splittings of excitonic levels are. Moreover, the theory [36] predicted a strong confinement-induced anisotropy of the electron effective Landé g factor in zinc-blende-based QW's and superlattices. A remarkable difference between the transverse and longitudinal g factor components has been observed on GaAs/Al$_x$Ga$_{1-x}$As and GaAs/AlAs QW structures respectively under optical orientation of free electrons in a tilted magnetic field [37–39] and in resonant spin-flip Raman scattering [40] as well as on Ga$_x$In$_{1-x}$As/InP QW's in ODMR experiments [41]. In [42] we have developed a theory of the electron g factor in QW's in an external electric field applied parallel to the structure growth direction, z. Experimentally, the electric field effect on the Zeeman splitting has been studied in GaAs/Al$_{0.35}$Ga$_{0.65}$As QW's by means of quantum beats spectroscopy in an in-plane magnetic field $B \perp z$ and electric fields $F \parallel z$ up to $9 \cdot 10^4$ V/cm [43]. Moreover, in [41] the ODMR signals were detected on samples with a built-in electric field arising due to one-sided p-modulation doping. The electric field reduced the overlap integrals between the electron and hole wavefunctions and, thus, caused an increase in the radiative recombination lifetime by two orders of magnitude.

Taking into account the D_{2d} point symmetry of an unbiased quantum-well structure grown along the direction [001], the effective 2×2 Hamiltonian for the lowest conduction subband $e1$ can be presented as a sum of two contributions

$$\mathcal{H}_c = E_c^0 + \frac{\hbar^2 k^2}{2M} + CF(k_x B_y - k_y B_x) \tag{5.30}$$
$$+ \eta F(\sigma_x k_y - \sigma_y k_x) + \frac{1}{2} g_\parallel \mu_B \sigma_z B_z + \frac{1}{2} g_\perp \mu_B (\sigma_x B_x + \sigma_y B_y),$$
$$\mathcal{H}_c' = \lambda(k_x B_x - k_y B_y) + \beta(\sigma_x k_x - \sigma_y k_y) + \zeta F(\sigma_x B_y + \sigma_y B_x). \tag{5.31}$$

Here $\alpha = x,y,z$ are the principal symmetry axes [100], [010] and [001], $k = (k_x, k_y)$ is the electron in-plane wave vector, σ_α are the Pauli spin matrices, $\mathbf{B}$ is the external magnetic field, μ_B is the Bohr magneton, F is the electric field assumed to be parallel to the growth direction, and the parameters $M, C, \eta, g_\parallel, g_\perp, \lambda, \beta$ and ζ are F^2-dependent. In addition to the conventional parabolic dispersion, we retain in equations (5.30, 5.31) spin-dependent and spin-independent terms linear in $\mathbf{k}$ and $\mathbf{B}$. The first contribution $\mathcal{H}_c$ is axially symmetrical while $\mathcal{H}_c'$ arises due to the lack of an inversion center in the compositional materials. The Zeeman effect is described by the g factor tensor components $g_{xx} = g_{yy} \equiv g_\perp$, $g_{zz} \equiv g_\parallel$ and $g_{xy} = g_{yx} = \zeta F$. The electric-field-induced in-plane anisotropy of the g factor arising due to the nonzero off-diagonal components g_{xy}, g_{yx} has been predicted by Korenev and Kalevich [44] and recently observed by Hallstein et al. [43]. Here we concentrate on the calculation of the diagonal components $g_\perp$ and $g_\parallel$.

The electric field effect on the longitudinal and transverse g factor components in the lattice-matched $Ga_{0.47}In_{0.53}As/InP$ QW structures is demonstrated in Fig. 6. The band parameters used in the calculation are as follows: the band gap $E_g = 0.813$ eV, the spin-orbit splitting of the valence band $\Delta = 0.356$ eV, $2p_{cv}^2/m_0 = 25.5$ eV for bulk $Ga_{0.47}In_{0.53}As$, $E_g = 1.423$ eV, $\Delta = 0.108$ eV, $2p_{cv}^2/m_0 = 20.4$ eV for bulk InP, the valence band offset $\Delta E_v = 0.356$ eV. In order to take account of the contribution to the g factor from remote bands, we added constants of $\Delta g = -0.13$ and -0.15 to the bulk values of g respectively for the well and barrier materials in which case $g(Ga_{0.47}In_{0.53}As) = -4.5$ and $g(InP) = 1.2$. The dotted curve in Fig. 6b presents the electric-field dependence of $g_\perp$ for a single heterointerface. In this case the transverse g factor increases from the bulk value of -4.5 at zero field to values $g_\perp > -3$ at $F > 1.25$ mV/Å. Comparing Fig. 6a and Fig. 6b one concludes that the quantum confinement in unbiased QW's has a stronger influence on $g_\perp$ rather than on $g_\parallel$ in agreement with the results of Kowalski et al. [41]. In QW's, at low electric fields, the both g factor components can be approximated by a sum of field-independent and quadratic-in-F terms. At sufficiently high electric fields the influence of the second interface upon the electron confinement becomes negligible which explains why in Fig. 6b the curves $g_\perp(F)$ for the 150 Å and 200 Å QW's approach asymptotically the dotted curve. For the 100 Å QW, the similar asymptotic behavior takes place at the higher fields. The dotted curve in Fig. 6a is calculated in the one-band approximation by averaging the bulk g factor values $g(k_z) = g(0) + hk_z^2$ in the well and barrier (see for details [36]). Recall that in this approximation the electron g factor remains isotropic.

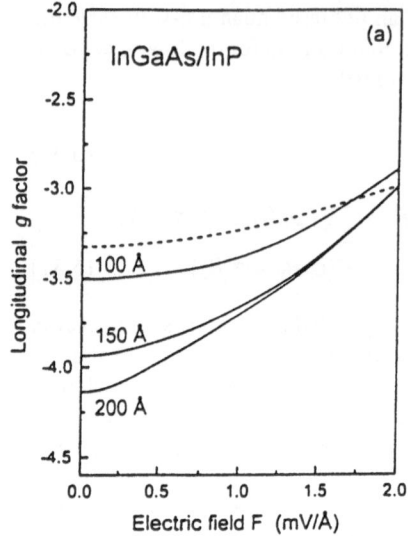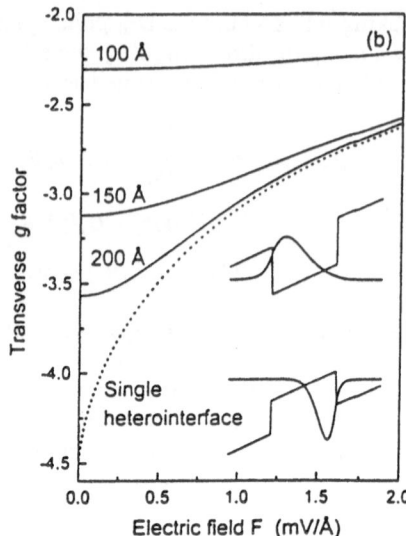

Figure 6 The longitudinal (a) and transverse (b) electron g factor as a function of the electric field applied to $Ga_{0.47}In_{0.53}As/InP$ QW structures with the well width of 100 Å, 150 Å and 200 Å. The dotted curve in (b) presents the transverse g factor for a biased single heterointerface. The dashed curve in (a) is calculated for a 100 Å QW in the single-band approximation, other curves are calculated in the Kane model. Inset: The schematic representation of the electron and heavy-hole envelope functions in a biased QW. [42]

Fig. 7 shows the electric-field dependence of the transverse g factor for a GaAs/ $Al_{0.35}Ga_{0.65}As$ QW structure. The parameters used are $E_g = 1.52$ eV, $\Delta = 0.34$ eV, $2p_{cv}^2/m_0 = 28.9$ eV for bulk GaAs, and $E_g = 1.94$ eV, $\Delta = 0.32$ eV, $2p_{cv}^2/m_0 = 26.7$ eV (dashed and solid curves) and 24.7 eV (dotted curves) for the barrier material, the band offset $\Delta E_v : \Delta E_c = 2 : 3$. The contribution from remote bands is taken into account by adding a constant of $\Delta g = -0.12$ to the values of $g_{\parallel}$, $g_{\perp}$ calculated in the Kane model. The difference between dashed and dotted curves in Fig. 7 illustrates the sensitivity of $g_{\parallel,\perp}$ to the barrier interband matrix element. Note that, for the above two values of p_{cv}, the g factor in the bulk barrier material is equal respectively to 0.57 and 0.67 and, for a 100 Å QW GaAs/$Al_{0.35}Ga_{0.65}As$, $g_{\perp}(F = 0) = -0.17$ and -0.13. The calculated field-induced contribution $g_{\perp}(F = 0.56 \text{ meV/Å}) - g_{\perp}(0) = 0.007$ is in a reasonable agreement with the measured value ~ 0.01 [43].

6 Nonlinear Optics of Quantum Microcavities

A natural way to enhance the light-matter interaction is to exploit band-edge resonances by tuning the incident-light frequency to the exciton-resonant spectral region.

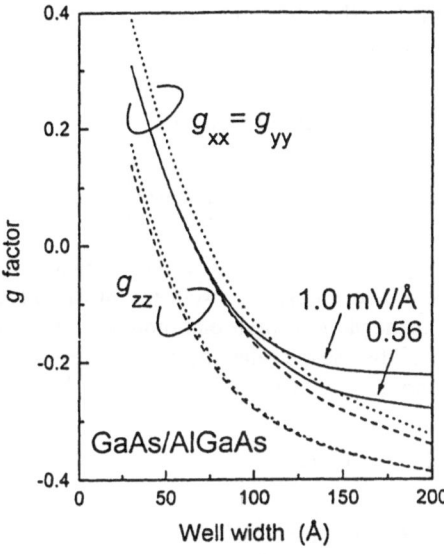

Figure 7
The electron g factor as a function of the GaAs/Al$_{0.35}$Ga$_{0.65}$As QW width. Dashed (dotted) curves represent the transverse and longitudinal g factor components in an unbiased QW calculated assuming that the bulk value of the g factor in the barrier material equals to 0.57 (0.67). The transverse g factor calculated for a biased QW with the electric field $F = 0.56$ and 1.0 meV/Å and for $g(\text{Al}_{0.35}\text{Ga}_{0.65}\text{As}) = 0.57$ is shown by solid curves. [42]

The quantum confinement of excitons in semiconductor nanostructures leads to the further increase in the resonant optical response. The photon confinement in microcavities with embedded QW's has opened a way for additional considerable enhancement of exciton-photon coupling [45–50].

The quantum microcavities can be particularly interesting for nonlinear optics since the nonlinear response is characterized by a stronger dependence upon the coupling constant. In [51], we have developed a theory of degenerate four-wave mixing in a semiconductor planar microcavity containing a QW between the cavity mirrors. A sequence of two coherent pulses with the wavevectors k_1 and k_2 creates a third-order dielectric polarization which serves as a source for a new light wave propagating in the direction $2k_2 - k_1$. The study of four-wave mixing near exciton resonances provides much information on the exciton dynamics and loss of coherence as well as on nonlinear mechanisms of exciton-photon interaction in semiconductor heterostructures [11,52–60]. The first experiments on four-wave mixing response of quantum wells embedded in GaAs/AlGaAs Fabry-Pérot microcavities have been recently reported [61,62].

We consider a multilayered heterostructure grown by using four compositional materials labelled as A,B,C,D and consisting of N_1 pairs of the C/B distributed Bragg reflector (DBR), an active region A with a quantum well D in its center and N_r pairs of the B/C DBR grown on the substrate A. The optical properties of the DBR's, or mirrors, are characterized by the amplitude reflection and transmission coefficients: r_{mj}, r'_{mj} (reflection from the left, $j = l$, and right, $j = r$, mirror for the light incident respectively from the active layer and from the external medium, vacuum or substrate) and t_{mj}, t'_{mj} (the same but for transmission through the mirror j).

The basic equations describing dynamics of the photonic mode in the microcavity and the exciton in the QW can be written as

$$\left[\frac{\partial}{\partial t} + i(\omega_0 - \bar{\omega}) + \Gamma\right] \bar{P}(t) = i\eta\Gamma_0\bar{E}(t) + \bar{F}_{NL}(t), \qquad (6.32)$$

$$\left[\frac{\partial}{\partial t} + i(\omega_c - \bar{\omega}) + \gamma_c\right] \bar{E}(t) = \bar{\Gamma}\left[\frac{i}{\eta}\bar{P}(t) + \frac{t'_{ml}}{2}\bar{E}_0(t)\right].$$

Here $\bar{\omega}$ is the current frequency of the light pulses, ω_0 and ω_c are the exciton and bare-photon-mode resonance frequencies, Γ and Γ_0 are the exciton non-radiative and radiative damping rates in a single QW, γ_c is the photon-mode damping rate determining the cavity linewidth, it is a sum, $\gamma_{cl} + \gamma_{cr}$, of the escape rates through the left and right mirrors, the parameter $\bar{\Gamma}$ is related to γ_{cj} and r_{mj} by [50]

$$\gamma_{cj} = \frac{1}{8}\left(1 - |r_{mj}|^2\right)\bar{\Gamma}, \qquad (6.33)$$

$\bar{P}(t)$ is the slowly-varying amplitude of the excitonic polarization averaged over the QW width L, $\bar{E}(t)$ is the electric field in the center of the QW, $\bar{E}_0(t)$ is the slowly-varying amplitude of the initial radiation. It consists of two pulses and, say, for Lorentzian pulses is given by

$$E_{0,1} \exp\left(-|t - t_1|/\tau_p\right) \exp\left(ik_{1,x}x\right) + E_{0,2} \exp\left(-|t - t_2|/\tau_p\right) \exp\left(ik_{2,x}x\right),$$

if (x,z) is the plane of incidence and $k_j = (k_{jx}, 0, k_{jz})$. Usually in four-wave mixing experiments the k_1 and k_2 vectors make small angles with the sample normal z and one may ignore an effect of the z component of the electric field and neglect angle dependencies of the reflection, transmission and nonlinear mixing coefficients. We omit everywhere the exponential functions $\exp(ik_{jx}x)$ bearing in mind that E_1, E_2 and E_3 describe the amplitudes of the three waves k_1, k_2 and $k_3 = 2k_2 - k_1$ independently detected. Other notations introduced above are as follows

$$\eta = \frac{\varepsilon_b}{2\pi} \frac{1}{kL}, \qquad (6.34)$$

$k = (\omega_0/c)n_c$ and the difference between n_c^2 and the background dielectric constant, ε_b, of the QW material is neglected. While deriving Eqs. (6.32) we considered thin enough QW's satisfying the condition $kL \ll 1$. The term $\bar{F}_{NL}(t)$ is a nonlinear part giving rise to the third-order polarization and consisting of the so-called anharmonic-oscillator-like, P^3, and two-level-like, P^2E, contributions [54,55,60,63]

$$\bar{F}_{NL}(t) = i|\bar{P}(t)|^2\left[\beta_1\bar{P}(t) + \beta_2\bar{E}(t)\right], \qquad (6.35)$$

where β_1 and β_2 are constant coefficients, and of the biexcitonic contribution. If the nonlinear term is neglected the transmission coefficient from vacuum to the substrate is given by [64]

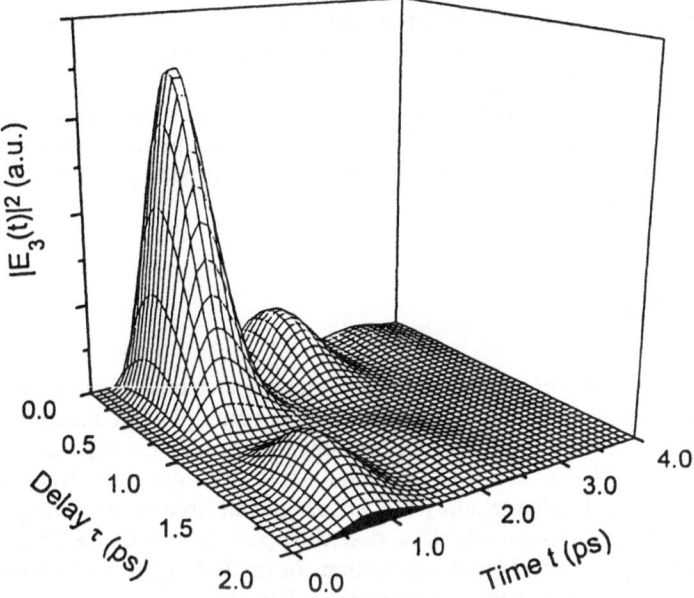

Figure 8 The calculated time-resolved $2k_2 - k_1$ signal for different time delays between the light pulses. The set of parameters used in the calculation is given in the text. The real time t is referred to the peak of pulse 2. The pattern reveals the damped oscillatory behavior in both τ- and t-dependencies. [51]

$$T = \left| \frac{2\sqrt{\gamma_{cl}\gamma_{cr}}\,(\omega - \omega_0 + i\Gamma)}{(\omega - \omega_c + i\gamma_c)(\omega - \omega_0 + i\Gamma) - \omega_R^2} \right|^2, \qquad (6.36)$$

where $2\omega_R$ is the so-called Rabi splitting between the exciton-polariton modes in the microcavity.

Fig. 8 illustrates the calculated evolution of time-dependent self-diffracted signal, $|E_3(t)|^2$, with increasing the interpulse delay for the resonant case $\bar{\omega} = \omega_0 = \omega_c$ and for the two-level-like mechanism of nonlinearity. We used the $Al_{0.2}Ga_{0.8}As$ λ-cavity parameters of Ref. [45] with the distributed Bragg reflectors comprising 24 $Al_{0.4}Ga_{0.6}As/$ AlAs stacks on the cavity front side (air interface) and 33 stacks on the substrate side. The refractive indices are as follows: $n_{ext,l} = 1$, $n_{ext,r} = 3.63$, $n_1 = 3.17$, $n_2 = 3.45$, $n_c = 3.54$. The calculated values of γ_c and $\bar{\Gamma}$ are 0.51 ps^{-1} and 117 ps^{-1}. A single QW was assumed to be embedded in the center of the active layer and values of $\Gamma = 1.0\,ps^{-1}$ and $\Gamma_0 = 0.05\,ps^{-1}$ were chosen for the exciton non-radiative and radiative damping rates. In this case the Rabi splitting $2\omega_R$ equals to 4.8 ps^{-1}. The signal in Fig. 8 exhibits quantum beats between the exciton and cavity modes, and a significant beat-like modulation is exposed in the dependencies on both t and τ. For the P^2E- and P^3-like nonlinearities the four-wave-mixing time-intergrated signal consists of monotonous and monoharmonic oscillating components. The both decay exponentially as a function of the delay time with the exponent determined by a sum of the photon and exciton damping rates. For the biexciton-mediated nonlinearity, the oscillating part is a damped multiharmonic signal containing overtones of $2\omega_R$ and the biexciton binding energy [51].

Kavokin et al. [65] have studied exciton-polaritons in microcavities with embedded quantum wires. The resonant diffraction of light at the long-period grating of quantum wires has been shown to be responsible for a unique four-mode polaritonic dispersion in the microcavity.

7 Conclusion

We have analyzed effects of the dielectric constant mismatch on the reflection and transmission spectra of the resonant Bragg and quasi-Bragg multiple-QW structures, showed that the long-range exchange interaction modifies significantly the fine structure of excitons localized in QW's or confined in quantum dots, demonstrated advantages of the optical orientation technique to study the anisotropic electron-hole exchange interaction in type II GaAs/AlAs superlattices, described peculiarities of the Zeeman effect in heterostructures, and considered implications of the Rabi splitting on the degenerate four-wave mixing in quantum microcavities. The exchange interaction within the exciton, electron and hole g factors and the polariton effect are extremely sensitive both to the structure geometry and to the exciton envelope function shape which makes the spectroscopic methods very effective for the characterization of semiconductor microstructures.

Acknowledgments

The author is grateful to S. V. Goupalov, A. A. Kiselev, A. V. Platonov and M. N. Tkachuk for useful discussions. Support from INTAS grant 93-3657 and Volkswagen Foundation is gratefully acknowledged.

Bibliography

[1] E.L. Ivchenko, in *Excitons*, edited by E.I. Rashba and M.D. Sturge (North-Holland, Amsterdam, 1982), p. 141.

[2] L.C. Andreani and F. Bassani, Phys. Rev. B **41**, 7536 (1990); L.C. Andreani, F. Tassone, and F. Bassani, Solid State Commun. **77**, 641 (1991).

[3] E.L. Ivchenko, Sov. Phys. Solid State **33**, 1344 (1991).

[4] E.L. Ivchenko, V.P. Kochereshko, A.V. Platonov, D.R. Yakovlev, A. Waag, W. Ossau, and G. Landwehr, Phys. Solid State **39**, N11 (1997).

[5] E.L. Ivchenko, A.I. Nesvizhskii, and S. Jorda, Phys. Solid State **36**, 1156 (1994); Superlatt. Microstruct. **16**, 17 (1994).

[6] Y. Merle d'Aubigné, A. Wasiela, H. Mariette, and A. Shen, in: *Proceeding of the 22nd International Conference of the Physics of Semiconductors*, Vancouver 1994, edited by D.J. Lockwood (World Scientific, Singapore, 1995) p. 1201.

[7] V.P. Kochereshko, G.R. Pozina, E.L. Ivchenko, D.R. Yakovlev, A. Waag, W. Ossau, G. Landwehr, R. Hellmann, and E.O. Göbel, in *Proceeding of the 22nd International Conference of the Physics of Semiconductors*, Vancouver 1994, edited by D.J. Lockwood (World Scientific, Singapore, 1995) p.1372; Superlatt. Microstruct. **15**, 471 (1994).

[8] V.A. Kosobukin and M.M. Moiseeva, Phys. Solid State **37**, 2036 (1995).

[9] D.R. Yakovlev, G.R. Pozina, V.P. Kochereshko, A. Waag, W. Ossau, and G. Landwehr, JETP Letters **61**, 628 (1995).

[10] T. Stroucken, A. Knorr, P. Thomas, and S.W. Koch, Phys. Rev. B **53**, 2026 (1996).

[11] M. Hübner, J. Kuhl, T. Stroucken, A. Knorr, S.W. Koch, R. Hey, and K. Ploog, Phys. Rev. Lett. **76**, 4199 (1996).

[12] Y. Merle d'Aubigne, A. Wasiela, H. Mariette, and T. Dietl, Phys. Rev. B **54**, 14003 (1996).

[13] G.E. Pikus and G.L. Bir, Sov.Phys.-JETP **33**, 108 (1971); **35**, 174 (1972).

[14] S.V. Goupalov, E.L. Ivchenko, and A.V. Kavokin, Proc. Int. Symposium 'Nanostructures: Physics and Technology', St. Petersburg 1996, p. 322; Superlatt. Microstruct., to be published.

[15] D. Gammon, E.S. Snow, B.V. Shanabrook, D.S. Katzer, and D. Park, Phys. Rev. Lett. **76**, 3005 (1996).

[16] E. Blackwood, M.J. Snelling, R.T. Harley, S.R. Andrews, and C.B.T. Foxon, Phys. Rev. B **50**, 14246 (1994).

[17] S.V. Goupalov, E.L. Ivchenko, and A.V. Kavokin, to be published.

[18] G. Bastard and J.Y. Marzin, Solid State Commun. **91**, 39 (1994).

[19] M. Nirmal, D.J. Norris, M. Kuno, M.G. Bawendi, Al.L. Efros, and M. Rosen, Phys. Rev. Lett. **75**, 3728 (1995); Al.L. Efros, M. Rosen, M. Kuno, M. Nirmal, D.J. Norris, and M.G. Bawendi, Phys. Rev. B **54**, 4843 (1996).

[20] M. Chamarro, M. Dib, C. Gourdon, P. Lavallard, O. Lublinskaya, and A.I. Ekimov, Phys. Rev. B **53**, 1336 (1996).

[21] U. Woggon, F. Gindele, O. Wind, and C. Klingshirn, Phys. Rev. B **54**, 1506 (1996).

[22] G.E. Pikus and E.L. Ivchenko, in *Excitons*, edited by E.I. Rashba and M.D. Sturge (North-Holland, Amsterdam, 1982), p. 205.

[23] R. Planel and C. Benoit a la Guillaume, in *Optical Orientation*, edited by F. Meier and B.P. Zakharchenya (North-Holland, Amsterdam, 1984), p. 353.

[24] E.L. Ivchenko, Pure & Appl. Chem. **67**, 463 (1995).

[25] E.L. Ivchenko and G.E. Pikus, Superlattices and Other Heterostructures. Symmetry and Optical Phenomena, Springer-Verlag, 1995.

[26] W.A.J.A. van der Poel, A.L.G.J. Severens, and C.T. Foxon, Optics Commun. **76**, 116 (1990).

[27] H. W. van Kesteren, E. C. Cosman, W. A. J. A. van der Poel, and C. T. Foxon, Phys. Rev. B **41**, 5283 (1990).

[28] S. Permogorov, A. Naumov, C. Gourdon, and P. Lavallard, Solid State Commun. **74**, 1057 (1990).

[29] E.L. Ivchenko, V.P. Kochereshko, A.Yu. Naumov, I.N. Uraltsev, and P. Lavallard, Superlatt. Microstruct. **10**, 497 (1991).

[30] C. Gourdon and P. Lavallard, Phys. Rev. B **46**, 4644 (1992).

[31] C. Gourdon, D.Yu. Rodichev, P. Lavallard, G. Bacquet, and R. Planel, J. de Physique IV, v.3, Coll.n°5, Suppl. JPII (Proc. 3rd Int. Conf. Optics of Excitons in Confined Systems, Montpellier 1993), p.183 (1993).

[32] C. Gourdon, I.V. Mashkov, P. Lavallard, and R. Planel, Proc. of ICPS-23, Berlin 1996, to be published.

[33] I. L. Aleiner and E. L. Ivchenko, Pis'ma Zh. Eksp. Teor. Fiz. **55**, 662 (1992) [JETP Letters **55**, 692 (1992)]; E. L. Ivchenko, A. Yu. Kaminski, and I. L. Aleiner, Zh. Eksp. Teor. Fiz. **104**, 3401 (1993) [JETP **77**, 609 (1993)].

[34] E.L. Ivchenko, A.Yu. Kaminski, and U. Rössler, Phys. Rev. B **54**, 5852 (1996).

[35] R.I. Dzhioev, H.M. Gibbs, E.L. Ivchenko, G. Khitrova, V.L. Korenev, M.N. Tkachuk, and B.P. Zakharchenya, Proc. 23rd Int. Symp. Compound Semicond. (St. Petersburg, Russia, 1996) Inst. Phys. Conf. Ser. No. 155, ch. 2, p. 173; Phys. Rev. B **57** (1997).

[36] E.L. Ivchenko and A.A. Kiselev, Sov. Phys. - Semicond. **26**, 827 (1992).

[37] E.L. Ivchenko, V.P. Kochereshko, I.N. Uraltsev, and D.R. Yakovlev, In *High Magnetic Fields in Semiconductor Physics III*, ed. by G. Landwehr, Springer Ser. Solid-State Sci., Vol. 101 (Springer, Berlin, Heidelberg 1992) p. 533.

[38] V.K. Kalevich and V.L. Korenev, JETP Lett. **56** 253 (1992).

[39] P. Le Jeune, D. Robart, X. Marie, T. Amand, M. Brousseau, J. Barrau, V. Kalevich, and D. Rodichev, Semicond. Sci. Technol. (1997).

[40] A.A. Sirenko, T. Ruf, K. Eberl, M. Cardona, A.A. Kiselev, E.L. Ivchenko, and K. Ploog, Proc. 12th Int. Conf. on the Application of High Magnetic Fields in Semiconductor Physics, (Würzburg 1996), ed. G. Landwehr and W. Ossau, World Scientific, 1996, p. 561.

[41] B. Kowalski, P. Omling, B.K. Meyer, D.M. Hofmann, C. Wetzel, V. Härle, F. Scholz, and P. Sobkowicz, Phys. Rev. B**49**, 14786 (1994).

[42] E.L. Ivchenko, A.A. Kiselev, and M. Willander, Solid State Commun. **102**, 375 (1997).

[43] S. Hallstein, M. Oestreich, W.W. Rühle, and K. Köhler, Proc. 12th Int. Conf. on the Application of High Magnetic Fields in Semiconductor Physics (Würzburg 1996), ed. G. Landwehr and W. Ossau, World Scientific, 1996, p. 593.

[44] V.K. Kalevich and V.L. Korenev, JETP Lett. **57** 571 (1993).

[45] C. Weisbuch, M. Nishioka, A. Ishikawa, and Y. Arakawa, Phys. Rev. Lett. **69**, 3314 (1992).

[46] R. Houdre, R.P. Stanley, U. Oesterle, M. Ilegems, and C. Weisbuch, Phys. Rev. B **49**, 16761 (1994).

[47] T.B. Norris, J.-K. Rhee, C.-Y. Sung, Y. Arakawa, M. Nishioka, and C. Weisbuch, Phys. Rev. B **50**, 14 663 (1994).

[48] L.C. Andreani, V. Savona, P. Schwendimann, and A. Quattropani, Superlatt. Microstruct. **15**, 453 (1994); V. Savona, L.C. Andreani, P. Schwendimann, and A. Quattropani, Solid State Commun. **93**, 733 (1995).

[49] S. Pau, G. Björk, J. Jacobson, H. Cao, and Y. Yamamoto, Phys. Rev. B **51**, 7090, 14 437 (1995).

[50] E.L. Ivchenko, M.A. Kaliteevski, A.V. Kavokin, and A.I. Nesvizhskii, J. Opt. Soc. Am. B **13**, 1061 (1996).

[51] Y. Fu, M. Willander, E.L. Ivchenko, and A.A. Kiselev, JETP Lett. **64**, 795 (1996); Phys. Rev B **55**, 9872 (1997).

[52] L. Schultheis, A. Honold, J. Kuhl, K. Köhler, and C.T. Tu, Phys. Rev. B **34**, 9027, (1986).

[53] K. Leo, M. Wegener, J. Shah, D.S. Chemla, E.O. Göbel, T.C. Damen, S. Schmitt-Rink, and W. Schäfer, Phys. Rev. Lett. **65**, 1340, (1990).

[54] M. Wegener, D.S. Chemla, S. Schmitt-Rink, and W. Schäfer, Phys. Rev. A **42**, 5675 (1990).

[55] S. Schmitt-Rink, S. Mukamel, K. Leo, J. Shah, and D.S. Chemla, Phys. Rev. A **44**, 2124 (1991).

[56] O. Carmel and I. Bar-Joseph, Phys. Rev. B **47**, 7606 (1993).

[57] S. Bar-Ad, I. Bar-Joseph, G. Finkelstein, and Y. Levinson, Phys. Rev. B **50**, 18375 (1994).

[58] J. Erland and I. Balslev, Phys. Rev. A **48**, R1765 (1993).

[59] J. Erland, K.-H. Pantke, V. Mizeikis, V.G. Lyssenko, and J.M. Hvam, Phys. Rev. B **50**, 15047 (1994).

[60] E.L. Ivchenko and A.I. Nesvizhskii, Pros. Int. Symp. "Nanostructures: Physics and Technology", St. Petersburg 1996, p.58.

[61] R. Shimano, S. Inouye, M. Kuwata-Gonokami, T. Nakamura, M. Yamanishi, and I. Ogura, Jpn. J. Appl. Phys. **34**, L817 (1995).

[62] F. Quochi, G. Bongiovanni, A. Mura, S. Gürtler, C. Dill, R. Houdre, and J.L. Staenli, Proc. 23nd Int. Conf. Phys. Semicond., Berlin 1996, ed. by M. Scheffler and R. Zimmermann, World Scientific, p. 3151.

[63] A. Schulze, A. Knorr, and S.W. Koch, Phys. Rev. B **51**, 10601 (1995).

[64] E.L. Ivchenko and A.V. Kavokin, JETP Lett. **62**, 710 (1995).

[65] A.V. Kavokin, E.L. Ivchenko, M.R. Vladimirova, M.A. Kaliteevki, and S.V. Goupalov, Superlatt. Microstruct., to be published.

Fano Resonances in the Optical Absorption of Low-dimensional Semiconductors

S. Glutsch

Friedrich-Schiller-Universität Jena,
Institut für Festkörpertheorie und Theoretische Optik,
Max-Wien-Platz 1, 07743 Jena

Abstract: Fano interference is a general phenomenon in the optical absorption of low-dimensional semiconductors, and is more the rule than the exception. The Fano interference is caused by the coupling of excitons belonging to higher sub-bands (Landau levels) to continua of lower subbands (Landau levels) via Coulomb interaction. We formulate conditions for Fano interference, identify low-dimensional semiconductors in which Fano resonances should be observed, and give an overview of theoretical and experimental results. The general theory is illustrated by calculating optical spectra of quantum-well wires.

1   Introduction

Fano interference is referred to as quantum mechanical coupling of a discrete state to a continuum of states. Fano [1] introduced a simple, exactly solvable model that was able to explain the phenomena observed in the optical spectra of rare gases. The resulting line shape is asymmetric and has a natural line width, i.e., an intrinsic inhomogeneous broadening that stems solely from the interaction of the discrete state with the continuum.

In semiconductors, the coupling of discrete and continuous states may result from electron-electron interaction, electron-phonon interaction, impurities, or specific band-gap alignments. Fano resonances appear in the transport properties [2,3], photocurrent measurements [4], Raman spectra [5], intersubband absorption [6], and interband absorption [7–10].

Quantum-size effects in low-dimensional semiconductors have been studied extensively in the last two decades [11], and are a vital part of today's semiconductor physics. The possibility of Fano interference was recognized long ago [12]. In 1989, Chu and Chang predicted a Fano resonance in quantum wells for the weakly allowed e1–hh3 transition [13], but the comparison with experiments was difficult because of the large inhomogeneous broadening which was introduced by the fabrication process. Only five years later, this Fano resonance was indeed observed by Oberli *et al.* [14] in asymmetric double quantum wells. In the same year, Willcox and Whittaker demonstrated numerically that allowed optical transition belonging to higher-order subbands should also show Fano interference [15]. These predictions were confirmed in experiments by Simmonds and co-workers [16] and by Bellani *et al.* [17]. According to numerical calculations, one-dimensional semiconductors (quantum wires) should reveal pronounced Fano resonances [18–20], but, until now, there has been no clear experimental evidence.

Besides the structural or geometrical confinement, magnetic fields can be used to force an electron-hole pair into a one-dimensional motion. In 1994, Fano resonances were both predicted theoretically and measured experimentally in Gallium Arsenide in a magnetic field [21].

The time-integrated four-wave-mixing signal of magnetic-field-induced Fano resonances, as studied by Siegner and co-workers [22], was found to show an anomalous behavior, in comparison with the time-resolved signal. Very recently, interesting results on the interplay of magnetic and geometrical confinement in wide quantum wells have been found by Bar-Ad *et al.* [23]. Since Fano resonance in the optical absorption of semiconductors is a relatively new subject, there is definitely more work necessary in order to accomplish a thorough understanding, especially in nonlinear optical spectroscopy.

The purpose of this paper is to show that Fano resonances are a general feature of the optical spectra in low-dimensional semiconductors and that the necessary coupling originates from the Coulomb interaction between electrons and holes. After a brief review of the Fano model in Sec. 2, we give in Sec. 3 a systematic description of the general mechanism of Fano interference in low-dimensional semiconductors that can be applied to various low-dimensional structures. We also identify possible candidates for Fano interference and give an overview of the present stand of theory and experiment. An explicit calculation for a one-dimensional structure is presented in Sec. 4, and a summary is given in Sec. 5.

2   Review of the Fano Model

The focus of this section is to briefly review the Fano model [1]. Based on simple algebra, we deduce the analytical form of the line shape. We then discuss the dependence of the line shape on the parameters of the model, the absorption of the discrete transition and the continuum, and the coupling matrix elements.

We consider a matrix A of the form

$$A = \begin{pmatrix} a_{11} & a_{12} & a_{13} & a_{14} & \cdots & a_{1n} \\ a_{21} & a_{22} & 0 & 0 & \cdots & 0 \\ a_{31} & 0 & a_{33} & 0 & \cdots & 0 \\ a_{41} & 0 & 0 & a_{44} & \cdots & 0 \\ \vdots & \vdots & \vdots & \vdots & \ddots & \vdots \\ a_{n1} & 0 & 0 & 0 & \cdots & a_{nn} \end{pmatrix} .$$

This matrix is sparse, i.e., only $O(n)$ elements are different from zero. The inversion can be performed by Gaussian elimination, and the elements of the inverse matrix $B = A^{-1}$ can be given in closed form, namely,

$$b_{1k} = \frac{a_{1k}}{a_{kk}} b_{11} \; ; \; b_{j1} = -\frac{a_{j1}}{a_{jj}} b_{11} \; ; \; b_{jk} = \frac{\delta_{jk}}{a_{jj}} + \frac{a_{j1} a_{1k}}{a_{jj} a_{kk}} b_{11} \quad (j,k = 2,\ldots,n)$$

$$b_{11} = \frac{1}{a_{11} - \sum\limits_{j=2}^{n} \dfrac{a_{1j} a_{j1}}{a_{jj}}} . \tag{2.1}$$

The optical susceptibility of a quantum-mechanical system is defined as ($\hbar =$ prefactors $= 1$):

$$\chi(\omega) = \sum_{\lambda} \frac{|\langle \varphi | \Phi_\lambda \rangle|^2}{\mathcal{E}_\lambda - (\omega + i0^+)} . \tag{2.2}$$

Here, $|\varphi\rangle$ is the initial state and 0^+ is a positive infinitesimal. The final states $|\Phi_\lambda\rangle$ and the corresponding transition energies $\mathcal{E}_\lambda$ are eigenstates and eigenvalues of a Hamiltonian $\hat{H}$, according to

$$\hat{H} | \Phi_\lambda \rangle = \mathcal{E}_\lambda | \Phi_\lambda \rangle \; ; \; \| \Phi_\lambda \| = 1 .$$

In a general base set, expression (2.2) writes

$$\chi(\omega) = \langle \varphi | [\hat{H} - (\omega + i0^+)]^{-1} | \varphi \rangle , \tag{2.3}$$

Up to prefactors, the optical absorption α is given by the imaginary part of χ. By virtue of Dirac's identity,

$$\frac{1}{\omega \pm i0^+} = \frac{P}{\omega} \mp i\pi\delta(\omega) \tag{2.4}$$

(P – principal value), it holds that

$$\alpha(\omega) \propto \sum_{\lambda=0}^{\infty} |\langle \varphi | \Phi_\lambda \rangle|^2 \delta(\omega - \mathcal{E}_\lambda) . \tag{2.5}$$

The Fano model describes the coupling of a discrete state with energy E_0 to a continuum with energies $E \in [E_{min}, E_{max}]$. The Hamiltonian and the initial state take the form

$$\hat{H} = \begin{pmatrix} E_0 & V_1^* & V_2^* & V_3^* & \cdots \\ V_1 & E_1 & 0 & 0 & \cdots \\ V_2 & 0 & E_2 & 0 & \cdots \\ V_3 & 0 & 0 & E_3 & \cdots \\ \vdots & \vdots & \vdots & \vdots & \ddots \end{pmatrix} \; ; \; |\varphi\rangle = \begin{pmatrix} \varphi_0 \\ \varphi_1 \\ \varphi_2 \\ \varphi_3 \\ \vdots \end{pmatrix}. \tag{2.6}$$

The continuum is represented by the discrete set $\{E_j\}$ $(j > 0)$ and the quantum mechanical coupling is mediated by the matrix elements V_j.

In order to perform the continuous limit, we introduce a density of states D such that

$$\sum_{j=1}^{\infty} f(E_j) = \int_{E_{min}}^{E_{max}} dE\, D(E)\, f(E)$$

for any function f that is integrable in the interval $[E_{min}, E_{max}]$ and, furthermore, we define the quantities

$$V_j = \frac{v(E_j)}{\sqrt{D(E_j)}} \; ; \; \varphi_j = \frac{\psi(E_j)}{\sqrt{D(E_j)}}$$

for $j > 0$.

For simplicity, we assume that v, φ_0 and ψ are real functions. A possible generalization will be discussed later. With the result (2.1) for the inverse matrix, the explicit form of the optical susceptibility (2.3) is

$$\chi(\omega) = \frac{(\varphi_0 - G - \frac{1}{2}i\Delta)^2}{E_0 - F - \omega - \frac{1}{2}i\Gamma} + K + \frac{1}{2}i\Lambda, \tag{2.7}$$

with the abbreviations

$$F(\omega) = P\int dE\, \frac{|v(E)|^2}{E - \omega} \; ; \qquad \frac{1}{2}\Gamma(\omega) = \pi|v(\omega)|^2$$

$$G(\omega) = P\int dE\, \frac{v(E)\,\psi(E)}{E - \omega} \; ; \qquad \frac{1}{2}\Delta(\omega) = \pi v(\omega)\,\psi(\omega) \tag{2.8}$$

$$K(\omega) = P\int dE\, \frac{|\psi(E)|^2}{E - \omega} \; ; \qquad \frac{1}{2}\Lambda(\omega) = \pi|\psi(\omega)|^2.$$

Introducing the dimensionless parameters

$$\varepsilon = \frac{E_0 - F - \omega}{\frac{1}{2}\Gamma} \; ; \; q = \frac{\varphi_0 - G}{\frac{1}{2}\Delta},$$

and taking into account the fact that $\Gamma\Lambda = \Delta^2$, the ratio A between $\alpha(\omega)$ and the continuum absorption $\alpha_{cont}(\omega)$ takes the simple form

$$A = \frac{\alpha(\omega)}{\alpha_{cont}(\omega)} = \frac{\mathrm{Im}\chi(\omega)}{\frac{1}{2}\Lambda} = \frac{(q+\varepsilon)^2}{1+\varepsilon^2}. \tag{2.9}$$

In Fig. 1, the function $A(\varepsilon,q)$ is plotted versus the normalized energy ε for various line-shape parameters $q = 0, 0.5, 1, 2$, and 3. It holds that $A(\varepsilon,q) = A(-\varepsilon, -q)$.

If Γ, Δ, and Λ are taken as constant in the energy range of interest, then A describes the line shape of the optical transition as function of the normalized energy ε. It is worthwhile to note that both the energy of the discrete state E_0 and the wave function φ_0 are modified by an admixture with continuum states, expressed by the principal-value integrals F and G, respectively. Importantly, the phase of the interaction, $\arg\left(v\psi\varphi_0^{-1}\right)$, which is contained in the parameter q, has an impact on the line shape and determines whether the slow rise is on the high- or low-energy side of the resonance.

From the invariance of the expression (2.3) against unitary transformation it is evident that the result (2.9) with the definitions (2.8) holds true for complex functions v, φ_0, and ψ, provided that $v\psi\varphi_0^{-1}$ is a real function.

We mention that the true optical susceptibility of the Fano model fulfills the sum rule,

$$\int_{-\infty}^{+\infty} d\omega\,\mathrm{Im}\chi(\omega) = \pi|\varphi_0|^2 + \pi\int_{E_{min}}^{E_{max}} dE\,|\psi(E)|^2,$$

which is apparent from Eqs. (2.2), (2.4), and the completeness of the eigenstates $|\Phi_\lambda\rangle$. This does no longer hold true for Eq. (2.9), because some parameters have been taken as constants.

It is worthwhile to discuss some special cases. For $\psi = 0$ and $v = 0$, the absorption of the discrete state should be recovered. This corresponds to the limit $q \to \infty$ in Eq. (2.9). However, the natural line shape does not approach a Lorentzian form, as Fig. 1 might suggest. Instead, as the coupling v approaches zero, the natural line width Γ drops to zero and the line shape goes over into a Dirac function. The line shape of the real physical system is then completely determined by homogeneous and inhomogeneous broadening. Even if there is coupling, the Fano resonance may not be recognized for small Γ. For $\psi = 0$ and $v \neq 0$, we observe a single line, but this time the transition energy is shifted by the amount F due to coupling to an invisible continuum. For $\varphi_0 = 0$ and $v = 0$, we obtain $q \neq 0$ and $\Gamma = 0$ so that $\alpha(\omega)/\alpha_{cont}(\omega) \equiv 1$. If an optically active continuum is coupled to an invisible Lorentzian line, i.e., $\varphi_0 = 0$, $v \neq 0$, the Fano interference is manifested as a dip ($q = 0$) at the energy $E_0 - F$ of the discrete transition.

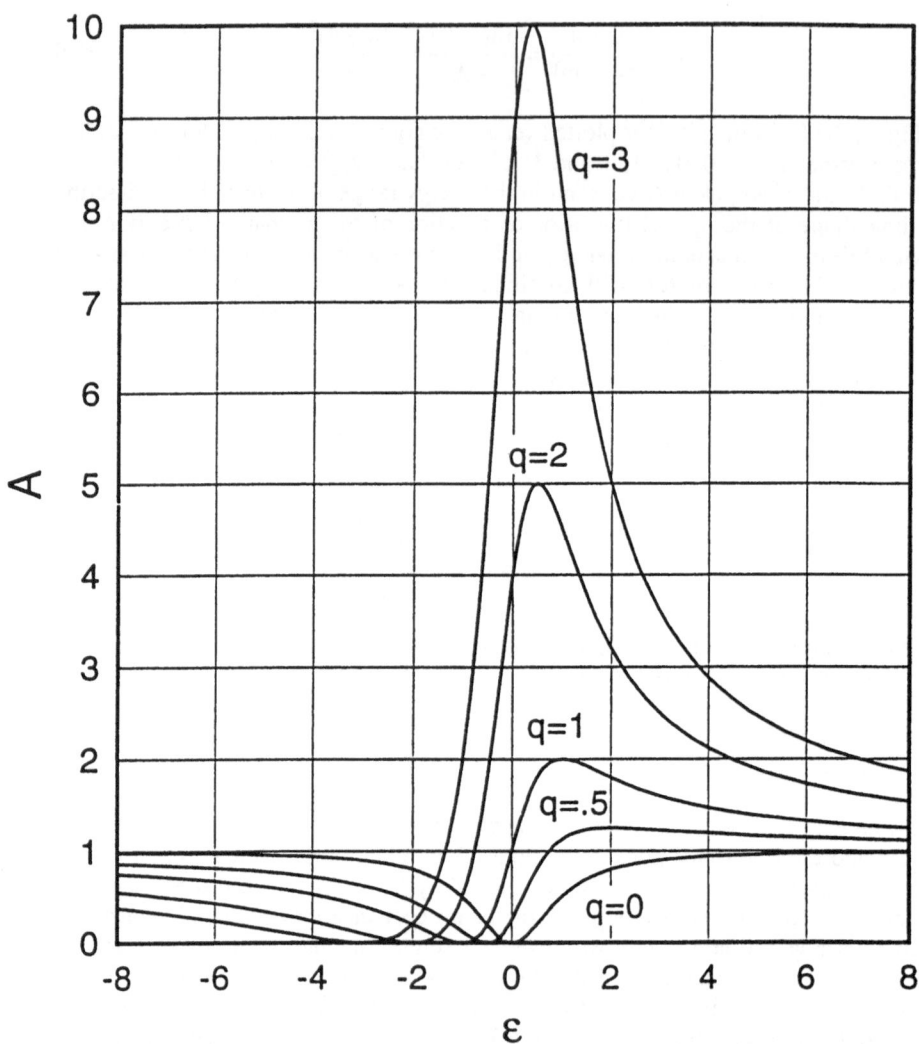

Figure 1 Natural line shape for different values of q. (Reverse the scale of abscissas for negative q.)

3   Fano Resonances as a General Feature of the Optical Absorption in Low-dimensional Semiconductors

After these preparations we go on to low-dimensional structures and show how the electron-hole Hamiltonian that governs the optical absorption can be brought into the form (2.6). Then we formulate criteria for Fano interference and present a list of systems for which Fano interference has either been predicted theoretically or has been observed.

Within the effective-mass approximation, the dynamics of the interband transitions in the vicinity of the fundamental gap can be described by means of the semiconductor Bloch equations in real space. Linearization in the optical field then yields the linear optical susceptibility, which can be presented as a generalized Elliott formula,

$$\chi(\omega) = \frac{|\mu_{cv}|^2}{\varepsilon_0} \frac{1}{\Omega} \sum_\lambda \frac{\left| \int d^3 R\, \Phi_\lambda(R,R) \right|^2}{\mathcal{E}_\lambda - \hbar(\omega + i\gamma)}. \tag{3.10}$$

Here, μ_{cv} is the interband dipole matrix element, $\varepsilon_0 = 8.85419 \times 10^{-12}\, As/(Vm)$ is the vacuum dielectric constant, and Ω is the normalization volume (area, length). The dephasing constant $\gamma > 0$ replaces the positive infinitesimal in Eq. (2.2).

The Hamiltonian

$$\begin{aligned}
\hat{H} = \;& \frac{1}{2m_e} \left| \frac{\hbar}{i} \nabla_e + e\, A(r_e) \right|^2 + U_e(r_e) - e\, E \cdot r_e \\
& + \frac{1}{2m_h} \left| \frac{\hbar}{i} \nabla_h - e\, A(r_h) \right|^2 + U_h(r_h) + e\, E \cdot r_h \\
& - \frac{e^2}{4\pi\varepsilon_0\varepsilon |r_e - r_h|^2}
\end{aligned} \tag{3.11}$$

with eigenvalues $\mathcal{E}_\lambda$ and eigenfunctions Φ_λ describes the motion of an electron-hole pair with masses $m_{e,h}$ under confinement potentials $U_{e,h}$ and an external electric and magnetic fields. The Coulomb potential is screened by the static dielectric constant ε. We assume normalization for discrete solutions and Dirac normalization for the continuous spectrum. In the latter case, the λ-sum in Eq. (3.10) has to be replaced by an integral.

Usually, the Hamiltonian (3.11) reveals symmetries so that the eigenvalue problem can be reduced to less than six coordinates. The remaining degrees of freedom are divided into directions of confinement $\vec{\eta}$ and free motion $\vec{\xi}$. Then Eqs. (3.10) and (3.11) take the general form

$$\chi(\omega) = \frac{1}{\varepsilon_0} \sum_\lambda \frac{\left| \int dv_{\vec{\eta}} \int dv_{\vec{\xi}}\, \mu_{\vec{\eta}}^*(\vec{\eta})\, \mu_{\vec{\xi}}^*(\vec{\xi})\, \Phi_\lambda(\vec{\eta},\vec{\xi}) \right|^2}{\mathcal{E}_\lambda - \hbar(\omega + i\gamma)} \tag{3.12}$$

$$\hat{H} = \hat{T}_{\vec{\xi}} + \hat{T}_{\vec{\eta}} + U(\vec{\eta}) + V(\vec{\eta},\vec{\xi}).$$

Here, $\hat{T}_{\vec{\xi}}$ and $\hat{T}_{\vec{\eta}}$ are the operators of the kinetic energy. The potential consists of two contributions, the confinement potential U and the Coulomb potential V with the properties

$$V(\vec{\eta},\vec{\xi}) < 0 ; \quad \lim_{|\vec{\xi}|\to\infty} V(\vec{\eta},\vec{\xi}) = 0.$$

Without Coulomb interaction, confined and free motion are separated and the eigenvalue problem for the confined direction reads

$$\left[\hat{T}_{\vec{\eta}} + U(\vec{\eta})\right] \phi_n(\vec{\eta}) = E_n \phi_n(\vec{\eta})$$
$$\left[\hat{T}_{\vec{\eta}} + U(\vec{\eta})\right] \phi_E(\vec{\eta}) = E \phi_E(\vec{\eta})$$

for the discrete spectrum E_n and the above-barrier continuous spectrum E.

In order to take into account Coulomb interaction, we define Coulomb matrix elements

$$V_{nn'}(\vec{\xi}) = \int dv_{\vec{\eta}}\, \phi_n^*(\vec{\eta})\, V(\vec{\eta},\vec{\xi})\, \phi_{n'}(\vec{\eta}).$$

If the off-diagonal Coulomb matrix elements ($n \neq n'$) are neglected, the eigenfunctions of the Hamiltonian (3.12) can be factorized according to $\Phi_\lambda(\vec{\eta},\vec{\xi}) = \Phi_{n\nu,E}(\vec{\eta},\vec{\xi}) = \phi_n(\vec{\eta})\, \psi_{n\nu,E}(\vec{\xi})$, where

$$\left[E_n + \hat{T}_{\vec{\xi}} + V_{nn}(\vec{\xi})\right] \psi_{n\nu,E}(\vec{\xi}) = \mathcal{E}_{n\nu,E}\, \psi_{n\nu,E}(\vec{\xi}).$$

The eigenvalues $\mathcal{E}_{n\nu}$ represent exciton Rydberg series with energies below the subband edge E_n, whereas $\mathcal{E}_{nE} = E_n + E$, $E \in (0,\infty)$ is the continuous spectrum associated with the n-th subband pair.

Fano interference is expected if (i) an exciton of a higher subband pair n' is degenerate with the continuum of a lower subband pair n, i.e.,

$$\mathcal{E}_{nE} = \mathcal{E}_{n'\nu} ; \tag{3.13}$$

(ii) the continuum is optically allowed, i.e.,

$$\int dv_{\vec{\eta}}\, \mu_{\vec{\eta}}^*(\vec{\eta})\, \phi_{n'}(\vec{\eta}) \neq 0 ; \quad \int dv_{\vec{\xi}}\, \mu_{\vec{\xi}}^*(\vec{\xi})\, \psi_{n' E}(\vec{\xi}) \neq 0 ; \tag{3.14}$$

and (iii) the coupling matrix element

$$v_{nE\,n'\nu} = \int dv_{\vec{\xi}}\, \psi_{nE}^*(\vec{\xi})\, V_{nn'}(\vec{\xi})\, \psi_{n'\nu}(\vec{\xi}) \tag{3.15}$$

is different from zero.

We mention that there is also coupling when the states that belong to the subbands n and n' in Eq. (3.15) are both discrete or both continuous. Since those contributions are energetically off-resonant, they do not qualitatively change the optical spectrum. The quantitative changes in the binding energies and oscillator strengths are however noticeable so that the simple Fano model fails to give accurate quantitative results [18,21].

In order to observe Fano resonances, the dimension of the semiconductor has to be less than three; because electrons and holes in bulk semiconductors can freely move in all three space directions and no subbands are formed. On the other hand, since continuum states are needed for Fano interference, the dimension of the semiconductor must be larger than zero, otherwise the spectrum would be fully discrete. Hence, Fano resonances should be found in one- and two-dimensional semiconductors.

The reduction of dimensionality can either be caused by geometric confinement or by magnetic field. A magnetic field reduces the degrees of freedom by two in the plane perpendicular to the field direction [24]. Bulk semiconductors in magnetic fields are thus one-dimensional systems [21] and excitons in quantum wells become zero-dimensional when a perpendicular magnetic field is applied [25]. The situation is more complicated when the direction of the magnetic field is perpendicular to geometrical confinement directions. For example, quantum wires in perpendicular magnetic field retain their one-dimensional nature [20].

In order to fulfill the condition (3.13), there should be at least two subband pairs. To see a pronounced Fano effect, the lowest-order exciton, $\nu = 1$, of the excited subband pair n' should be situated above the continuum edge E_n.

The conditions (3.14) and (3.15) are usually fulfilled if all symmetries are properly taken into account in the formulation of the Hamiltonian (3.12) and the boundary conditions. An exception to this rule are systems with electron-hole symmetry caused by nearly perfect geometrical confinement or magnetic field. In this case, forbidden optical transitions may be coupled to allowed ones, leading to Fano resonances with a line-shape parameter $q = 0$, also called "window resonances" [26], that are manifested as dips in the continuum.

Practically, however, those transitions often become weakly allowed because of leakage of wave functions through the barriers or non-resonant Coulomb coupling. Then q is small in magnitude but different from zero.

If the confinement potential U is periodic, the spectrum of the confined motion is continuous and gives rise to minibands. These continua also contribute to Fano interference, but the effect is rather small [27].

After these general considerations we are able to identify low-dimensional semiconductors in which Fano interference should be observable. A selection of those structures, for which theoretical or experimental results have already been published, is shown in Table 1. Apparently, there are more theoretical results than experimental ones. This is due to the fact that large inhomogeneous broadening is found in semiconductor structures and a high quality of the interfaces is required. Due to technological progress in the fabrication of quantum wells, Fano resonances have been observed at the e1–hh3

Table 1 Low-dimensional semiconductors for which Fano interference is expected: theoretical and experimental results.

Dimension	System	Theory	Experiment
2	quantum well	[13,15,28–30]	[14,16,17]
	quantum well $+ \vec{E}_\perp$	[31]	
	superlattice	[27,32]	
	superlattice $+ \vec{E}_\perp$	[33]	[40]
1	bulk $+ \vec{B}$	[17,21,34]	[21]
	superlattice $+ \vec{B}_\perp$	[27]	
	quantum wire	[18,19,30]	
	quantum wire $+ \vec{B}_\perp$	[20]	

and e2–hh2 transitions. The realization of lateral confinement in one-dimensional systems is much more difficult, and, up until now, there has been no convincing experimental evidence for Fano resonances in quantum wires, although the coupling of different subbands due to Coulomb interaction is stronger than in two-dimensional semiconductors. In bulk semiconductors in a magnetic field, the confinement is not geometrical and thus does not show roughness on an atomic scale.

A particular challenge is the numerical treatment of Fano resonances and, in turn, Fano resonances are a signature for high numerical accuracy. Numerical calculations are essentially based upon four methods: (i) iterative diagonalization (Lanczos algorithm) [13,29], (ii) solution of a sparse linear set of equations (Coulomb Green's function) [15,19,33], (iii) the solution of an initial-value problem (equation-of-motion method) [31], and the solution of a multichannel scattering problem [20].

Currently, there is much interest in T-shaped quantum wires [35,36], because these structures are produced completely by molecular-beam epitaxy and have larger confinement and a higher quality than etched quantum wires. However, in most cases, there is only one subband pair [36] and the condition (3.13) is not fulfilled.

In principal, quantum wells and superlattices with in-plane magnetic fields should show Fano resonances, but the two-particle problem has not yet been solved. Even the solution for individual electrons and holes is very complicated [37]. Superlattices in parallel magnetic fields have been studied by Dignam and Sipe [38], but their analysis was limited to discrete states.

4 Example

The purpose of this section is to illustrate how the abstract theoretical considerations of the last section are applied to a specific structure. As an example, we choose a flat quantum-well wire, where the well thickness d is much smaller than the wire width b.

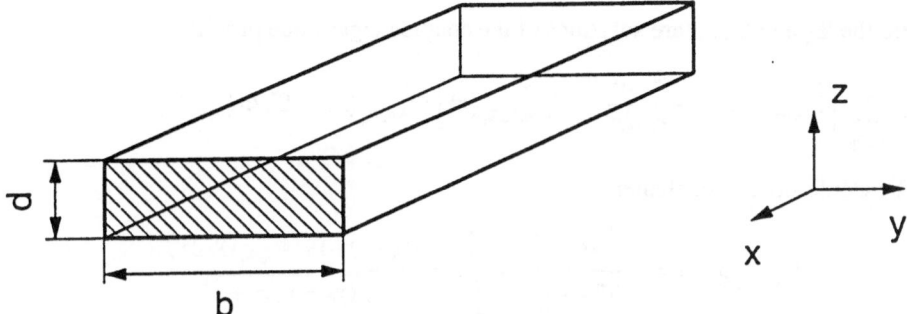

Figure 2 Geometry of the quantum-well wire, and notations.

A sketch of the sample is shown in Fig. 2. For simplicity, we assume infinitely high barriers. If x is the wire direction and y the direction of the lateral confinement, the symbols introduced in the last section are identified as follows:

$$\vec{\eta} = (y_e, y_h) \; ; \; dv_{\vec{\eta}} = dy_e \, dy_h \; ; \; \mu_{\vec{\eta}}(y_e, y_h) = \delta(y_e - y_h)$$

$$\vec{\xi} = x = x_e - x_h \; ; \; dv_{\vec{\xi}} = dx \; ; \; \mu_{\vec{\xi}}(x) = \delta(x)$$

$$T_{\vec{\eta}} = -\frac{\hbar^2}{2m_e} \frac{\partial^2}{\partial y_e^2} - \frac{\hbar^2}{2m_h} \frac{\partial^2}{\partial y_h^2} \; ; \; T_{\vec{\xi}} = -\frac{\hbar^2}{2m} \frac{\partial^2}{\partial x^2} \; ; \; \frac{1}{m} = \frac{1}{m_e} + \frac{1}{m_h}$$

$$U(y_e, y_h) = \begin{cases} -\dfrac{\hbar^2 \pi^2}{2mb^2} & \text{for } (y_e, y_h) \in [0, b] \times [0, b] \\ +\infty & \text{elsewhere} \end{cases}$$

$$V(y_e, y_h, x) = -\frac{e^2}{4\pi\varepsilon_0 \varepsilon \sqrt{(y_e - y_h)^2 + x^2}}$$

The constant $\hbar^2\pi^2/(2mb^2)$ has been subtracted from the potential so that the continuum starts at $\hbar\omega = 0$.

The eigenvalues and eigenfunctions for the confined motion are

$$E_{n_e, n_h} = \frac{\hbar^2\pi^2(n_e^2 - 1)}{2m_e b^2} + \frac{\hbar^2\pi^2(n_h^2 - 1)}{2m_h b^2}$$

$$\phi_{n_e n_h}(y_e, y_h) = \sqrt{\frac{2}{b}} \sin(\frac{n_e \pi y_e}{b}) \sqrt{\frac{2}{b}} \sin(\frac{n_h \pi y_h}{b}).$$

In the basis of the functions ϕ_{n_e, n_h}, the optical susceptibility is written as

$$\chi(\omega) = \frac{|\mu_{cv}|^2}{\varepsilon_0} \sum_{\lambda} \frac{\left| \sum_{n=1}^{\infty} \Phi_{\lambda nn}(0) \right|^2}{E_\lambda - \hbar(\omega + i\gamma)},$$

where the $\mathcal{E}_\lambda$ and $\Phi_{\lambda nn'}$ are solutions of the coupled eigenvalue problem

$$\sum_{n'_e n'_h} \left[E_{n_e n_h} \delta_{n_e n_h} - \frac{\hbar^2}{2m} \frac{d^2}{dx^2} + V_{n_e n_h n'_e n'_h}(x) \right] \Phi_{\lambda n'_e n'_h}(x) = \mathcal{E}_\lambda \Phi_{\lambda n_e n_h}(x). \qquad (4.16)$$

The Coulomb matrix elements

$$V_{n_e n_h n'_e n'_h}(x) = -\frac{e^2}{4\pi\varepsilon_0 \varepsilon} \int_0^b dy_e \int_0^b dy_h \frac{\phi_{n_e n_h}(y_e, y_h) \phi_{n'_e n'_h}(y_e, y_h)}{\sqrt{(y_e - y_h)^2 + x^2}}$$

are nonzero for $(n_e + n'_e + n_h + n'_h) \bmod 2 = 0$ so that Eqs. (4.16) separate into those for even-parity [$(n_e + n'_e) \bmod 2 = 0$] and odd-parity [$(n_e + n'_e) \bmod 2 = 1$] states. On the other hand, because of equal electron and hole wave functions (4.16), optical transitions are allowed only for $n_e = n_h$.

All quantities are expressed in Hartree units, $2E_B = me^4/[(4\pi\varepsilon_0 \varepsilon)^2 \hbar^2]$ and $a_B = 4\pi\varepsilon_0 \varepsilon \hbar^2/(me^2)$ for energies and lengths, respectively. For Gallium Arsenide it holds that $2E_B = 9.4\,\text{meV}$ and $a_B = 12\,\text{nm}$. We choose a wire width $b = 1.2$ and an effective-mass ratio $m_h/m_e = 7$, which is about that for GaAs. A small homogeneous broadening $\hbar\gamma = 0.01$ is introduced.

The optical susceptibility was calculated according to Eq. (2.3), in the basis of sub-band pairs for the motion in the confinement (y) direction with an energy cutoff of 24. The wire (x) direction was discretized using 10,000 finite elements on the interval $[0, 1000]$.

The results of different approximations for the Coulomb interaction are shown in Fig. 3. The imaginary part of the optical susceptibility on a logarithmic scale is plotted versus the transition energy $\hbar\omega$. The transitions are classified according to $(n_e, n_h; \nu)$, where (n_e, n_h) is the subband pair and $\nu \geq 1$ denotes the exciton quantum number.

First, we completely neglect the Coulomb interaction (Fig. 3a). The optical absorption is then proportional to the density of states. The subbands start at energies $\pi^2(n^2 - 1)/(2mb^2)$. The $\omega^{-1/2}$ singularities at the subband edges are removed by the finite homogeneous broadening. The smoothness of the curve indicates that the continuum was properly discretized.

When we allow for Coulomb interaction *within* each subband pair, i.e., $n_e = n'_e$, $n_h = n'_h$, we end up with a spectrum that is a superposition of spectra of individual subbands (Fig. 3b). The subband edges turn into Rydberg series of excitons, each exciton having a Lorentzian line shape. In contrast to the free-particle case, the subband continua are nearly constant in the region of interest and the singularities at the subband edges disappear [39].

The situation changes drastically when we take into account Coulomb coupling *between* all optically allowed subbands, i.e., $n_e = n_h$, $n'_e = n'_h$ (Fig. 3c). Then the $(2,2;1)$ exciton turns into a Fano resonance by coupling to the $(1,1)$ continuum. The low-energy side shows a slow decrease and a dip is observed at the high-energy side. This indicates that the line-shape parameter q is negative. The same effect takes place for each subband pair (n,n) with $n > 1$.

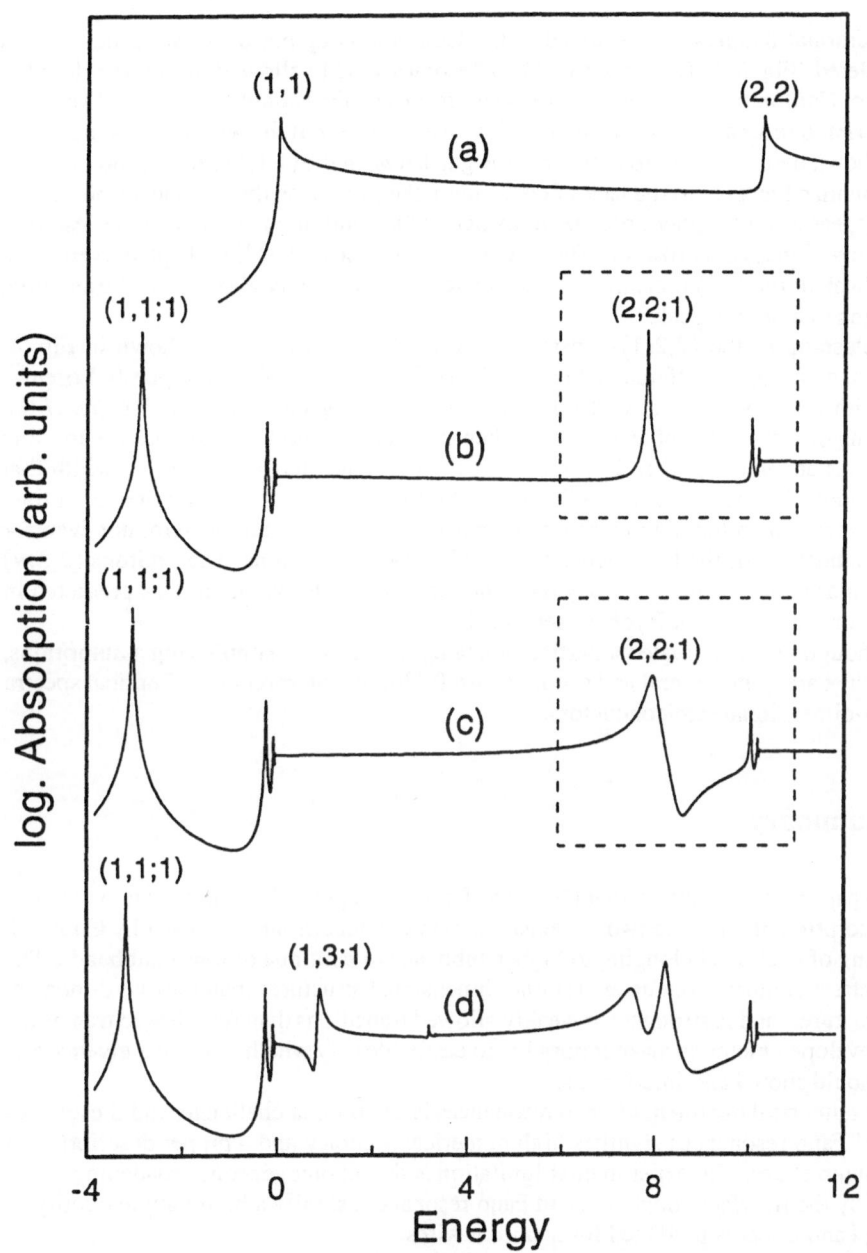

Figure 3 Calculated absorption of the quantum-well wire of width $b = 1.2$ for different approximations of the electron-hole interaction: a) neglect of the Coulomb interaction, b) intrasubband interaction, c) intra- and intersubband interaction for optically allowed subband pairs, and d) interaction of all subband pairs.

Additional features are observed if the Coulomb coupling of all subband pairs is considered (Fig. 3d). The transition (1,3) becomes weakly allowed, which is solely the result of Coulomb interaction. Due to quantum-mechanical coupling to the (1,1) continuum, a Fano resonance appears, this time with a positive q. If the wire width is increased (not shown here), this exciton falls into the gap between the (1,1;1) exciton and the (1,1) continuum edge. Due to the lack of continuum, the line shape then remains Lorentzian. The appearance of higher order excitons below the continuum onset is a manifestation of center-of-mass quantization. The interpretation of the (2,2;1) line shape is even more complicated: the (2,2) transition is degenerate with the weakly allowed (1,5) transition, resulting in a double peak.

A close-up of the (2,2;1) transitions (dashed boxes in Fig. 3) is shown in Fig. 4. Without coupling of different subband pairs (left), the excitons remain purely Lorentzian, with a line width that is determined by the homogeneous broadening $\hbar\gamma$. If the coupling of different subband pairs is taken into account, each exciton has a Fano line shape. As apparent for the (2,2;1) exciton, the resonance has a natural line width that is considerably larger than the full width at half maximum $2\hbar\gamma$ of the Lorentzian line. The absorption on the high-energy side does not drop completely to zero, not even for $\gamma = 0^+$, because of the low-energy tails of higher-order excitons. The excitons $(2,2;\nu)$ are similar to each other, i.e., they have the same height-to-width ratio, as predicted in Fano's original work for Rydberg series [1].

Although we have treated a particular example with some simplifying assumptions, the results are quite general and should be useful for the interpretation of optical spectra of low-dimensional semiconductors.

5   Summary

In this paper we have shown that Fano interference is a general phenomenon in the optical absorption of one- and two-dimensional semiconductors and is caused by Coulomb coupling of excitons belonging to higher subbands to continua of lower subbands. The Fano effect is more pronounced for one-dimensional structures than for two-dimensional structures, and is stronger for weakly allowed transitions than for allowed transition. The developed methods make it possible to easily identify whether or not a given structure should show Fano interference.

The numerical treatment of Fano resonances is a particular challenge, and the appearance of Fano resonances signifies high numerical accuracy and a proper description of continuum states. The experimental limitation is the inhomogeneous broadening introduced by the interface roughness and Fano resonances signify a high sample quality. A strong Fano effect is predicted for quantum wires.

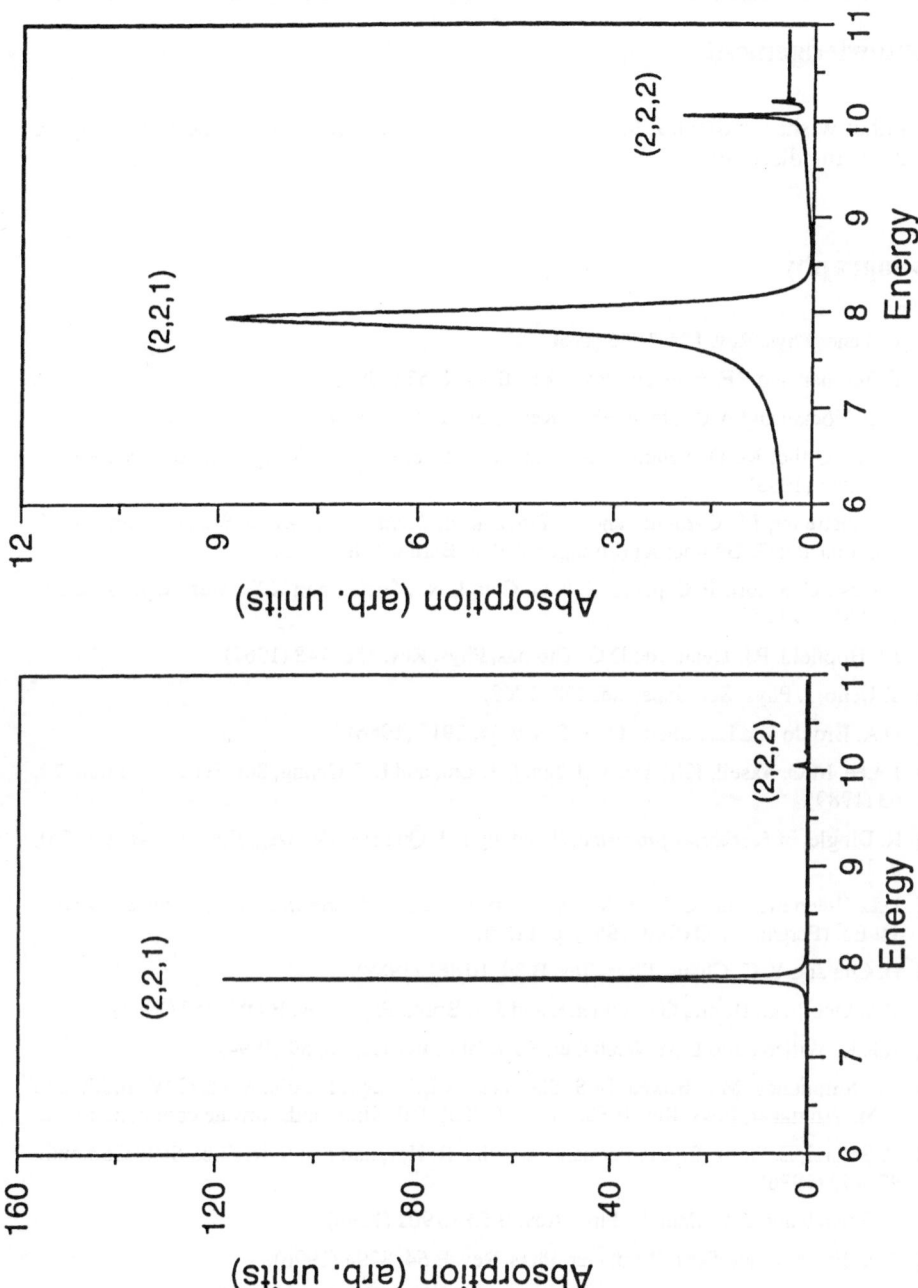

Figure 4 Close-up of the (2,2;1) resonance (dashed boxes in Fig. 3) without intersubband interaction (left) and with intersubband interaction (right).

Acknowledgement

The author wishes to thank J. M. Baker, F. Bechstedt, K. Hannewald, and J.-M. Wagner for interesting discussions.

Bibliography

[1] U. Fano, Phys. Rev. **124**, 1866 (1966).

[2] E. Tekman and P.F. Bagwell, Phys. Rev. B **48**, 2553 (1993).

[3] J. U. Nöckel and A.D. Stone, Phys. Rev. B **50**, 17415 (1994).

[4] T. dell'Ortho, M. Di Ventra, J. Almeide, C. Coluzza, and G. Margaritondo, Phys. Rev. B **52**, 2265 (1995).

[5] G. Abstreiter, M. Cardona, and A. Pinczuk, in *Light scattering of Solids IV*, ed. by M. Cardona and G. Güntherodt (Springer-Verlag, Berlin, 1984), p. 5.

[6] J. Faist, C. Sirtori, F. Capasso, S.-N.G. Chu, L.N. Pfeiffer, and K.W. West, Optics Letters **21**, 985 (1996).

[7] J.J. Hopfield, P.J. Dean, and D.G. Thomas, Phys. Rev. **158**, 748 (1967).

[8] T. Ueno, J. Phys. Soc. Japan **26**, 438 (1969).

[9] D.A. Broido and L.J. Sham, Phys. Rev B **34**, 3917 (1986).

[10] J. Lee, M.O. Vasell, K.T. Tsu, G.J. Jan, C.P. Liu, and L.F. Chang, Sol. State. Commun. **72**, 63 (1989).

[11] R. Dingle, in *Festkörperprobleme XV*, ed. by H.J. Queisser (Vieweg, Braunschweig, 1975), p. 21.

[12] D.L. Greenaway and G. Harbeke, *Optical properties and band structure of semiconductors*, 1st Ed. (Pergamon, Oxford, 1968), p. 113 ff.

[13] H. Chu and Y.-C. Chang, Phys. Rev. B **39**, 10861 (1989).

[14] D.Y. Oberli, G. Böhm, G. Weimann, and J.A. Brum, Phys. Rev. B **49**, 5757 (1994).

[15] A.R.K. Willcox and D.M. Whittaker, Superlatt. Microstr. **16**, 59 (1994).

[16] P.E. Simmonds, M.J. Birkett, M.S. Skolnick, W.I.E. Tagg, P. Sobkowicz, G.W. Smith, and D.M. Whittaker, Phys. Rev. B **50**, 11251 (1995); P.E. Simmonds, private communication.

[17] V. Bellani, E. Pérez, S. Zimmermann, L. Viña, R. Hey, and K. Ploog, Solid State Commun. **97**, 459 (1996).

[18] S. Glutsch and D.S. Chemla, Phys. Rev. B **53** 15902 (1996).

[19] A.N. Forshaw and D.M. Whittaker, Phys. Rev. B **54**, 8794 (1996).

[20] M. Graf, P. Vogl, and A.B. Dzyubenko, Phys. Rev. B **54**, 17003 (1996).

[21] S. Glutsch, U. Siegner, M.-A. Mycek, and D.S. Chemla, Phys. Rev. B **52**, 17009 (1994).

[22] U. Siegner, M.-A. Mycek, S. Glutsch, and D.S. Chemla, Phys. Rev. Lett. **74**, 470 (1995).

[23] S. Bar-Ad, P. Kner, M.V. Marquezini, S. Mukamel, and D.S. Chemla, Phys. Rev. Lett. **78**, 1363 (1997).

[24] L.P. Gor'kov and I.E. Dzyaloshinskii, Sov. Phys. JETP **26**, 449 (1968) [Zh. Eksp. Theor. Fiz. **53**, 717 (1967)].

[25] C. Stafford, S. Schmitt-Rink, and W. Schäfer, Phys. Rev. B **41**, 10 000 (1990).

[26] H. Friedrich, *Theoretical Atomic Physics* (Springer, New York, 1991).

[27] S. Glutsch, P. Lefebvre, and D.S. Chemla, Phys. Rev. B **55**, 15786 (1997).

[28] J.A. Brum and D.Y. Oberli, Journal de Physique II, Vol. 3, Colloque C5, 191 (1993).

[29] R. Winkler, Phys. Rev. B **51**, 14 395 (1995).

[30] S. Glutsch, D.S. Chemla, and F. Bechstedt, Phys. Rev. B **51**, 16 885 (1995).

[31] S. Glutsch, F. Bechstedt, and D.S. Chemla, in *Proceedings of the 23rd International Conference on the Physics of Semiconductors, Berlin, Germany, 1996*, ed. by M. Scheffler and R. Zimmermann (World Scientific, Singapore, 1996), p. 1987.

[32] K. Maschke, P. Thomas, and E.O. Göbel, Phys. Rev. Lett. **67**, 2646 (1991).

[33] D.M. Whittaker, Europhys. Lett. **31**, 55 (1995).

[34] W. Becker, B. Gerlach, T. Hornung, and R.G. Ulbrich, in *Proceedings of the 18th International Conference on the Physics of Semiconductors, Stockholm, Sweden, 1986*, ed. by O. Engström (World Scientific, Singapore, 1987), p. 1713.

[35] W. Wegscheider L.N. Pfeiffer, M.M. Dignam, A. Pinczuk, K.W. West, S.L. McCall, and R. Hull, Phys. Rev. Lett. **71**, 4071 (1993).

[36] W. Langbein, H. Gislason, and J.M. Hvam, Phys. Rev. B **54**, 14 595 (1996).

[37] J.C. Maan, in *Application of High Magnetic Fields in Semiconductor Physics. Proceedings of the International Symposium, Grenoble, France, 1982*, ed. by G. Landwehr (Springer-Verlag, Berlin 1983), p. 163.

[38] M.M. Dignam and J.E. Sipe, Phys. Rev. B **45**, 6819 (1992).

[39] T. Ogawa and T. Takagahara, Phys. Rev. B **44**, 8138 (1991).

[40] C. Holfeld et al., to be published.

Ultrafast Oscillations in the Time-Resolved Emission of a Resonantly Excited Semiconductor Microcavity

Martin Koch

Bell Laboratories, Lucent Technologies, Holmdel NJ 07733, USA
Present address: Physics department, University of Munich,
80799 Munich, Germany.

Abstract: We investigate the linear and nonlinear emission dynamics of a semi-conductor microcavity which is resonantly excited using 100 fs pulses. In the time-resolved emission two different types of oscillations are observed. In the low density limit we observe normal mode oscillations which reflect an oscillatory energy transfer between the excitonic and the photonic subsystem. The temporal pattern which is formed by these oscillations in a nonlinear four-wave mixing experiment shows several features which are characteristic of the mixed photon-exciton nature of the semiconductor microcavity system. The second type of oscillations is observed at high excitation densities and arises from an interference between the laser mode and an oblique mode defined by the pump pulse.

1 Introduction

The first to realize that spontaneous emission of an absorber can be enhanced or suppressed by placing the absorber inside a small microresonator was Purcell in 1946 [1]. Since then many elaborated theories on this fundamental effect have been developed, accompanied by numerous experiments on atoms in the microwave and in the optical regime and lately also on semiconductor microcavities [2]. The work on microcavities is not only of general interest since it enables to study basic concepts of light-matter interaction but also important for technological reasons. The latter aspect includes the possibility to achieve ultra-low laser thresholds and record modulation bandwidths in vertical cavity surface emitting lasers (VCSEL's) [3–6] and to control the emission wavelength of polymer light emitting diodes [7,8].

A microcavity commonly consists of two highly reflecting mirrors facing each other with a small spacing d. For photons which are in between these boundaries the two mirrors represent nearly infinitely high barriers. Similar to the textbook problem of a "particle in a box", where electronic confinement concentrates the electronic density of states into discrete levels, the photonic density of states inside the cavity is concentrated into discrete modes at frequencies $\omega_n = nc\pi/d$ (with n=1,2,3,...). As a result of this

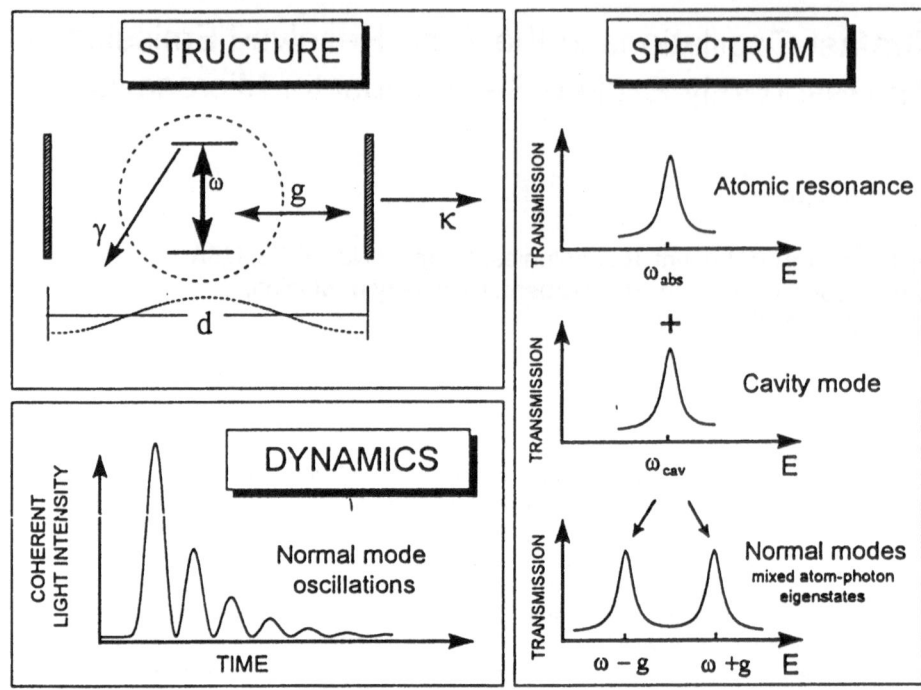

Figure 1   Upper left: schematic drawing of a two-level absorber with a homogeneous linewidth of γ in a λ cavity with a cavity lifetime of $1/\kappa$. The coupling strength between absorber and cavity is characterized by the coupling frequency g. Lower left: linear dynamics of the strongly coupled absorber-cavity system after pulsed excitation. Right: spectrum of the strongly coupled absorber-cavity system.

quantization only photons with a frequency corresponding to these modes are allowed to enter the structure in the surface normal direction, i.e. in the direction perpendicular to the mirror planes. Once inside the cavity the photons do not remain there for infinite times but escape from the structure with a constant rate κ. Moreover, if the space between the mirrors is not empty but contains an optical absorber intracavity photons can get absorbed (see the schematic drawing in the upper left of Fig. 1). The absorbed energy is either reemitted at some later times or is lost for the system due to dissipation. Dissipative channels include electronic dephasing, spectral relaxation and spontaneous emission into directions other than that of the surface normal.

Depending on the ratio between the total dissipation rate γ, the inverse cavity lifetime κ and the coupling frequency g between the absorber and the intracavity field two regimes are distinguished: if the dissipation exceeds the coupling frequency, i.e., if $g \ll (\kappa, \gamma)$ the composite absorber-cavity system is in the weak coupling limit. In this regime spontaneous emission is an irreversible process. It is simply enhanced or suppressed as compared to free space depending on the cavity frequency being tuned in or out of resonance with the absorbers transition frequency. In the strong coupling limit, in contrast, when $g \gg (\kappa, \gamma)$ radiative decay becomes reversible. A photon emitted by the absorber remains inside the cavity long enough to get reabsorbed, subsequently reemitted, and so on. This coherent energy transfer between the photonic and the electronic subsystem reflects itself in pronounced oscillations in the light intensity coherently emitted by the system after a pulsed excitation (Fig. 1, lower left) [9–11]. Accordingly, the spectrum of the system which is observed in a transmission or reflection measurement splits into two Lorentzian shaped lines (right part of Fig. 1). These lines are called normal modes and correspond to the two mixed absorber-photon eigenstates of the composite system [12–15]. In the case of a semiconductor microcavity (SMC) the split states are also called "cavity polariton" [16]. This terminology is based on the similarities between the exciton-cavity system and a bulk polariton. Since a variation in the angle of incidence results in a blueshift of the cavity mode the cavity polariton shows a distinct dispersion curve [16].

In general it is possible to switch between these two regimes of coupling. The detailed switching mechanism, however, depends on the physical nature of the absorber inside the cavity. If the cavity contains Rydberg atoms the switching is accomplished by increasing the average number of atoms N inside the cavity for a given time [17]. Without affecting γ and κ this increases the effective coupling rate $g = g_0$ where g_0 is the coupling rate for a single atom and is given by $g_0 = \mu \sqrt{\omega_c} \mid \Psi(r = 0) \mid^2 / 2\hbar\varepsilon_0 L$. Here μ denotes the atom's dipole moment, ω_c the cavity resonance, L the cavity length and Ψ the cavity mode function at the position $r = 0$ of the absorbing atom. In the very extreme a few or even a single atom can be sufficient to cause a significant normal mode splitting [14]. The splitting is then also called vacuum Rabi splitting and is described by the fully quantized Jaynes-Cummings model [18].

For a SMC which is the system investigated in the present article switching between the weak and the strong coupling regime is accomplished by varying the light intensity that illuminates the cavity. A cavity initially in the strong coupling limit can be switched to the low coupling limit using intense laser irradiation. With increasing laser power the collective excitonic dipole moment decreases due to bleaching caused by phase space filling and screening of the Coulomb interaction [19]. Accordingly, the coupling frequency g of the system decreases. Additionally, γ is enhanced due to carrier-carrier scattering. Thus, the regime of $g \ll (\kappa, \gamma)$ where the normal mode splitting has collapsed is easily attained; especially after resonant and pulsed laser irradiation.

It is important to note that in semicondutor microcavities a strong coupling limit in the sense that one single electron-hole pair causes a vacuum Rabi splitting with a single photon can not be realized at present. This is because the damping rate of the excitonic polarization exceeds the coupling frequency of a single electron-hole pair by far. Yet, if a SMC contains sufficient "absorber material" a normal mode splitting can be observed [16,20]. Due to the large number of electron-hole pairs and photons involved a quantization of the intracavity field is not necessarily required to model a SMC. Depending on the particular problem addressed a SMC may alternatively be described semiclassically [15] or even purely classically [13]. Because of a variety of many-particle effects, however, the interaction of a semiconductor with light is more complicated than the light-matter interaction of an isolated atom. Consequently, the simple atomic model which assumes two state atoms (see Fig. 1) fails for a quantitative description of a SMC. Yet, since a semiclassical atomic model reflects some structural aspects of the coupled exciton-cavity system it may lead to qualitative results and, hence, can help to enlighten part of the underlying physics. In fact, it can be quite instructive due to its simplicity and transparency. It is therefore chosen as basis for the calculations that accompany the experimental results presented below.

In this article we investigate the emission dynamics of a semiconductor microcavity after it is resonantly excited by short laser pulses. By varying the intensity of the excitation pulses both regimes of coupling can be encountered.

We first concentrate on the linear dynamics of the strongly coupled exciton-cavity system after excitation with a single weak pump pulse (section 3). The dynamics which is dominated by normal mode oscillations is monitored by upconverting the light intensity emitted form the structure into the direction of the reflected pump pulse. The particular issue addressed in section 3 is the effect of inhomogeneous line broadening (IB) on the frequency and the amplitude of the oscillations. We find that the oscillation frequency slightly increases for moderate values of IB. Yet, as IB exceeds a certain value the normal mode oscillations are heavily damped and their frequency drastically decreases. These findings are contrasted with the case of homogeneous line broadening where no increase in the oscillation frequency is observed.

Since the linear dynamics of a strongly coupled SMC is basically indistinguishable from that of purely electronic multi-level systems it does not necessarily indicate the mixed exciton-photon nature of the system. As was shown recently, however, the physical nature of a multi-level system can be concluded from its nonlinear dynamics observed in a time-resolved four-wave mixing (FWM) experiment [21,22]; a special type of pump and probe experiment where two excitation pulses are employed. In section 4 we therefore perform a time-resolved FWM experiment to monitor the nonlinear response of the SMC system. Like the linear dynamics the nonlinear dynamics is dominated by normal mode oscillations. Yet, the observed oscillation pattern significantly differs from that of electronic multi-level systems and displays several features which we find to be characteristic of the essential mixed electron-photon nature of the SMC system. In addition, we present spectrally resolved FWM measurements which indicate the presence of a coherent energy transfer between the intracavity field and a biexcitonic transition.

Section 5 focuses on the regime of weak coupling when the cavity is heavily pumped with a single intense pulse. Under these conditions the normal mode splitting has collapsed and laser action takes place. Again the cavity emission is dominated by pronounced oscillations. This time, however, the oscillations are due to an interference between the surface normal mode into which lasing is expected and an oblique pump mode.

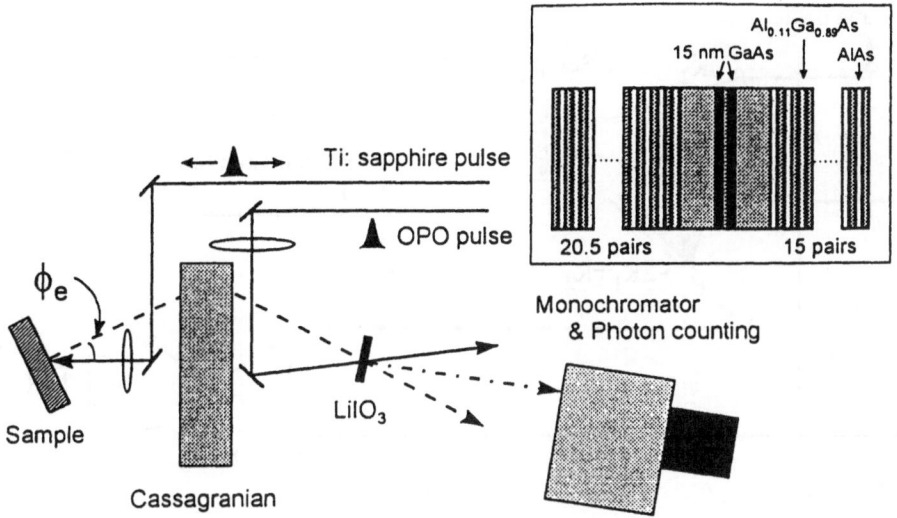

Figure 2 Experimental setup for time-resolved detection of the linear cavity emission via two-color upconversion. The inset shows the sample structure.

2 Sample Geometry and Experimental Setups

The sample investigated throughout this article is a planar microcavity structure grown by molecular beam epitaxy. The structure which is schematically shown in the inset of Fig. 2 comprises two 15 nm GaAs quantum wells, which are separated by a 10 nm $Al_{0.3}Ga_{0.7}As$ barrier and are placed at the antinode of a λ cavity that has an $Al_{0.3}Ga_{0.7}As$ spacer layer. The top and bottom mirrors consist of 15 and 20.5 pairs of $Al_{0.11}Ga_{0.89}As$-AlAs quarter-wave stacks. At 10 K, the temperature at which all experiments are performed, a cavity linewidth of 1 meV is measured. This value corresponds to a Q factor of 1540 which demonstrates the high quality of the cavity part of the structure. The cavity resonance can be tuned either by varying the angle of incidence [16] or, alternatively, for a fixed incidence angle by varying the position on the sample. The latter becomes possible since the spacer layer is typically slightly wedge-shaped due to a non-uniform epitaxial growth. If the cavity resonance is tuned around the energy of the 1s-heavy hole exciton the typical anticrossing behavior [16,20] with a minimum splitting of 4.2 meV is observed.

Two different experimental setups are employed. In section 3 and 5 we use only one pump beam and investigate the linear emission of the cavity emitted in either the direction of the reflected pump beam or in the direction of the surface normal (into which lasing is expected). A schematic diagram of the setup is shown in Fig. 2. The sample, which is inside a cryostat (not shown) is optically excited under a small external angle ϕ_e. The pump pulses which are provided by a Ti:sapphire laser are focused down to a spot size of 15 μm (FWHM). They have a duration of 100 fs and a spectral

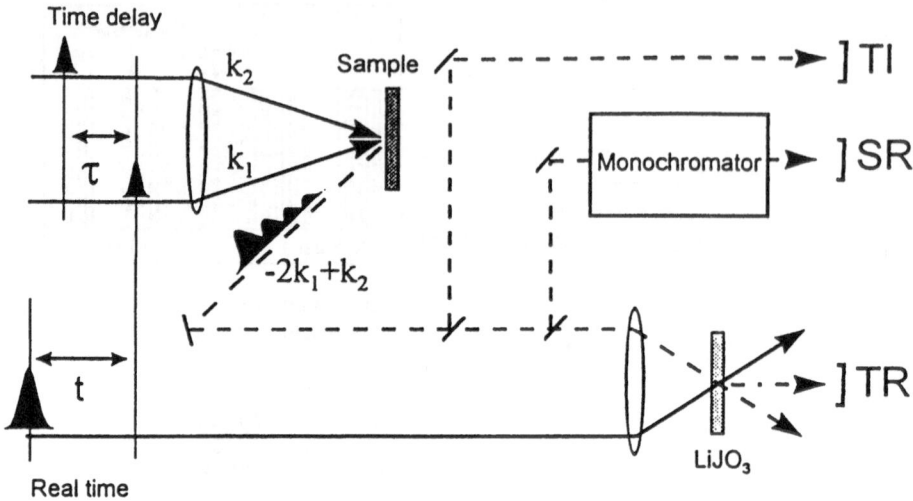

Figure 3   Experimental setup for a nonlinear four-wave-mixing experiment with time-integrated (TI), spectrally resolved (SR) and time-resolved detection (TR).

width of 17 meV (FWHM). The emission from the cavity is collected by a cassagranian and focused onto a $LiIO_3$ crystal where it is temporally resolved by upconversion [23] using synchronized pulses of an optical parametric oscillator. To suppress undesired scattered light at wavelengths other than the signal wavelength the upconverted signal is sent through a monochromator and is finally detected via single photon counting. The direction in which the emission is analyzed can be selected by varying the sample orientation.

The second experimental technique employed is nonlinear FWM in the self-diffraction configuration. This technique has been successfully employed in the last years to study various coherent phenomena in semiconductors and semiconductor nanostructures. The FWM setup is shown in Fig. 3. Two laser pulses with wavevectors k_1 and k_2 that are delayed with respect to each other (time delay τ) are focused onto the sample (spot size 100 μm) under an angle of 2.5 degrees with respect to the surface normal. While the first pulse sets up a macroscopic first order polarization $P^{(1)}(t)$ in the material, the second delayed pulse acts twofold: in a first step, it interacts with the polarization left over from the first pulse, creating a population of excitons. In a second step it generates a third order polarization $P^{(3)}(t,\tau)$, which acts as a source for the nonlinear signal. In a simple picture the FWM signal is proportional to the polarization $P^{(1)}(\tau)$ in the sample at the time of arrival of the second pulse [22]. In this sense the second pulse probes the decay of the macroscopic polarization due to phase destroying scattering events. In the presence of a microcavity it additionally samples the oscillatory revival of $P^{(1)}$ due to the energy exchange with the intracavity field.

Since our sample is opaque the FWM experiment is performed in a reflection geometry, i.e. the signal is detected into a direction corresponding to $-2k_1 + k_2$ [24]. The FWM signal can be analyzed either time-integrated (TI) (upper right), spectrally resolved (SR) by means of a monochromator or time-resolved (TR) by upconverting it in a LiIO$_3$ crystal using a third strong reference pulse (lower right). For a more detailed discussion on the FWM technique we refer to Ref. [22].

3 Linear Dynamics in the Strong Coupling Limit: the Role of Inhomogeneous Broadening

For the experimental investigation of the linear response we choose a sample position at which the cavity mode is close to resonance with the transition of the 1s quantum well exciton. The cavity emission following the excitation with a weak resonant laser pulse is temporally resolved as described above; the direction in which the emission is analyzed is that of the reflected pulse. As can be seen from Fig. 4 the cavity emission is dominated by pronounced oscillations with a period of 935 fs. This value is in excellent agreement with the normal mode splitting of 4.4 meV that can be observed in the spectrum of the reflected pulse (see inset of Fig. 4). Based on the experiences gained on atomic microcavities [2] we conclude that the oscillations are normal mode oscillations and arise from a reversible energy transfer between the macroscopic polarization inside the

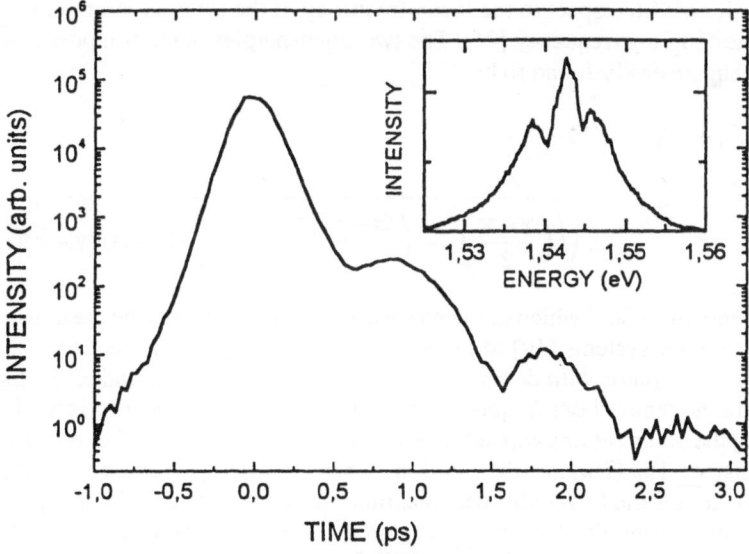

Figure 4   Time resolved cavity emission into the direction of the reflected pulse. Inset: corresponding spectrum. The two dips correspond to the normal modes of the system.

quantum wells and the intracavity field. The overall decay of the signal in Fig. 4 is relatively fast; over five orders of magnitude within 3 ps. The fast decay is consistent with the reflected pulse spectrum in the inset from which a normal mode linewidth of 2 meV is obtained; a value much larger than the cavity linewidth.

In imitation of the concept of linewidth averaging known from homogeneously broadened transitions (see below, Eq. 3.2) it can be concluded that the excitonic transition in our sample is considerably inhomogeneously broadened. This demonstrates that despite of the progresses which have been made in epitaxial crystal growth many samples still suffer from IB resulting from long range disorder. It is therefore desirable to understand the influence of IB on the dynamics of the coupled exciton-microcavity system. Since it would be extremely costly to grow a whole set of samples with various amounts of IB we address this problem theoretically for the moment using a semiclassical atomic model.

We start with the Maxwell-Bloch equations in the linear response regime which describe the intracavity field $\alpha(t)$ and the polarization $p(t)$ of a two level absorber which represents the exciton positioned inside the cavity. In a rotating frame [25] the equations read:

$$
\begin{aligned}
\dot{\alpha}(t) &= (-\kappa + i\Theta)\alpha(t) + g \cdot p(t) + \kappa\varepsilon(t) \\
\dot{p}(t) &= (-\gamma + i\Delta)p(t) - g \cdot \alpha(t)
\end{aligned}
\tag{3.1}
$$

As defined in the introduction κ and γ are the damping constants of the cavity and the two level absorber, respectively. Θ and Δ are the detunings of the cavity and the optical transition with respect to the mean frequency of the external laser field $\varepsilon(t)$ and g denotes the coupling frequency [26]. The two eigenenergies of the composite exciton-photon system are easily found to be:

$$
\lambda_{\pm} = -\left(\frac{\gamma+\kappa}{2}\right) + i\left(\frac{\Theta+\Delta}{2}\right)
$$
$$
\pm \sqrt{\left(\frac{\gamma-\kappa}{2}\right)^2 - \left(\frac{\Theta-\Delta}{2}\right)^2 - g^2 + \frac{i}{2}[(\Theta-\Delta)(\gamma-\kappa)]} \tag{3.2}
$$

We now extend this model which has been successfully used to describe the dynamics of atomic microcavity systems [10] to incorporate Gaussian inhomogeneous broadening. Instead of one oscillator with detuning Δ we assume N oscillators (index i) which are detuned from the central laser frequency by Δ_i. The Gaussian distribution is modeled by weightening the polarizations with a factor w_i, with $\sum_i w_i = 1$ to keep the total oscillator strength constant. The Gaussian distribution is characterized by a broadening parameter b where $2b\sqrt{2ln2}$ is the full-width-half-maximum of the distribution. For simplicity the coupling frequency and the dephasing rate g are assumed to be identical for all excitons. Altogether, equations (3.1) take the form [27]:

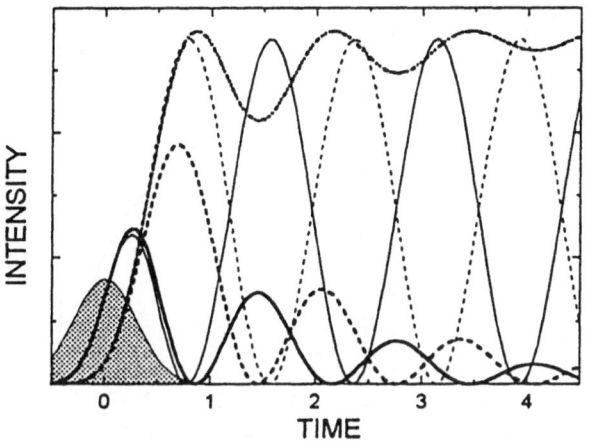

Figure 5
Squared magnitudes
of the intracavity
field (solid lines) and
the macroscopic po-
larization (dashed
lines) for an unbroad-
ened (thin lines) and
an inhomogeneously
broadened ($b = 1.2$)
system (thick lines).
g is 2 and κ, γ, Θ, and
Δ are set to zero. The
dashed-dotted line
shows the total po-
larization. The exci-
tation pulse is shown
as a gray area.

$$\dot{\alpha}(t) = (-\kappa + i\Theta)\alpha(t) + g\sum_i w_i \cdot p_i(t) + \kappa\varepsilon(t)$$

$$\dot{p}_i(t) = (-\gamma + i\Delta_i)p_i(t) - g \cdot \alpha(t)$$

(3.3)

Fig. 5 shows the dynamics of an unbroadened and an inhomogeneously broadened system; the dephasing rate γ, the cavity damping rate κ and the detunings are set to zero for simplicity. Since experimentally an intensity is detected we plot here the squared magnitudes of the intracavity field (solid lines) and the macroscopic polarization $P^{(1)}(t) = \sum_i w_i p_i(t)$ (dashed lines). The external pump pulse is shown as a gray area. In the absence of any broadening the observed normal mode oscillations are undamped like expected for a non-dissipative system (thin lines). The intensity of the intracavity field has a π-phase shift with respect to the squared macroscopic polarization; i.e. at the position of a minimum in $|\alpha(t)|^2$ we find a maximum in $|P^{(1)}(t)|^2$ and vice versa [28].

In the presence of IB, in contrast, $|\alpha(t)|^2$ and $|P^{(1)}(t)|^2$ are heavily damped (Fig. 5, thick lines). The damping is a consequence of a destructive interference between the individual polarizations within the Gaussian distribution. The interference leads to a reduction of the macroscopic polarization which is the only driving force for the intracavity field after the pump pulse is switched off. Since there is no real dissipation in the system ($\kappa = \gamma = 0$) the energy is not lost as for a damped system but "stuck" in the excitonic part of the composite system. This effect, due to which the numerical curves show so called collapse and revivals [29] for later times [30], can be visualized by plotting the quantity $\sum_i |w_i p_i(t)|^2$ ("total polarization") (Fig. 5, dashed-dotted line).

Another feature that is apparent from Fig. 5 is the higher oscillation frequency of the broadened system. A systematic study, the result of which is shown in the left plot of Fig. 6, reveals that the oscillation frequency first increases with IB, but quickly collapses as the broadening parameter b exceeds the coupling energy g. This behavior which is in agreement with recent calculations in the frequency domain [31,32] is somewhat

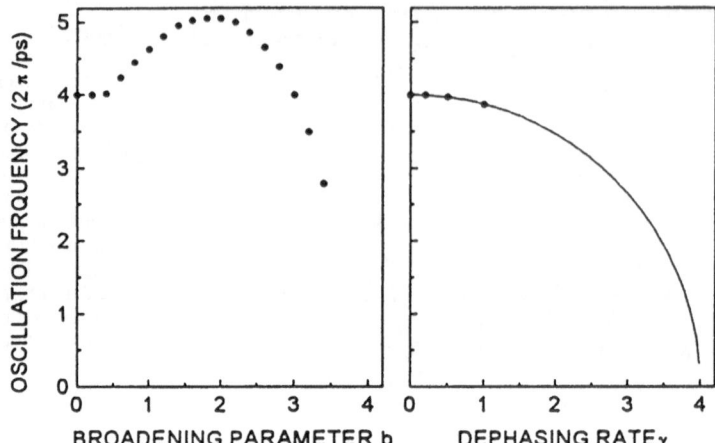

Figure 6   Left part: Oscillation frequency as a function of the broadening parameter b. Right part: Collapse of the normal mode splitting with increasing homogeneous broadening. The dots represent values extracted from the numerical calculations, the solid line is obtained from equation (3.2) for κ, Θ, $\Delta = 0$.

surprising, because for a homogeneously broadened system the normal mode splitting simply collapses as γ is increased. However, it can be intuitively understood on the basis of equation (3.1) where the oscillation frequency is given by the absolute value of the square root expression. With κ, γ, and Θ being zero this expression reduces to $\sqrt{-(\Delta/2)^2 - g^2}$. In the absence of detuning and any broadening the oscillation frequency is simply $2g$. In the case of a Gaussian distribution, in contrast, most of the oscillators have a finite Δ which, altogether, results in a higher value for the oscillation frequency. As discussed above an IB results in a strong damping of the macroscopic polarization. In this sense an IB acts similar to a homogeneous line broadening caused by an irreversible dephasing of the polarization. As the dephasing rate g is increased no increase in the oscillation frequency is observed. Instead the normal mode splitting steadily collapses (right part of Fig. 6). In the case of IB the damping induced collapse at some point overcompensates for the enhanced normal mode splitting resulting from the "collective detuning". This overcompensation finally leads to a drastic decrease in the oscillation frequency.

It is interesting to note that the phase shift between $|P^{(1)}(t)|^2$ and $|\alpha(t)|^2$ is no longer π as for the unbroadened case but somewhat larger. This can be seen in Fig. 5 where the minimum in $|\alpha(t)|^2$ in the vicinity of $t = 2$ does not coincide with the corresponding maximum in $|P^{(1)}(t)|^2$. These phase shifts can be understood by a careful discussion on the basis of the equations of motion. A detailed discussion addressing these points will be given elsewhere [33].

It will be interesting to compare the above results obtained with a simple atomic model to calculations based on a more elaborated theory including a continuum of band states and many-particle effects [34].

4   Nonlinear Dynamics in the Strong Coupling Limit: Observation of a Characteristic Oscillation Pattern

In section 3 we made the reasonable assumption that the oscillations reflect a reversible coherent energy transfer between the material polarization and the intracavity field. Yet, there is a purely classical point of view to that problem. An alternative description which does not include the concept of coupling is a Fabry-Perot resonator containing a Lorentzian absorber [13]. Thus the existence of a coherent energy transfer mediated by a coupling can not unambiguously be concluded from the linear dynamics. Furthermore, the dynamics displayed in Fig. 4 shows no features that are unique to the SMC system. Atomic, molecular and semiconductor spectroscopy is full of examples for systems that, despite of having a completely different nature, show an identical linear behavior; i.e. two peaks in the spectrum and cosine-like oscillations in the time domain. Simple examples for those systems include an electronic 3-level system in which case the oscillations in the linear emission reflect the motion of a quantum mechanical wave packet and a 2×2-level system where they result from a polarization interference [21,22]. It was shown, however, that a 3-level and a 2×2-level system display a different oscillation pattern in their nonlinear response. Investigating the nonlinear dynamics, e.g. in a TR-FWM experiment, can therefore help to enlighten the physical nature of the system under investigation [21,22]. In view of the findings we focus in the following on the nonlinear response of the strongly coupled exciton-cavity system. We find the TR-FWM signal to be dominated by pronounced normal mode oscillations. The oscillation pattern, however, is rather complex and very different from that of a set of coupled electronic oscillators (e.g. an electronic 3-level system). The nonlinear dynamics also differs from that of an exciton in free space which does not exhibit any oscillations in its dynamics. By comparing the experimental results with model calculations on the basis of the Maxwell-Bloch equations we find the complex oscillation pattern to result from the fact that in the presence of a microcavity the field driving the polarization is not the field of the short external pulse but the oscillating intracavity field which persists for much longer times. In other words, the observed behavior reflects the mixed character of the coupled photon-electronic state in the microcavity. Our results also bare implications for the case of a bulk-polariton. Because of the strong analogies between a cavity-polariton and a bulk polariton a TR-FWM experiment on the latter system should lead to similar results.

For the FWM experiments the cavity is tuned slightly below the heavy hole resonance to minimize complications due to mixing between the cavity mode and the light hole transition. The normal mode splitting under these conditions amounts to 6.1 meV as determined from the FWM spectrum [35]. The TR-FWM signal for time delays τ of 100 and 170 fs is shown in Fig. 7. The signal exhibits oscillations with a period of 0.68 ps in agreement with the splitting obtained from the spectrum. The oscillations which significantly vary in amplitude as the time delay τ is varied are superimposed onto a slowly varying envelope of nearly 3 ps FWHM. In the absence of a cavity this envelope is given by an interplay between electronic dephasing, many-particle Coulomb

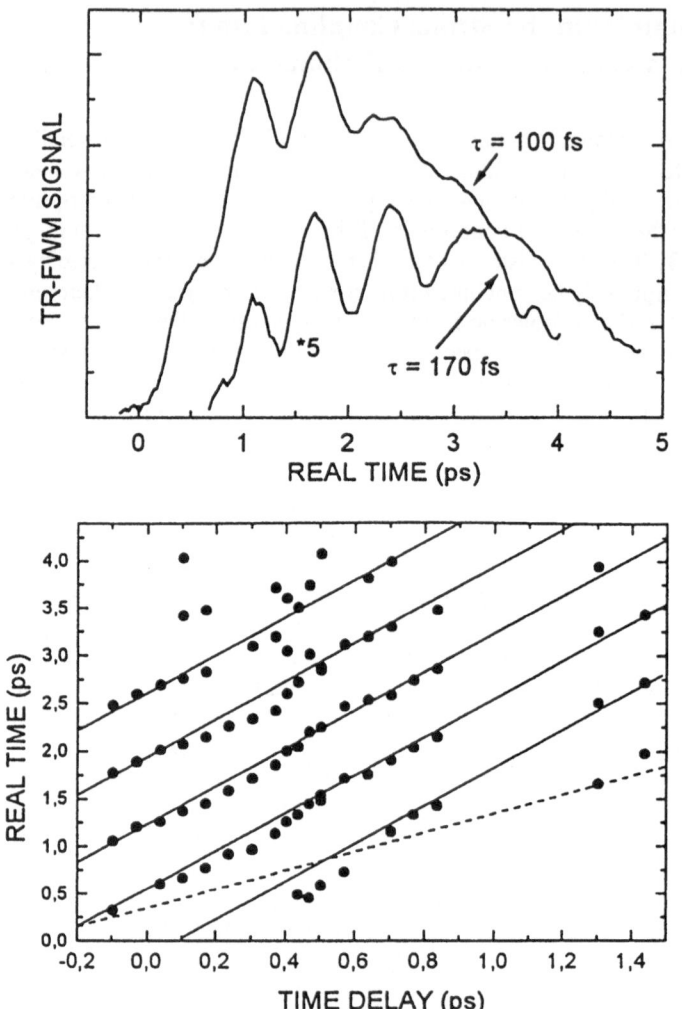

Figure 7
Time-resolved
FWM signal for
time delays τ of
100 and 170 fs.

Figure 8 Positions of the signal minima plotted versus *t* and τ.

effects and disorder induced inhomogeneous broadening [36–38]. It can be expected
that the presence of the microcavity leads to additional effects. This, however, should
be confirmed by model calculations including all of the above effects.

Here, we concentrate on the observation and the modeling of the signal oscillations.
In particular, we are interested in the oscillation pattern, i.e., in how the oscillation
minima shift in real time *t* as the time delay τ is varied. The pattern extracted from a set
of transients is shown in Fig. 8 where we plot the positions of the oscillation minima
as dots versus *t* and τ. The minima clearly follow a slope of $t = 2\tau$ (indicated by the

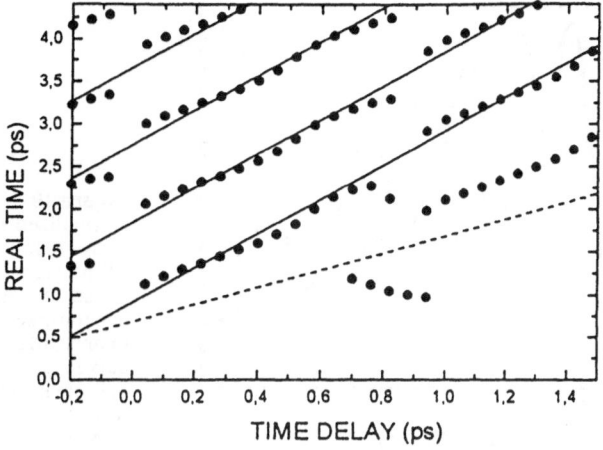

Figure 9
Positions of the minima in $|\alpha^{(2k_2-k_1)}|^2$ for Θ = -1.5 meV, Δ = 0 meV, g = 2 meV, and practically no dissipation ($1/\kappa = 1/\gamma = 1/\xi =$ 100 000 ps).

solid lines), as opposed to the $t = \tau$-slope (dashed line) known from a purely electronic system of coupled oscillators. A close inspection reveals the minima to perform slight undulations around the $t = 2\tau$-slope. Furthermore a crossover behavior suggests itself in the vicinity of $\tau = 0.5$ ps.

To understand the origin of this behavior analytical calculations have been performed [39]. To this end equations (3.1) have been extended by a third equation describing the dynamics of an excitonic density n (with a lifetime ξ). In free space the FWM intensity is proportional to the third order polarization. In the presence of a microcavity, however, we have to go one step further and iteratively calculate an intracavity field $\alpha^{(2k_2-k_1)}$ associated with the direction $2k_2 - k_1$ [24]. Part of this field leaks out of the cavity and gives rise to the FWM signal. By assuming δ-pulse excitation it becomes possible to derive an analytical expression for $\alpha^{(2k_2-k_1)}$ [39]. If all detunings and damping rates are for simplicity set to zero one ends up with an expression which (apart from prefactors) is essentially a superposition of four terms each of which describes a different (t,τ)-behavior. One among these terms describes a $t = 2\tau$-behavior. This term differs from the others in that it linearly increases with time t [40]. After one normal mode oscillation at the latest it exceeds all others and gives the leading contribution to the oscillation pattern. Following the colculation reveals that the terms $t = 2\tau$-behaviour finally results from the phase conjugations [41]. Note that in the case of an inhomogeneously broadened 2-level system the phase conjugation leads to the so called "photon echo", i.e. to a sharp emission spike at $t = 2\tau$ [22,42–44]. The terms increase is caused by the fact that in the presence of a cavity the polarization is not directly driven by $\varepsilon(t)$ as in free space but by the intracavity field α which is nonzero for much longer times. Furthermore α is periodically revived due to its energy exchange with the absorber. In some sense the effect of the microcavity on the nonlinear response can be seen in analogy to the Coulomb exchange terms known from polymers [45] and semiconductors [36]. In the absence of dephasing these terms which renormalize energy and field cause an infinite signal rise [37]. In order to visualize the analytical results numerical

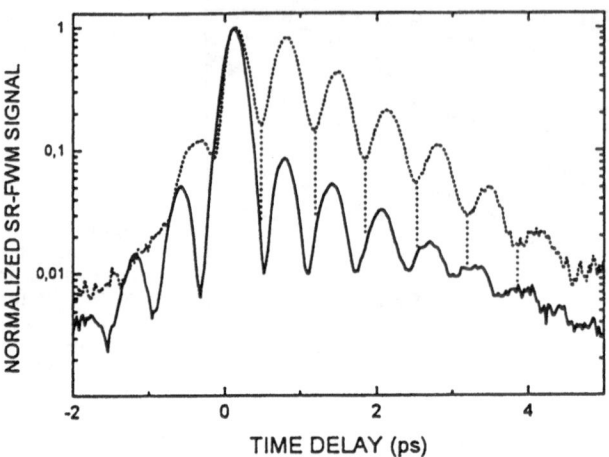

Figure 10
Normalized spectrally resolved FWM signal detected at the exciton-like resonance (dashed curve) and at the cavity-like resonance (solid curve).

calculations assuming 100 fs-pulses have been performed. For the case of no detuning and practically no dissipation the oscillation minima form a series of almost straight lines with a $t = 2\tau$-slope. In accordance with the analytical result only very faint undulations around this slope are observed. They arise from the combined action of all other terms since at times at which the first term is zero some others are not, which then cause slight shifts of the signal minima (and maxima) with respect to the $t = 2\tau$-lines. If, however, a slight detuning is introduced to better match the experimental conditions the undulations become more pronounced. This can be seen from Fig. 9 where we plot the (t,τ)-pattern obtained numerically for a cavity detuning of -1.5 meV. In addition to the undulations, a crossover behavior suggests itself in the vicinity of $\tau = 0.9$ ps repeating itself in τ with the oscillation frequency. The introduction of dephasing tends to enlarge the undulation behavior around the $t = 2\tau$-slope (depending on the exact choice of γ and ξ) which is then observed for even larger detunings. The additional introduction of an IB or a moderate local field does not change the (t,τ)-structure significantly [39]. Altogether, the atomic model reproduces several key features of the experimental data and, hence, gives a good qualitative description of the coupled exciton-cavity system [46].

As mentioned above, the oscillation pattern observed in time-resolved FWM clearly differs from that of an electronic 3-level system [21]. This finding seems to contradict spectrally resolved FWM experiments which have been performed recently [47–49]. The signature of a 3-level system in SR-FWM are two spectral peaks, the intensity of which oscillates in phase as the time delay τ is varied [50]. The measurements presented in [48] indicate exactly this behavior. Yet, measurements with improved signal to noise ratio which are shown in Fig. 10 reveal a clear phase shift between the excitonic-like mode and the cavity-like mode [35]. It is reasonable to assume that this phase shift which is not observed for a 3-level system is a signature characteristic of the mixed exciton-photon character of the SMC system; i.e. the spectral counterpart to the

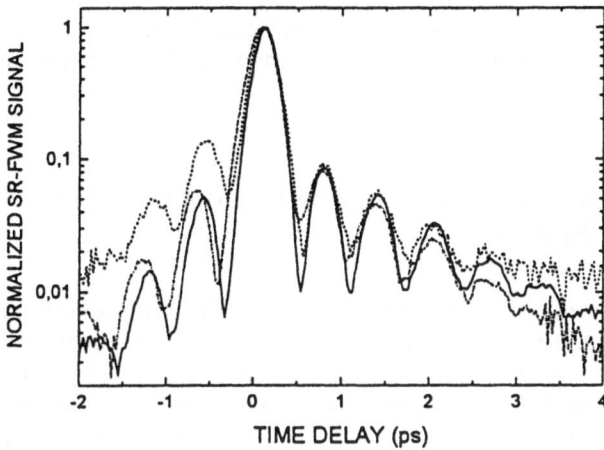

Figure 11 Spectrally resolved FWM signal detected at the high energy side (806.3 nm, dash-dotted) the peak (806.9 nm, solid) and the low energy side (807.3 nm, dashed) of the cavity-like resonance. The curves are normalized for better comparison.

complex $t = 2\tau$-oscillation pattern observed in the time domain. Model calculations to clarify this point are currently under way.

There are, however, other interesting features which can be observed in Fig. 10. Noteworthy is a considerable enhancement of the cavity-like resonance with respect to the exciton-like mode around $\tau = 0$. Furthermore, the signal to negative times is different for the two modes. While the exciton-like mode is unmodulated for times earlier than -800 fs the cavity-like resonance exhibits pronounced oscillations.

The different character of the oscillations to negative times as compared to the oscillations for $\tau > 0$ ps becomes even more obvious when the detection wavelength is varied around the cavity-like resonance (Fig. 11). In contrast to the oscillations for positive time delays the oscillations to negative times exhibit a pronounced phase shift as the detection wavelength is scanned across the resonance. Simultaneously there are severe changes in the relative signal strength. For a quantum well not embedded in a microcavity it was found that biexcitonic effects can lead to a FWM signal to negative times [51–53]. This suggests that the oscillations to negative times arise from a coherent energy transfer between the intracavity field and a biexcitonic transition. To confirm this, however, further spectrally resolved measurements with various polarizations should be performed [53,54].

As discussed above the linear response of the exciton-cavity system is very sensitive to the irradiating laser power [15,55]. An increase in pump intensity easily leads to a bleaching of the excitonic transition and therefore to a collapse of the normal mode splitting. This switching to the weak coupling limit can also be observed in the nonlinear response. For high laser intensities the FWM spectrum reduces to a single peak positioned energetically between the two normal modes observed at low intensities. Accordingly, the TR-FWM signal exhibits no oscillations. The signal maximum shifts with a $t = \tau$-slope over a large τ-range (not shown). Yet, as we will see in the next section there are also ultrafast oscillations in the cavity emission if the structure is heavily pumped. The oscillations reported in the following are different from normal mode oscillations. They arise from an interference between two cavity modes corresponding to two different angles.

5 Linear Dynamics in the Weak Coupling Limit: Ultrafast Oscillations in the Laser Emission

In this section we investigate the cavity emission for the case of intense resonant pumping. Like section 3 we excite the sample under a small angle while we tune the central laser frequency to match for the oblique cavity mode (pump mode). In contrast to section 3 we detect the emission in the direction of the surface-normal, the direction into which lasing is expected (laser mode). The time resolved emission in this direction is depicted in Fig. 12. Unlike for the case of nonresonant optical pumping, which has been investigated by a variety of groups [56–59] the emission instantaneously follows the excitation by the pump pulse which is marked by the high peak around 0 ps. For nonresonant pumping the signal onset is delayed since the carrier distribution first has to transfer energy to the lattice. For resonant excitation, in contrast, only a momentum relaxation is required [60]; a process which is extremely fast at high densities.

The emission dynamics in Fig. 12 is again dominated by pronounced oscillations which are superimposed onto a nearly exponential decay. The period of the oscillations slightly increases in time, from 510 fs between the first two peaks to over 700 fs for later oscillations. Interestingly, it does not significantly depend on the pump intensity [61]. Although there is some indication for a laser threshold with increasing pump power [62] the oscillation frequency remains basically unaffected. This can be seen in the inset of Fig. 12 where we show a typical plot of the signal maxima versus pump intensity.

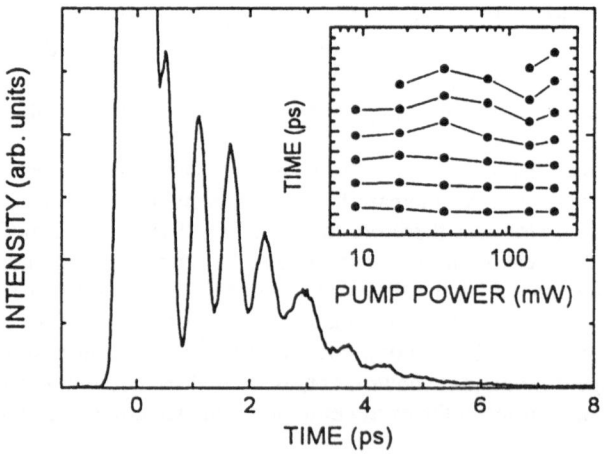

Figure 12
Time resolved emission in the direction of the surface-normal. Inset: Positions of the oscillations maxima for different pump powers (the lines are guide to the eye).

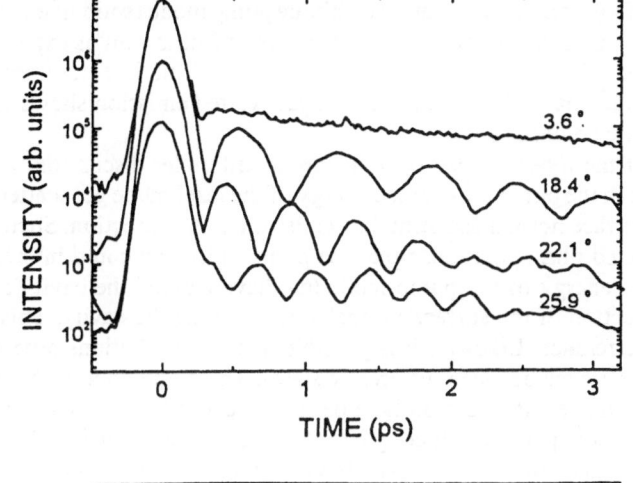

Figure 13
Time resolved emission in the direction of the surface-normal for different pump angles ϕ_e. The curves are vertically displaced for clarity.

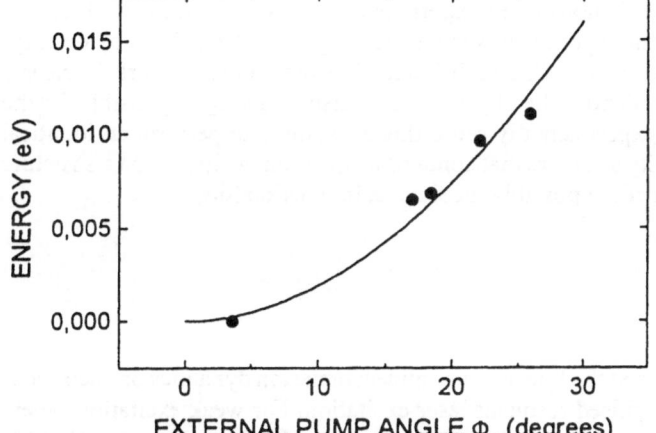

Figure 14
Experimentally observed oscillation frequency multiplied by Planck's constant (dots) and the energy difference between pump mode and laser mode (solid line) as a function of the external pump angle ϕ_e.

Although a full model predicts ultrafast relaxation oscillations up to 1.4 THz [58] (a value considerably exceeding the 100 GHz limit predicted by a simple rate equation model [5]) the insensitivity of the oscillation frequency to the pump power excludes the possibility of the signal modulations being relaxation oscillations. Furthermore, the relaxation oscillations predicted in [58] are much less pronounced than the oscillations reported here.

In order to understand the origin of the oscillations we measure the surface-normal emission for different pump angles ϕ_e. The results which are shown in Fig. 13 clearly demonstrate a strong dependence of the oscillation frequency on pump angle. To further clarify this dependence we plot in Fig. 14 the mean oscillation frequency multiplied by Planck's constant h as function of ϕ_e. A comparison with the energy difference that is expected between the surface normal mode and the oblique pump mode (solid line in the inset) shows that the oscillation frequency is in excellent agreement to what is expected for an interference between the pump mode and the laser mode. This interpretation is further supported by the fact that the spectrum of the upconverted emission shows two peaks resulting from the two modes [63].

The physical nature of the interference phenomenon is still under discussion. One possibility is that photons in the pump pulse at the energy of the laser mode get scattered into the direction of the surface normal and stimulate emission in this direction. Such an emission would have a fixed phase relationship to the pump pulse and could interfere with pump photons at the energy of the pump mode that have entered the cavity and are scattered into the direction of the surface normal while leaving the cavity. This is a simple polarization interference. Likewise it is possible that the oscillations reflect a quantum mechanical interference between different k-states. The pump photons have a finite in-plane wavevector and excite electron-hole pairs in specific k-states with finite k. In contrast, the electron-hole pairs which couple to the laser mode are near $k = 0$. A coupling between these k-states through many-body Coulomb interaction is expected theoretically [19] and was demonstrated experimentally at low densities [64]. At the high densities of the present experiment, such a coupling may be mediated by density-polarization product terms in the Coulomb interaction [65]. Further work is clearly needed to unambiguously identify the physical mechanism drawing responsible for the oscillations. One way to experimentally settle this question is to perform an identical experiment on a microcavity structure that contains no quantum wells; i.e. on a structure free of any a medium that could possibly mediate an interaction [66].

6 Summary

In summary we have analyzed the linear and nonlinear emission dynamics of a semiconductor microcavity under pulsed resonant laser excitation. For weak excitation where the sample is in the strong coupling regime the emission is dominated by normal mode oscillations. We have shown that inhomogeneous broadening leads to a strong damping of these oscillations and of the overall signal. The damping is caused by a destructive interference between the individual oscillators within the inhomogeneously broadened line. Simultaneously inhomogeneous broadening leads to a slight increase of the oscillation frequency.

Normal mode oscillations observed in the nonlinear response show a complex pattern. A comparison with calculations on the basis of the Maxwell-Bloch equations has shown that the complex pattern is characteristic of the mixed exciton-photon nature of the semiconductor microcavity system. Although the Maxwell-Bloch equations do not satisfactorily describe all aspects of the nonlinear dynamics (e.g. there are discrepancies between theory and experiment if the time-integrated four-wave-mixing signal is considered) they qualitatively reproduce some key features which are observed experimentally. We can record the fact that a qualitative description of the coupled exciton-cavity system is to some extent possible without considering any many-particle Coulomb effects.

If the cavity is heavily pumped the normal mode splitting collapses and oscillations are no longer observed; neither in four-wave-mixing nor in the direction of the reflected pump pulse. The laser emission perpendicular to the sample plane, in contrast, is strongly modulated under these conditions. We have shown these oscillations to arise from an interference between the laser mode and an oblique pump mode.

Acknowledgments

The author is grateful to J. Shah for innumerable discussions which have stimulated this work and for making the authors stay at Bell Laboratories so pleasant. Furthermore the author thanks T. Meier for performing the calculations presented in section 4, F. Löser for the collaboration on section 3, J. E. Cunningham for providing the sample, and M. Tsuchiya, H. Wang, F. Jahnke, and A. Knorr for fruitful discussions. The authors stay at Bell Laboratories was financially supported by the Alexander-von-Humboldt foundation.

Bibliography

[1]   E.M. Purcell, Phys. Rev. **69**, 681 (1946).

[2]   for an overview see: *Confined Electrons and Photons: New Physics and Applications*, edited by E. Burstein and C. Weisbuch (Plenum Press, New York, 1995); *Cavity Quantum Electrodynamics*, edited by P.R. Berman (Academic, Boston, 1994).

[3]   J.L. Jewell, K.F. Huang, K. Tai, Y.H. Lee, R.J. Fischer, S.L. McCall, and A.Y. Cho, Appl. Phys. Lett. **55**, 424 (1989).

[4]   H. Yokoyama, Science **256**, 66 (1992).

[5]   Y. Yamamoto, G. Björk, H. Heitmann, and R. Horowicz , in *Optics of Semiconductor Nanostructures*, Ed. by F. Henneberger, S. Schmitt-Rink, and E.O Göbel, Akademie Verlag, Berlin (1993).

[6]   E.F. Schubert, N.E.J. Hunt, M. Micovic, R.J. Malik, D.L. Sivco, A.Y. Cho, G.J. Zydzik, Science **265**, 943 (1994).

[7]   U. Lemmer, R. Hennig, W. Guss, A. Ochse, J. Pommerehne, R. Sander A. Greiner, R.F. Mahrt, H. Bässler, J. Feldmann, and E.O. Göbel, Appl. Phys. Lett. **66**, 1301 (1995).

[8]   N. Tessler, G.J. Denton, and R.H. Friend, Nature **382**, 695 (1996).

[9]   Y. Kaluzny, P. Goy, M. Gross, J. M. Raimond, and S. Haroche, Phys. Rev. Lett. **51**, 1175 (1983).

[10] R. J. Brecha, L. A. Orozco, M. G. Raizen, M. Xiao, and H.J. Kimble, J. Opt. Soc. Am. **B 12**, 2329 (1995).

[11] T. B. Norris, J.-K. Rhee, C.-Y. Sung, Y. Arakawa, M. Nishioka, and C. Weisbuch, Phys. Rev. **B 50**, 14663 (1994).

[12] J.J. Sanchez-Mondragon, N.B. Narozhny, and J.E. Eberly, Phys. Rev. Lett. **51**, 550 (1983).

[13] Y. Zhu, D.J. Gauthier, S.E. Morin, Q. Wu, H.J. Carmichael, and T.W. Mossberg, Phys. Rev. Lett **64**, 2499 (1990).

[14] R.J. Thompson, G. Rempe, and H.J. Kimble, Phys. Rev. Lett. **68**, 1132 (1992).

[15] F. Jahnke, M. Kira, S. W. Koch, G. Khitrova, E.K. Lindmar, T.R. Nelson, D.V. Wick, J.D. Berger, O. Lyngnes, H. M. Gibbs, K. Tai, Phys. Rev. Lett. **77**, 5257 (1996).

[16] R. Houdre, C. Weisbuch, R.P. Stanley, U. Oesterle, P. Pellandini, and M. Illegems, Phys. Rev. Lett. **73**, 2043 (1994).

[17] M. G. Raizen, R. J. Thompson, R. J. Brecha, H.J. Kimble, and H. J. Carmichael, Phys. Rev. Lett. **63**, 240 (1989).

[18] E. T. Jaynes and F.W. Cummings, Proc. IEEE **51**, 89 (1963).

[19] H. Haug and S.W. Koch, in: *Quantum Theory of the Optical and Electronic Properties of Semiconductors*, World Scientific, Singapore (1993).

[20] C. Weisbuch, M. Nishioka, A. Ishikawa, and Y. Arakawa, Phys. Rev. Lett. **69**, 3314 (1992).

[21] M. Koch, J. Feldmann, G. v. Plessen, E.O. Göbel, P. Thomas, and K. Köhler, Phys. Rev. Lett. **69**, 3631 (1992).

[22] M. Koch, G. v. Plessen, J. Feldmann, E.O. Göbel, Chem. Phys. **120**, 367 (1996).

[23] J. Shah, IEEE J. Quantum Electron **24**, 276 (1988).

[24] Because of multiple reflections inside the cavity, conceptually there is no difference between the FWM signals emitted into the forward and into the backward directions.

[25] L. Allen and J.H. Eberly, in: *Optical Resonance and Two-Level Atoms*, Dover Publications, New York (1975).

[26] Our notation is chosen in imitation of the one used by H.J. Kimble in [2].

[27] Note, that our model assumes the absence of any mutual interaction between the individual exciton polarizations; a reasonable assumption for very low excitation densities.

[28] We note that a full cycle in the unsquared magnitudes is completed only after two subsequent maxima in $|\alpha(t)|^2$.

[29] P. Meystre and M. Sargent, *Elements of Quantum Optics* **III**, Springer Verlag, Berlin (1990).

[30] Due to dissipation the phenomenon of collapse and revival can not be observed in a real SMC.

[31] S. Pau, G. Björk, H. Cao, E. Hanamura, and Y. Yamamoto, Solid State Comm. **98**, 781 (1996).

[32] R. Houdre, R.P. Stanley, and M. Illegems, Phys. Rev. **A 53**, 2711 (1996).

[33] F. Löser, M. Koch, J. Shah, unpublished.

[34] See the article by F. Jahnke et al. in this issue.

[35] Since the cavity is detuned with respect to the excitonic transition the mixing between the two modes is incomplete. The result is a weaker mode that has more the character of an exciton and a stronger mode that is more cavity-like [20,47].

[36] D.S. Kim, J. Shah, T.C. Damen, W. Schäfer, F. Jahnke, S. Schmitt-Rink, and K. Köhler, Phys. Rev. Lett. **69**, 2725 (1992).

[37] M. Lindberg, R. Binder, and S.W. Koch, Phys. Rev. A **45**, 1865 (1992).

[38] F. Jahnke, M. Koch, T. Meier, J. Feldmann, W. Schäfer, P. Thomas, S.W. Koch, E.O. Göbel, and H. Nickel, Phys. Rev. B **50**, 8114 (1994).

[39] M. Koch, J. Shah, and T. Meier, Phys. Rev. B, Rapid Comm. (Jan. 1998).

[40] This term will not grow for infinite times if damping is included.

[41] Y.R. Shen in: *The Principles of Nonlinear Optics*, Wiley & Sons, N.Y. (1984).

[42] T. Yajima and Y. Taira, J. Phys. Soc. Japan **47**, 1620 (1979).

[43] G. Noll, U. Siegner, S.G. Shevel, and E.O. Göbel, Phys. Rev. Lett. **64**, 792 (1990).

[44] M.D. Webb, S. T. Cundiff, and D.G. Steel, Phys. Rev. Lett. **66**, 934 (1991).

[45] T. Meier and S. Mukamel, Phys. Rev. Lett. **77**, 3471 (1996).

[46] The variations in the oscillation depth which have been observed in Fig. 7 are also reproduced.

[47] H. Wang, J. Shah, T.C. Damen, W.Y. Jan, J.E. Cunningham, M. Hong, and J. P. Mannaerts, Phys. Rev. **B51**, 14713 (1995).

[48] H. Wang, J. Shah, T.C. Damen, L.N. Pfeiffer, and J.E. Cunningham, phys. stat. sol (b) **188**, 381 (1995).

[49] P.V. Kelkar, V. Kozlov, A.V. Numikko, C.-C. Chu, D.C. Grillo, J. Han Hua, and R. L. Gunshor, *Quantum Electronics and Laser Science Conference 1996*, QThB6, Technical Digest, Anaheim (1996).

[50] V.G. Lyssenko, J. Erland, I. Balslev, K.-H. Panthke, B.S. Razbirin, and J.M. Hvam, Phys. Rev. B **48**, 5720 (1993).

[51] B.F. Feuerbacher, J. Kuhl, and K. Ploog, Phys. Rev. B **43**, 2493 (1991).

[52] E.J. Mayer, G.O. Smith, J. Kuhl, D. Bennhardt, T. Meier, A. Schluze, P. Thomas, S.W. Koch, R. Hey, and K. Ploog, in *Ultrafast Phenomena IX*, ed. by P.F. Barbara, W. H. Knox, G. A. Mourou, and A.H. Zewail, Springer-Verlag (1994).

[53] H. Wang, J. Shah, T.C. Damen, and L.N. Pfeiffer, Solid State Comm. **91**, 869 (1994).

[54] T.F. Albrecht, K. Bott, T. Meier, A. Schluze, M. Koch, S.T. Cundiff, J. Feldmann, W. Stolz, P. Thomas, S.W. Koch and E.O. Göbel, Phys. Rev. B **54**, 4436 (1996).

[55] J.-K. Rhee, D.S. Citrin, and T.B. Norris, Y. Arakawa, and M. Nishioka, Solid State Comm. **97**, 941 (1996).

[56] G. Pompe, T. Rappen, M. Wegener, Phys. Rev. B **51**, 7005 (1995).

[57] P. Michler, A. Lohner, W.W. Rühle, and G. Reiner, Appl. Phys. Lett. **66**, 1599 (1995).

[58] F. Jahnke, S.W. Koch, and K. Henneberger, Appl. Phys. Lett. **62**, 2313 (1993).

[59] M. Hilpert, M. Hofmann, C. Ellmers, M. Östereich, H.-C. Schneider, F. Jahnke, S.W. Koch, W.W. Rühle, H.D. Wolf, D. Bernklau, and H. Riechert, *Hot Carriers in Semiconductors X*, Berlin (1997).

[60] H. Wang, J. Shah, T.C. Damen, and L.N. Pfeiffer, Phys. Rev. Lett. **66**, 1599 (1995).

[61] On the assumption that the portion of the power absorbed in the sample can be seen from the difference between the incident and the reflected spectrum (about 7 %) we estimate a pump power of 100 mW to correspond to a carrier density of $2.4 \cdot 10^{13}$ cm^{-2}, a value which should be high enough to ensure lasing.

[62] M. Koch, J. Shah, H. Wang, T. C. Damen, W.Y. Yan, and J.E. Cunningham, *International Quantum Electronics Conference 1996*, Technical Digest, Sydney (1996).

[63] M. Tsuchiya, M. Koch, J. Shah, T.C. Damen, W.Y. Yan, and J.E. Cunningham, Appl. Phys. Lett. **71**, 1240 (1997).

[64] S.T. Cundiff, M. Koch, W.H. Knox, J. Shah, and W. Stolz, Phys. Rev. Lett. **77**, 1107 (1996).

[65] A. Knorr, private communication.

[66] Alternatively one could perform an experiment on a sample similar to ours at a spatial position where the cavity is tuned far below the band edge. Unfortunately, the present sample does not allow sufficient detuning.

Ultrashort Pulse Propagation and Excitonic Nonlinearities in Semiconductor Microcavities

F. Jahnke, M. Ruopp, M. Kira, S. W. Koch

Department of Physics and Material Sciences Center,
Philipps-University Marburg, Renthof 5, 35032 Marburg, Germany

Abstract: Excitonic normal-mode coupling in semiconductor microcavities is studied on the basis of coupled Maxwell-semiconductor Bloch equations. Excitonic nonlinearities due to phase space filling, plasma screening and dephasing are consistently included on a microscopic level. Results for the influence of inhomogeneous exciton broadening on the plasma saturation of normal-mode coupling as well as pulse propagation effects are presented.

1 Introduction

Radiatively coupled excitons in quantum wells and microcavities show many properties similar to atoms in the micromaser. For example, excitonic normal-mode coupling [1] resembles vacuum-field Rabi oscillations [2,3] and effects of radiative coupling between excitons in multiple quantum wells [4,5] are similar to superradiance of atoms [6]. The description of few atoms interacting with a high-finesse cavity mode requires a fully quantum-mechanical treatment which can be given, e.g., in terms of the Jaynes-Cummings model. Since in semiconductors (bulk or quantum wells) the light-matter interaction can no longer be attributed to a small number of atoms (comparable with the number of photons), a semi-classical analysis is possible [7] except for luminescence or photon correlation studies. Hence, reflection, transmission and absorption experiments in semiconductors have a different interpretation in comparison to the atomic case. Vacuum-field Rabi oscillations of a single excited atom can be viewed as periodic energy exchange between the excited-atom zero-photon state and the deexcited-atom one-photon state. However, excitonic normal-mode coupling in semiconductor microcavities is due to the interaction of the externally driven coherent exciton polarization and the cavity field similar to coupled classical oscillators. In addition, the saturation mechanisms in atoms (e.g., power broadening or local field effects) are completely different from those in semiconductors (phase space filling and Coulomb interaction between carriers).

Recent experiments on ultrafast pulse excitation of excitons in semiconductor microcavities [8–10] lead to the question how femtosecond laser pulses propagate in a semiconductor microcavity and how the exciton saturation affects the microcavity response. In the following, we investigate the pulse propagation by directly solving Maxwell's equations for the microcavity geometry. The influence of nonlinear exciton saturation is studied on the basis of a microscopic model which incorporates the influence of phase-space filling, screening of the Coulomb interaction as well as exciton dephasing due to carrier and polarization scattering. After summarizing our theory in the following section 2, we discuss the interplay of exciton dephasing and disorder-induced inhomogeneous broadening on the saturation of excitonic normal-mode coupling in section 2.2. Results for the propagation of femtosecond laser pulses in semiconductor microcavities are discussed in section 4.

2 Theory

We consider a semiconductor microcavity where quantum wells (QWs) are placed between distributed Bragg reflectors (DBRs) consisting of alternating quarter wavelength layers as schematically shown in Fig. 1. The mirror layers are non-absorbing in the considered wavelength region. Furthermore, we assume that only in-plane components of the QW polarization are excited (e.g., for the heavy hole transition of the lowest QW subband). When an externally applied field propagates in growth direction (i.e., orthogonal to the layers of the QWs or DBRs), we have to treat only the one-dimensional wave equation

$$\left[\frac{\partial^2}{\partial z^2} - \frac{n^2(z)}{c_0^2}\frac{\partial^2}{\partial t^2}\right] E(z,t) = -\mu_0 \frac{\partial^2}{\partial t^2} P(z,t), \qquad (2.1)$$

where $n(z)$ describes the refractive index profile of the microcavity. The source term of Eq. (2.1) contains the sum of the excitonic polarizations of the individual QWs. The z-dependence of the excitonic polarization for the i-th QW is determined by the confinement wave function $\xi_i(z)$,

$$P(z,t) = \sum_i P_{QW,i}(t)\,|\xi_i(z)|^2. \qquad (2.2)$$

Note that the wave equation (2.1) has to be solved for the entire microcavity structure in order to include both the feedback of the DBRs and the field components leaving the microcavity. However, with Eq. (2.2), the source contributions of the excitonic polarization are restricted to the QW positions. In a Bloch basis with the in-plane carrier momentum k, the macroscopic QW polarization $P_{QW,i}(t)$ follows from

$$P_{QW,i}(t) = \frac{1}{L^2} \sum_k d_{cv}^* \Psi_{k,i}(t) + \text{c.c.} \tag{2.3}$$

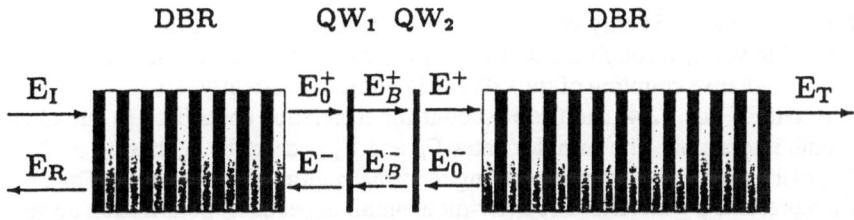

Figure 1 Schematic outline of a planar microcavity consisting of two quantum wells (QWs) between distributed Bragg reflectors (DBRs). The various field components propagate in forward (E^+) and backward (E^-) direction as well as incident (E_I), reflected (E_R) and transmitted field (E_T) are indicated.

where d_{cv} is the dipole matrix element. Coupled equations of motion for the coherently driven excitonic polarization $\Psi_{k,i}(t)$ and the subband occupation probability $f_{k,i}^{e,h}(t)$ for electrons and holes of the i-th QW have been derived within the second Born approximation [11]

$$\left[\hbar\frac{\partial}{\partial t} - \varepsilon_{k,i}^e(t) - \varepsilon_{k,i}^h(t)\right]\Psi_{k,i}(t) - \left[1 - f_{k,i}^e(t) - f_{k,i}^h(t)\right]\Omega_{k,i}(t)$$
$$= i\left[\Gamma_{k,i}(t)\Psi_{k,i}(t) + \frac{1}{L^2}\sum_{k'}\Gamma_{k,k',i}(t)\Psi_{k',i}(t)\right], \tag{2.4}$$

$$\hbar\frac{\partial}{\partial t}f_{k,i}^a(t) + \Omega_{k,i}(t)\Psi_{k,i}^*(t) - \Omega_{k,i}^*(t)\Psi_{k,i}(t)$$
$$= i\left\{\Sigma_{k,i}^{in,a}(t)[1 - f_{k,i}^a(t)] - \Sigma_{k,i}^{out,a}(t)f_{k,i}^a(t) + \Sigma_{k,i}^{pol,a}(t)\right\}. \tag{2.5}$$

The LHS of Eqs. (2.4) and (2.5) correspond to the well-known Hartree-Fock equations with phase-space filling nonlinearities, the renormalized free-particle energies $\varepsilon_{k,i}^a$ and the renormalized Rabi-energy $\Omega_{k,i}$,

$$\varepsilon_{k,i}^a(t) \;=\; \varepsilon_k^a - \frac{1}{L^2}\sum_{k'} V_{k-k'}\, f_{k',i}^a(t), \tag{2.6}$$

$$\Omega_{k,i}(t) \;=\; d_{cv}\, E_{QW,i}(t) + \frac{1}{L^2}\sum_{k'} V_{k-k'}\, \Psi_{k',i}(t), \tag{2.7}$$

where $V_{k-k'}$ is the QW matrix element of the Coulomb potential and ε_k^a are the free-carrier energies. The cavity field interacting with the i-th QW is given by

$$E_{QW,i}(t) = \int dz\, |\xi_i(z)|^2\, E(z,t). \tag{2.8}$$

In Eqs. (2.4) and (2.5) only polarization and occupation of the i-th QW are explicitly coupled. However, through the self-consistent calculation of the field $E_{QW,i}(t)$ from Eq. (2.1), radiative coupling of the QWs is included consistently.

The RHS of Eqs. (2.4) and (2.5) account for correlation contributions due to carrier Coulomb interaction. The complex rates Γ_k and $\Gamma_{k,k'}$ describe dephasing of the excitonic polarization and the corresponding renormalization of $\varepsilon_{k,i}$ and $\Omega_{k,i}$. The real part of Γ_k represents a generalized (carrier-momentum dependent) polarization decay rate. However, this decay rate is compensated to a large extent by the off-diagonal dephasing rate $\Gamma_{k,k'}$ which couples the polarization of various k-states [9]. The fact that the k-sum of the dephasing terms vanishes,

$$\sum_k \left[\Gamma_{k,i}(t)\Psi_{k,i}(t) + \frac{1}{L^2}\sum_{k'}\Gamma_{k,k',i}(t)\Psi_{k',i}(t) \right] = 0 \tag{2.9}$$

shows the microscopic nature of dephasing as an interference effect rather than irreversible decay of the individual Ψ_k.

Note that the correlation contributions include screening corrections of the Hartree-Fock terms described by Eqs. (2.6) and (2.7). The rates Σ_k contain the influence of carrier and polarization interaction on the occupation probabilities. For the detailed expressions of Γ and Σ see Ref. [11].

2.1 Light Propagation in Quantum Wells

For the solution of the light propagation problem, we neglect the QW extension in comparison to the light wavelength in Eqs. (2.2) and (2.8) using $|\xi_i(z)|^2 \simeq \delta(z - z_i)$ with the QW position z_i. On the other hand, the QW extension is included in the description of the electronic properties, especially in the QW matrix elements of the Coulomb potential. Note that the carrier confinement in QWs is due to the band-gap offset between the well and buffer material whereas the corresponding refractive index change is usually very small. Then, for QWs with constant refractive index of the well and buffer region, we obtain from Eq. (2.1) the solution

$$E(z,t) = -\frac{\mu_0 c}{2} \sum_i \frac{\partial}{\partial t} P_{\text{QW},i}(t - \frac{|z-z_i|}{c}) + E_0^{\pm}(t \mp \frac{z}{c}). \qquad (2.10)$$

The homogeneous solutions $E_0^{\pm}$ correspond to freely propagating waves incident on the QWs in forward and backward direction. The first term on the RHS of Eq. (2.10) describes the field which is radiated from the QWs. For a QW at $z = 0$, we find two contributions

$$\frac{\partial}{\partial t} P_{\text{QW}}(t - \frac{|z|}{c}) = \begin{cases} \frac{\partial}{\partial t} P_{\text{QW}}(t - \frac{z}{c}) & z > 0 \\[2mm] \frac{\partial}{\partial t} P_{\text{QW}}(t + \frac{z}{c}) & z < 0 \end{cases} \qquad (2.11)$$

describing the QW emission in forward direction for $z > 0$ as well as a wave propagating away from the QW in backward direction for $z < 0$. For a set of N QWs, the resulting transmitted and reflected field components, E^+ and E^-, follow from

$$E^+(z,t) = E_0^+(t - \frac{z}{c}) - \frac{\mu_0 c}{2} \sum_{i=1}^{N} \frac{\partial}{\partial t} P_{\text{QW},i}(t - \frac{z-z_i}{c}), \qquad (2.12)$$

$$E^-(z',t) = E_0^-(t + \frac{z'}{c}) - \frac{\mu_0 c}{2} \sum_{i=1}^{N} \frac{\partial}{\partial t} P_{\text{QW},i}(t + \frac{z'-z_i}{c}), \qquad (2.13)$$

where $z \leq z_1$ and $z' \geq z_N$. E_0^+ and E_0^- are the fields incident on the QW system from the LHS and RHS, respectively. The field at the j-th QW contains in addition to $E_0^{\pm}$ also the radiation from the other QWs

$$E_{\text{QW},j}(t) = E_0^+(t - \frac{z_j}{c}) + E_0^-(t + \frac{z_j}{c}) - \frac{\mu_0 c}{2} \sum_i \frac{\partial}{\partial t} P_{\text{QW},i}(t - \frac{|z_j - z_i|}{c}), \qquad (2.14)$$

which leads within a self-consistent solution to radiative coupling between the QWs. For QWs in a microcavity, as shown in Fig. 1, one always has fields E_0^+ and E_0^- applied on both sides of the QW system due to the presence of the mirrors. On the other hand, the emission of the QW system, E^+ and E^-, acts as incident waves for the DBRs together with the field(s) externally applied to the microcavity.

At this point it is instructive to discuss the radiative broadening of the QW transitions due to propagation effects, which is fully included in the above discussed treatment. We consider the simplest case of a single QW with an external field applied from the LHS, i.e., $E_0^+ = E_0$ and $E_0^- = 0$. After Fourier transforming E and P according to

$$E(t) = \int \frac{d\omega}{2\pi} e^{-i\omega t} E(\omega), \qquad (2.15)$$

we obtain from Eqs. (2.12) and (2.13) the complex reflection and transmission coefficients, $r = E^-/E_0$ and $t = E^+/E_0$, respectively,

$$r(\omega) = \frac{ik}{2} \frac{1}{\varepsilon_0 \varepsilon} \frac{P(\omega)}{E_0(\omega)}, \tag{2.16}$$

$$t(\omega) = 1 + r(\omega), \tag{2.17}$$

where $\omega = ck$, $c = c_0/n$, $c_0^2 = \frac{1}{\mu_0 \varepsilon_0}$ and $n^2 = \varepsilon$ have been used. Note that the excitonic susceptibility $\chi(\omega) = P(\omega)/E_{QW}(\omega)$ is defined with the field at the QW position, E_{QW}, that follows from

$$E_{QW}(\omega) = E_0(\omega) - \frac{\mu_0 c}{2} i\omega P_{QW}(\omega). \tag{2.18}$$

Using Eq. (2.18) to introduce the excitonic susceptibility in Eq. (2.16), we find for the QW reflectivity due to the excitonic polarization

$$r(\omega) = \frac{\frac{ik}{2} \frac{1}{\varepsilon_0 \varepsilon} \chi(\omega)}{1 - \frac{ik}{2} \frac{1}{\varepsilon_0 \varepsilon} \chi(\omega)}. \tag{2.19}$$

For the linear susceptibility of the 1s exciton,

$$\chi_{1s}(\omega) = -g \frac{|d_{cv}|^2}{\hbar\omega - E_{1s} + i\gamma}, \tag{2.20}$$

we obtain from Eq. (2.19) the analytical solution

$$R = |r|^2 = \frac{\Gamma^2}{(\hbar\omega - E_{1s})^2 + (\gamma + \Gamma)^2}. \tag{2.21}$$

Hence, the reflectivity spectrum of the exciton contains in addition to the dephasing γ also the radiative broadening

$$\Gamma = \frac{k}{2\varepsilon_0 \varepsilon} g |d_{cv}|^2, \tag{2.22}$$

which is a result of the broken translational symmetry of the QW in growth direction [12–15]. It was shown in Refs. [4,5] that optical coupling between several quantum wells can further modify this radiative exciton decay.

2.2 Light Propagation in Dielectric Mirrors

Within the DBRs, the wave equation (2.1) contains no source terms. Hence, we can describe the wave propagation in the DBRs by free (homogeneous) solutions that have to obey Maxwell's boundary conditions at all interfaces between the dielectric layers. Figure 2 shows two adjacent layers of the DBRs. The field in the l-th layer,

$$E_l(z,t) = E_l^+(z,t) + E_l^-(z,t) \tag{2.23}$$

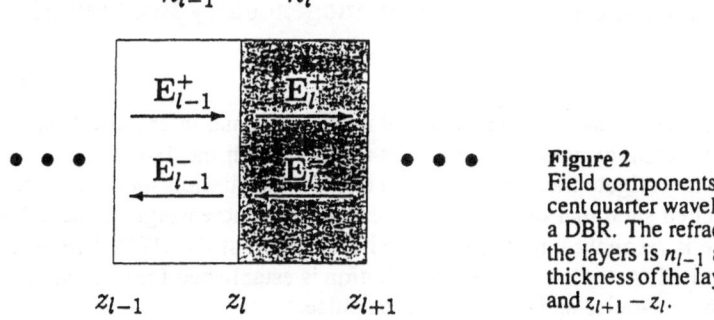

Figure 2
Field components in two adjacent quarter wavelength layers of a DBR. The refractive index of the layers is n_{l-1} and n_l and the thickness of the layers is $z_l - z_{l-1}$ and $z_{l+1} - z_l$.

contains components traveling in forward and backward direction,

$$E_l^+(z,t) = E_l^+(t - \frac{z - z_l}{c_l}), \qquad (2.24)$$

$$E_l^-(z,t) = E_l^-(t + \frac{z - z_{l+1}}{c_l}), \qquad (2.25)$$

where $c_l = c_0/n_l$ and n_l is the refractive index of the considered layer. Maxwell's boundary conditions require the continuity of $E(z)$ and $E'(z)$ at $z = z_l$ which leads to

$$E_{l-1}^+(z_l,t) + E_{l-1}^-(z_l,t) = E_l^+(z_l,t) + E_l^-(z_l,t), \qquad (2.26)$$

$$\frac{1}{c_{l-1}} \left[E_{l-1}^+(z_l,t) - E_{l-1}^-(z_l,t) \right] = \frac{1}{c_l} \left[E_l^+(z_l,t) - E_l^-(z_l,t) \right]. \qquad (2.27)$$

The definitions of E_l^+ and E_l^- in Eqs. (2.24) and (2.25) are choosen such that

$$E_l^+(z_l,t) = E_l^+(t) \quad E_{l-1}^+(z_l,t) = E_{l-1}^+(t - \Delta t_{l-1}) \qquad (2.28)$$

$$E_{l-1}^-(z_l,t) = E_{l-1}^-(t) \quad E_l^-(z_l,t) = E_l^-(t - \Delta t_l) \qquad (2.29)$$

with the propagation times $\Delta t_{l-1} = \frac{z_l - z_{l-1}}{c_{l-1}}$ and $\Delta t_l = \frac{z_{l+1} - z_l}{c_l}$. Hence, we can express the field components propagating away from the surface at $z = z_l$ in terms of incident waves with the matrix equation

$$\begin{pmatrix} E_l^+(t) \\ E_{l-1}^-(t) \end{pmatrix} = \hat{M}_l \begin{pmatrix} E_{l-1}^+(t - \Delta t_{l-1}) \\ E_l^-(t - \Delta t_l) \end{pmatrix} \qquad (2.30)$$

where

$$\hat{M}_l = \frac{1}{n_l + n_{l-1}} \begin{pmatrix} 2n_{l-1} & n_l - n_{l-1} \\ n_{l-1} - n_l & 2n_l \end{pmatrix} \qquad (2.31)$$

follows from Eqs. (2.26) and (2.27). With Eqs. (2.30) and (2.31) we can describe the propagation of time-dependent fields through the successive layers of the DBRs.

3 Normal-Mode Coupling with Inhomogeneously Broadened Quantum-Well Excitons

In this section we discuss the propagation of a weak external laser pulse through the semiconductor microcavity where the pulse is resonant with the 1s-exciton transition of the QWs. We investigate the changes of the cavity transmission spectrum when the QWs are pre-excited such that the 1s-exciton resonance is increasingly bleached. If an intense pre-pulse is resonant with the interband absorption of the QWs, free electrons and holes are excited. A thermal carrier distribution is established for a sufficient time delay between the pre-pulse and the weak probe pulse.

It is well-known that additional free carriers can efficiently bleach the excitonic resonances due to phase-space filling, screening of the Coulomb interaction and scattering. The resulting exciton saturation is attributed to a reduction of the oscillator strength and to excitation-induced dephasing. In Ref. [9] it has been shown that the exciton saturation in very good quality samples with weak inhomogeneous broadening is dominated by resonance broadening due to carrier and polarization interaction (excitation induced dephasing). In the following we study the interplay of this excitation-induced broadening and additional inhomogeneous broadening of the exciton resonance due to disorder.

For the numerical solutions, we consider a GaAs/AlAs microcavity structure with two DBRs of 99.5% reflectivity. In the $\frac{3}{2}\lambda$ spacer between the DBRs two QWs are placed at the static-field anti-nodes. The light propagation through the DBR mirrors can be evaluated using Eqs. (2.30) and (2.31). The resulting transmitted field from the first DBR and the reflected field from the second DBR serve as input quantities for Eqs. (2.12)-(2.14) describing the interaction of the QWs with the intra-cavity field. The excitation of the QWs and the excitonic QW polarization follows from Eqs. (2.3)-(2.8). For the investigation of intense ultrashort pulse propagation through a microcavity, as discussed e.g. in Ref. [16], the equations have to be solved self-consistently. However, for a weak pulse propagating through the pre-excited microcavity, the QW polarization Ψ_k remains in the linear regime (with respect to the weak probe pulse) and the probe pulse induced changes of the carrier occupation f_k^a can be neglected. Then f_k^a contains only the pre-excited carrier distributions which are described by Fermi-Dirac functions for the electrons and holes. Correspondingly, also the dephasing rates Γ_k and $\Gamma_{k,k'}$ in Eq. (2.4) as well as the screening properties of the Coulomb interaction in Γ and Σ, which are treated using the Lindhard formula, do not change during the pulse propagation.

Figure 3 shows the calculated imaginary part of the QW susceptibility for increasing plasma density. The homogeneous resonance broadening is only due to excitation-induced dephasing whereas background contributions are neglected. First we treat the case without additional inhomodeneous broadening. For a small density, the pronounced 1s-exciton resonance leads to well-separated normal-mode peaks in the cavity transmission spectrum (Fig. 4). The increasing plasma density results in strong broadening of the 1s-exciton transition which reduces the height of the normal-mode peaks. On the other hand, the normal-mode splitting is determined by the exciton oscillator strength.

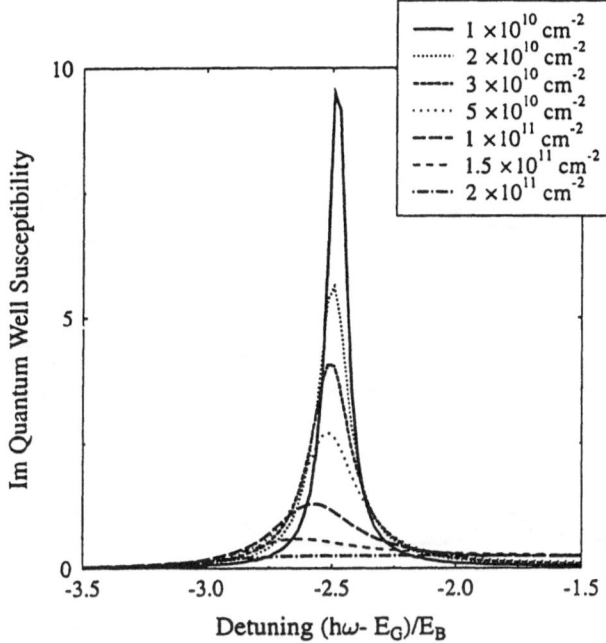

Figure 3
Imaginary part of the optical susceptibility for an 8 nm QW and various quasi-equilibrium electron-hole pair densities at the carrier temperature 77 K. The detuning is given relative to the unrenormalized band-gap E_G in terms of the 3d binding energy $E_B=4.2$ meV.

For the large splitting-to-linewidth ratio of the normal-mode spectrum in Fig. 4, we obtain only a very small reduction of the normal-mode splitting with increasing carrier density. As discussed in Ref. [9], we can therefore directly conclude that the exciton saturation is dominated by excitation-induced dephasing rather than by a reduction of the exciton oscillator strength. Only when the density dependent band-gap shrinkage becomes so large that the band edge is degenerate with the 1s-exciton and the cavity resonance, normal-mode coupling is no longer possible due to rapid dephasing of the continuum states and the single cavity resonance appears in the transmission spectrum. Note that the different dephasing of excitonic and interband states is fully included in our theory. For the calculation of the sequence of curves in Figs. 3 and 4 only the input density has been changed.

Earlier experiments on excitonic normal-mode coupling in semiconductor microcavities have been performed on samples where disorder-induced inhomogeneous broadening leads to a splitting-to-linewidth ratio smaller than one [8,17]. To model this situation we included in our calculations a Gaussian distribution of excitonic resonances with half-width $\Gamma_{inh} = 0.35E_B$ (Fig. 5) and $0.7E_B$ (Fig. 6). For increasing broadening (excitation- or disorder-induced) the normal-mode peaks start to overlap. This leads to a reduction of the normal-mode splitting even for unchanged exciton oscillator strength.

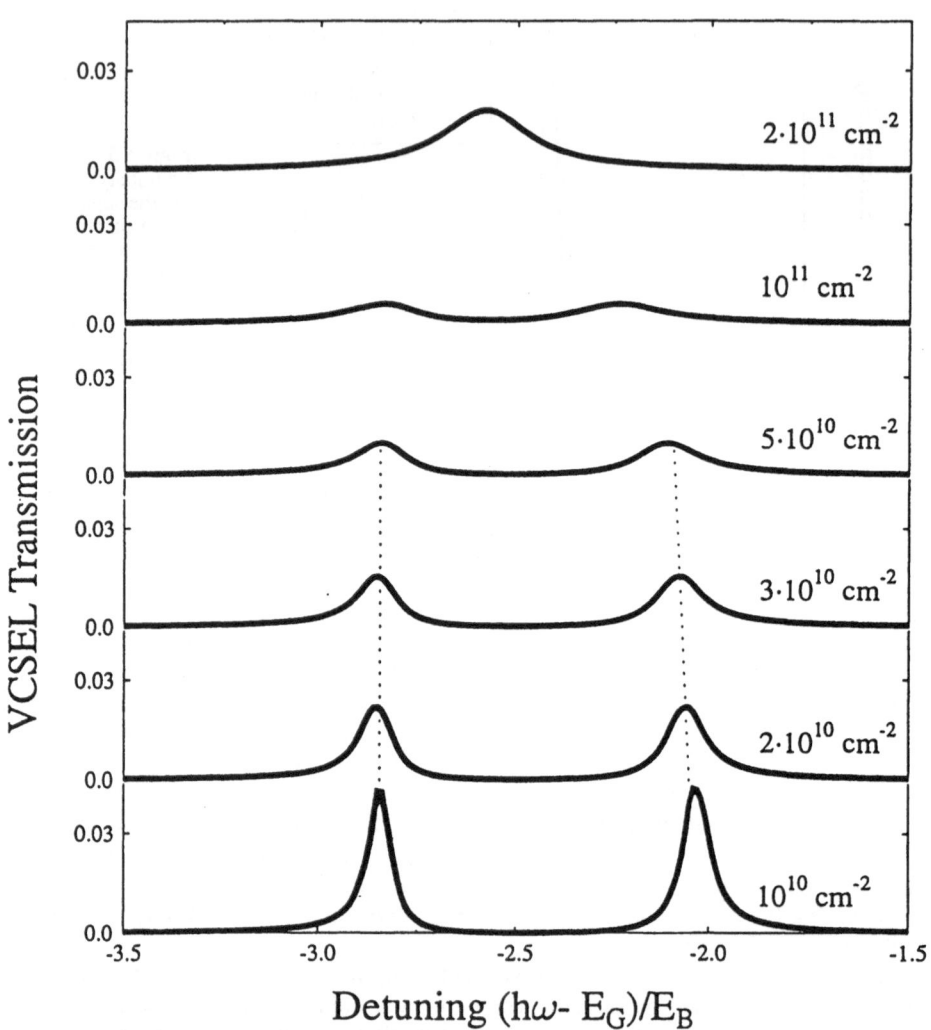

Figure 4 Calculated microcavity transmission for increasing plasma density and bleaching of the exciton according to Fig. 3. The cavity resonance is slightly tuned with respect to the exciton resonance to obtain equal-height normal-mode peaks.

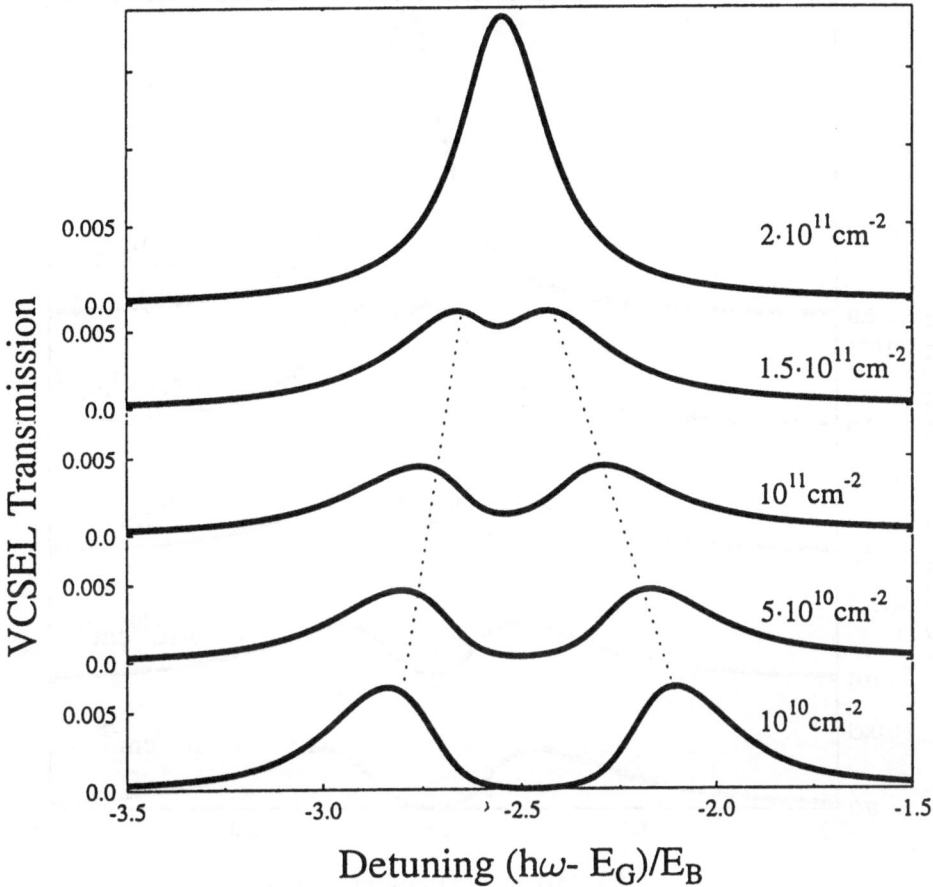

Figure 5 Microcavity transmission spectra for increasing plasma density and additional inhomogeneous broadening $\Gamma_{inh}=0.35\,E_B$.

The interpretation of experiments in samples with large disorder-induced inhomogeneous broadening is additionally complicated by the fact that the carrier density, which is necessary to obtain a comparable saturation of the excitonic resonance, increases with increasing inhomogeneous broadening. Starting from the QW exciton reflection or transmission spectrum at the carrier density 10^{10} cm^{-2}, we plot in Fig. 7 the density, at which the exciton peak height is reduced by 25% (dotted line), 50% (solid line) and 75% (dashed line). Since for larger additional inhomogeneous broadening Γ_{inh} the density-induced changes in the exciton spectrum are smaller, the corresponding saturation density increases.

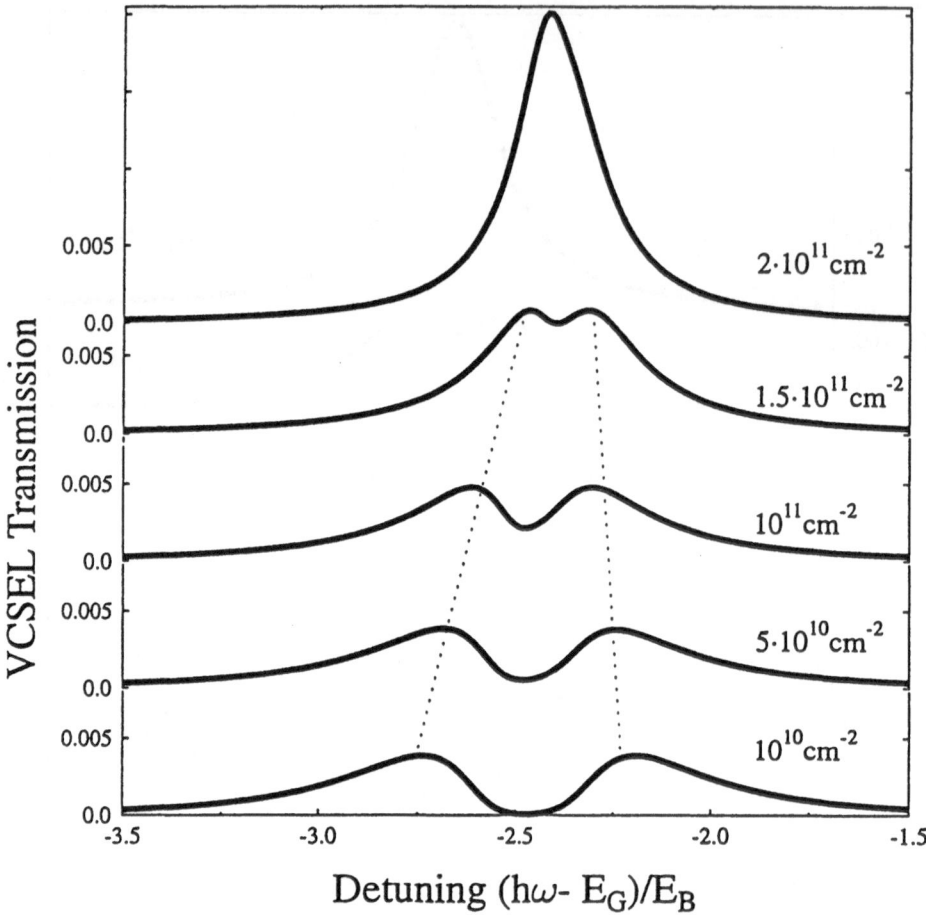

Figure 6   Same as Fig. 5 but with additional inhomogeneous broadening Γ_{inh}=0.7 E_B.

4   Ultrafast Pulse Propagation in Semiconductor Microcavities

The information obtained from our theory includes also the temporal and spatial evolution of the cavity field and the polarization dynamics. For the case of vanishing inhomogeneous broadening, we show in Fig. 8 the time-resolved intra-cavity field at the QW position, $|E_{QW,i}|$, and the corresponding QW polarization, $|P_{QW,i}|^2$, after 100 fs pulse excitation. The pulse is resonant with the 1s exciton and has it's temporal maximum at the front mirror at $t = 0$. For a small background plasma density of the pre-excited system, we obtain normal-mode oscillations describing the periodic energy exchange between the cavity-field and the QW polarization. Since these oscillations are the time-domain representation of the normal-mode splitting in Fig. 4, the oscillation period for

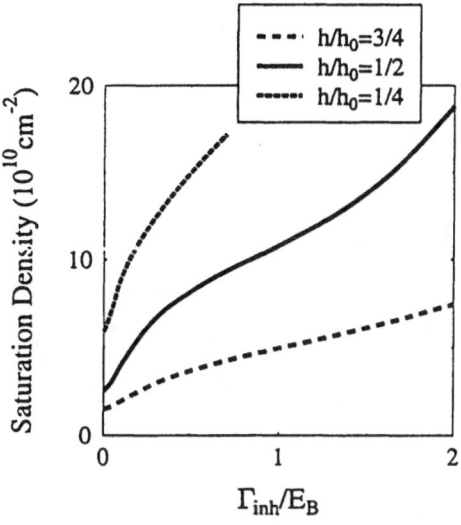

Figure 7
Carrier density at which the 1s-exciton peak-height is reduced to 3/4, 1/2 and 1/4 of the reference value at 10^{10} cm^{-2} and 77 K carrier temperature for increasing inhomogeneous broadening Γ_{inh} of the exciton spectrum.

carrier densities smaller than 5×10^{10} cm^{-2} is nearly unchanged and the decay of the cavity field is mainly determined by the cavity lifetime. The smaller normal-mode splitting for the density 10^{11} cm^{-2} in Fig. 4 leads to a larger oscillation period in Fig. 8. The faster decay of the cavity field under these conditions is only visible on a longer time scale. At very high carrier densities, the large excitation-induced dephasing leads to a rapid non-radiative polarization decay and the missing re-radiation of the QWs causes a strong damping of the cavity field.

For the considered system with two QWs at the static anti-node positions of a $\frac{3}{2}\lambda$ cavity, field and polarization for both QWs are identical since the spatial extension of the 100 fs pulse is much larger than the extension of the microcavity and the pulse spectrum fits entirely into the microcavity stop-band. For comparison, the thin solid line in Fig. 8 shows the field at the cavity anti-node position for the case without QWs. The field reaches it's maximum value on the time-scale of the pulse duration before it exponentially decays with the cavity lifetime.

From our calculations we also obtain the spatial field distribution inside and outside of the microcavity. As shown in Fig. 9, due to the high reflectivity of the microcavity most of the applied 100 fs pulse is reflected. For a background density 10^{10} cm^{-2} and vanishing inhomogeneous broadening, we obtain pronounced normal-mode oscillations in the reflected and transmitted field components of the microcavity emission. For times > 0.1 ps, the field inside the microcavity is distributed according to the the static pattern of the cavity mode. Only the height of the field pattern follows the normal-mode oscillations.

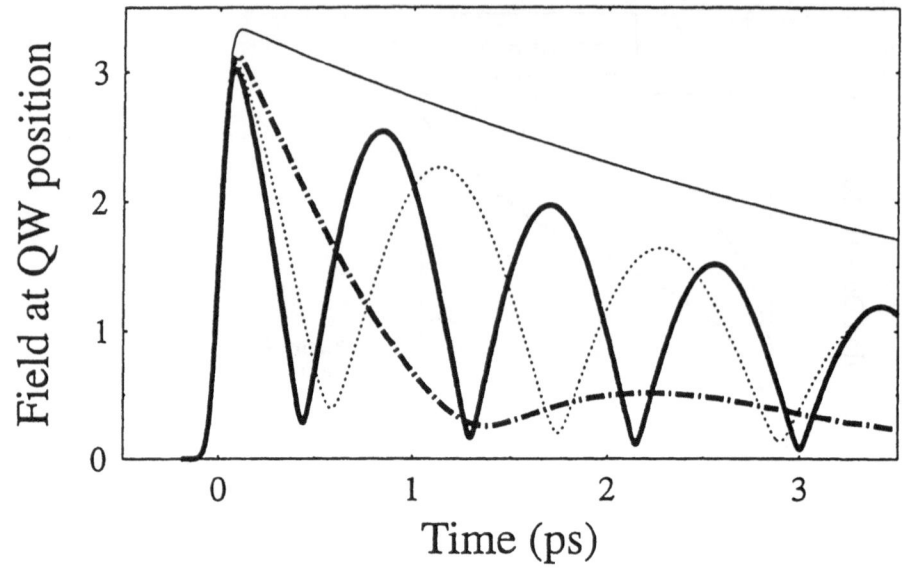

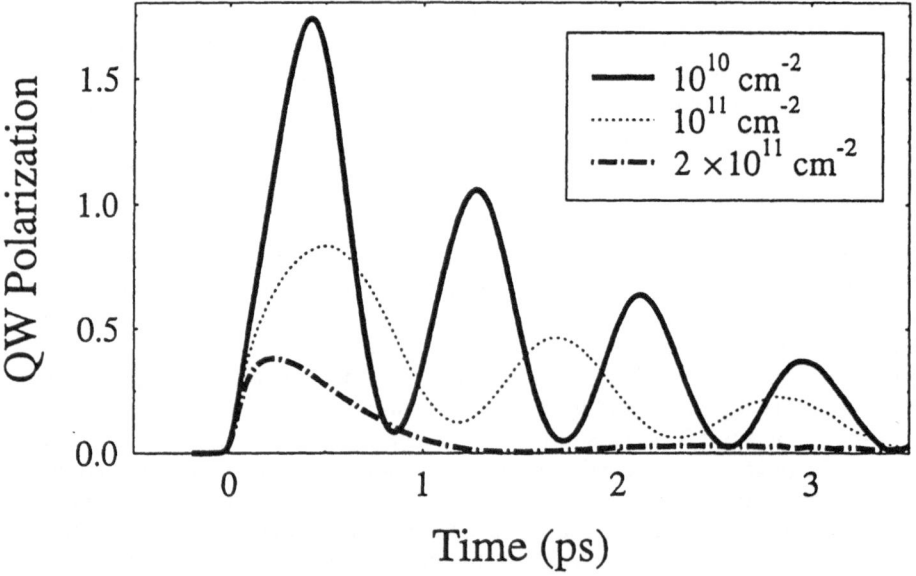

Figure 8 Cavity field at the QW position, $|E_{QW,i}|$, and QW polarization, $|P_{QW,i}|^2$ after excitation of the microcavity with a weak 100 fs pulse and various background plasma density at temperature 77 K. The thin solid line shows the cavity field for a microcavity without QWs.

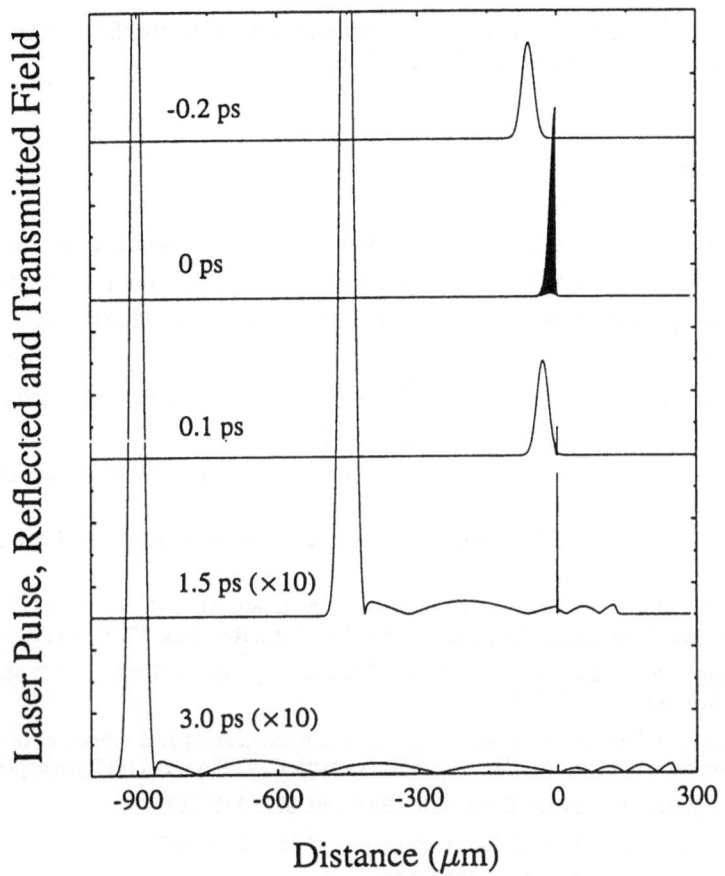

Figure 9 Spatial field distribution $|E(z,t)|$ for the propagation of a 100 fs laser pulse through a microcavity with two QWs. The microcavity is located at position zero.

In conclusion, we highlighted in this article some of the interesting many-body effects of exciton saturation and ultrashort pulse propagation in semiconductor microcavities. Due to the high-quality quantum wells inside the cavities, this system allows us to study details of the light-semiconductor interaction.

Acknowledgments

We would like to thank H.M. Gibbs, G. Khitrova and co-workers for ongoing experiment/theory collaborations and many stimulating discussions. This work was supported by the Deutsche Forschungsgemeinschaft through the Leibniz prize. We acknowledge a grant for CPU time at the Forschungszentrum Jülich. M. K. thanks the Commission of the European Communities for financial support.

Bibliography

[1] C. Weisbuch, M. Nishioka, A. Ishikawa, and Y. Arakawa, Phys. Rev. Lett. **69**, 3314 (1992).

[2] J.J. Sanchez-Mondragon, N.B. Narozhny, and J.H. Eberly, Phys. Rev. Lett. **51**, 550 (1983).

[3] R.J. Thompson, G. Rempe, and H.J. Kimble, Phys. Rev. Lett. **68**, 1132 (1992).

[4] D.S. Citrin, Phys. Rev. B **49**, 1943 (1994).

[5] T. Stroucken, A. Knorr, P. Thomas, and S.W. Koch, Phys. Rev. B **53**, 2026 (1996).

[6] M. Gross and S. Haroche, Phys. Rep. **93**, 301 (1982).

[7] Y. Zhu, D.J. Gauthier, S.E. Morin, Q. Wu, H.J. Carmichael, and T.W. Mossberg, Phys. Rev. Lett. **64**, 2499 (1990).

[8] J.-K. Rhee, D.S. Citrin, T.B. Norris, Y. Arakawa, and M. Nishioka, Solid State Commun. **97**(11), 941 (1996).

[9] F. Jahnke, M. Kira, S.W. Koch, G. Khitrova, E.K. Lindmark, T.R. Nelson Jr., D.V. Wick, J.D. Berger, O. Lyngnes, H.M. Gibbs and K. Tai, Phys. Rev. Lett. **77**, 5257 (1996).

[10] H. Wang, J. Shah, T.C. Damen, W.Y. Jan, J.E. Cunningham, M. Hong, and J.P. Mannaerts, Phys. Rev. B **51**, 14713 (1995).

[11] F. Jahnke, M. Kira and S.W. Koch, "Linear and Nonlinear Optical Properties of Excitons in Semiconductor Quantum Wells and Microcavities", Z. Phys. B. (1997), accepted.

[12] V.M. Agranovich and O.A. Dubovskii, JETP Lett. **3**, 223 (1966).

[13] M. Orrit, C. Aslangul, and P. Kottis, Phys. Rev. B **25**, 7263 (1982).

[14] E. Hanamura, Phys. Rev. B **38**, 1228 (1988).

[15] L.C. Andreani, F. Tassone, and F. Bassani, Solid State Commun. **77**, 641 (1990).

[16] O. Lyngnes, J.D. Berger, J. Prineas, S. Park, G. Khitrova, H.M. Gibbs, F. Jahnke, M. Kira and S.W. Koch, "Nonlinear Emission Dynamics from Semiconductor Microcavities in the Nonperturbative Regime", Solid State Commun. (1997), accepted.

[17] R. Houdré, J.L. Gibernon, P. Pellandini, R.P. Stanley, U. Oesterle, C. Weisbuch, J. O'Gorman, and B. Roycroft, Phys. Rev. B **52**, 7810 (1995).

Ultrafast Switching of Surface Emitting Semiconductor Microlasers

S.G. Hense, M. Elsässer, and M. Wegener

Institut für Angewandte Physik der Universität Karlsruhe
Kaiserstr. 12, D-76128 Karlsruhe, Germany

Abstract: Microcavities offer the unique possibility to study the interplay between a single quantized mode of the light field and the semiconductor material. Dynamic processes which are of interest for a high modulation bandwidth in information transfer can thus be investigated on a model system and eventually be applied in the design of actual devices. The novel switching mechanism presented in this article enables the modulation on a sub-picosecond timescale. It is based on the transient heating of the electronic distribution functions, and it is a many-body effect which is typical of semiconductor lasers. This behaviour will be demonstrated both on a model structure and on a commercially available electrically pumped device, and it will be compared with numerical simulations based on different models.

1 Introduction

The development of microlasers in general and especially of vertical cavity surface emitting lasers (VCSEL) [1–3] has clearly enhanced the already large application potential of semiconductor lasers. Employing them in optical communication and data processing necessitates a high modulation bandwidth, which up to now seemed to be intrinsically limited by relaxation oscillations. These oscillations stem from the energy transfer between the electronic and photonic system, characterized by a frequency up to several 10 GHz [4]. However, recent theoretical publications [5] have predicted the possibility to increase this value by one order of magnitude. Such ultrafast switching is based on energy transfer processes within the carrier system itself and can be achieved by pumping an already operating VCSEL with an additional pulse, which in our case is optical. According to the conventional rate equation description of semiconductor lasers, one should expect that this increase in carrier density is followed by an increase in laser emission. In sharp contrast to this, the laser will actually react to the perturbation with a switch-off on a sub-picosecond time-scale. The mechanism of this process, which is a unique feature of semiconductor lasers, is schematically depicted in Fig. 1: Before the arrival of the perturbation pulse (a), the carriers[1] exhibit their stationary

---

[1] For clarity, only electrons are shown, and spectral holes have been omitted.

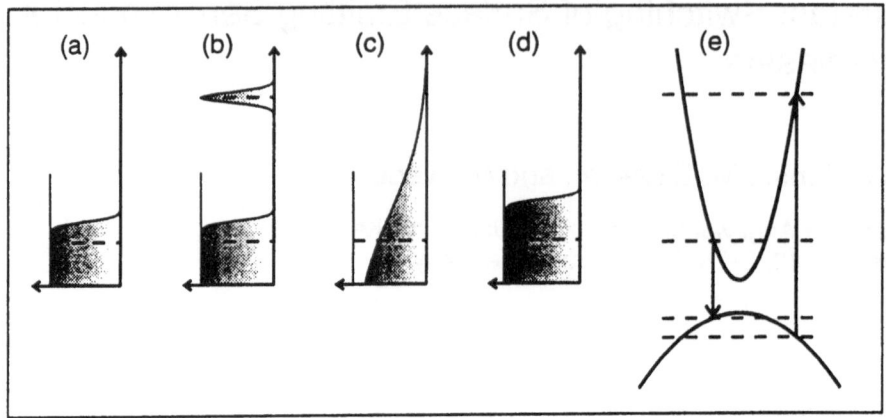

Figure 1   Schematic of the switching due to transient carrier heating.

distribution, characterized by a certain carrier density and mean kinetic energy. The additional carriers are excited at an energy above the average (b), so that the energy density in the carrier system is increased, which leads to hot and thus shallow distributions (c). Inversion at the spectral position of the cavity mode is no longer given, and the laser switches off. Due to phonon emission the distributions cool down again (d), inversion is recovered, and the laser switches back on. Fig. 1 also clarifies why this process cannot be explained by a simple four-level-model: Going from (b) to (c), the occupation probability decreases for all levels that correspond to the four-level-system. Consequently the sum of the occupations of these four levels will decrease as well. In the four-level-system, however, the sum of the occupations would have to be constant.

This article reviews both experiments and model calculations that have been performed in order to demonstrate and examine the suggested switching [6,7]. It is organized as follows: After giving some information about the experimental technique and the samples that have been examined, section 2 will give experimental proof for the switching and discuss its dependence on various parameters. In section 3 we will compare these experimental results with numerical simulations based on two different semiclassical models, which will be explained there as well. Finally, section 4 will summarize and compare both experimental and numerical results.

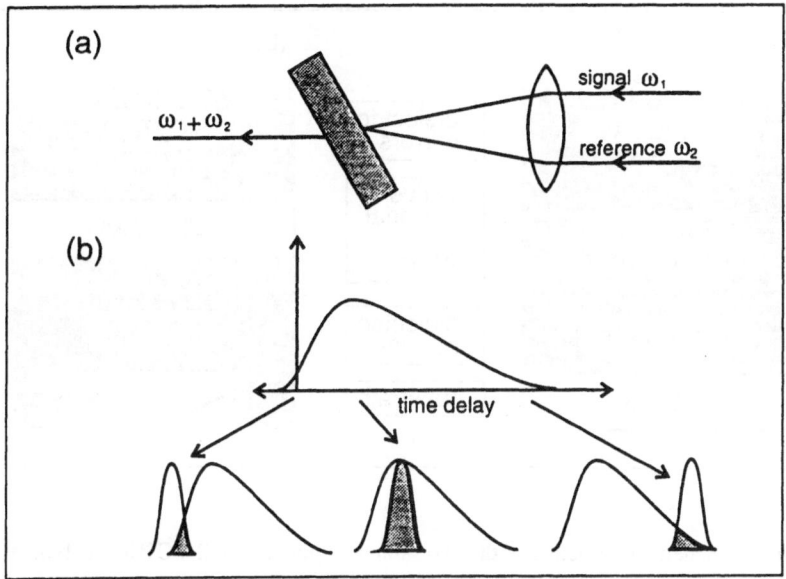

Figure 2 Schematic of an up-conversion experiment.

2   Experiments

2.1   Experimental Technique

The necessary temporal resolution in our experiments is achieved by using the so-called up-conversion technique. Both the emission of the sample and a fraction of the exciting laser pulse are focused onto a nonlinear crystal, where the sum-frequency of *signal* and *reference* is generated. Its intensity will be proportional to the part of the emission of the sample which arrives at the crystal together with the reference pulse, and by delaying the reference pulse it is thus possible to monitor different parts of the emission. An up-conversion scan consists in a systematic variation of the reference delay and the subsequent measurement of the sum-frequency signal, and it can mathematically be described by a convolution of the original signal with the reference pulse. Since the delay can typically be varied in steps of 8 fs, which is far shorter than the duration of the laser pulses, the limit in temporal resolution is given by the temporal width of the laser pulses (in our case 120 fs). A schematic depiction of the up-conversion technique is shown in Fig. 2.

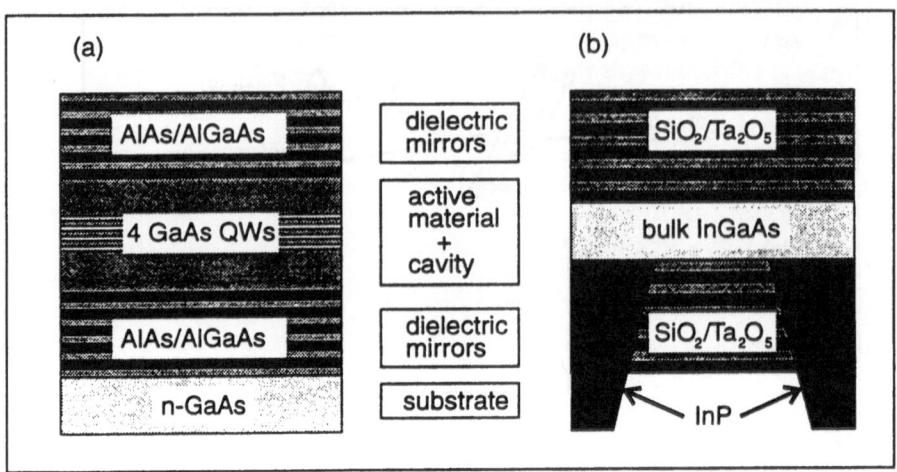

Figure 3 Schematic structure of the two samples: Quantum well VCSEL and bulk micro-laser.

2.2 Samples and Set-ups

When choosing appropriate samples for the investigation of the switching process, one has to keep in mind two aspects: On one hand, it is important to confirm that the effect is not a purely academic one, i.e. it is observable on actual devices. On the other hand, it will be useful to examine model structures as well, since they exhibit the characteristic features more clearly and thus may provide more information about the underlying mechanisms. In our case the crucial point will be the "filling factor", i.e. the ratio of the active volume to the size of the cavity: In spite of the high gain coefficient, the total amplification will be small for a quantum well laser, which results in rather long time constants, especially for the recovery of the emission subsequent to the switch-off. On the other hand, a bulk gain medium may have limited application possibilities due to its high threshold, but the large "filling factor" will result in much shorter time constants, and the switching effect will be observable much more clearly. According to the above considerations, experimental results on two different samples will be reported here [6,7]: The first one is a commercially available VCSEL device with four GaAs quantum wells as the active medium in an AlGaAs cavity. Its structure is shown in Fig. 3(a), and it represents a fairly advanced level in VCSEL technology. Technical details about such devices can e.g. be found in [8]. Mounted in an IC and held under ambient conditions, the sample is pumped by a stationary electrical current, delivering about 1 mW at 1.459 eV photon energy for a current of 5 mA. It is perturbed by an optical femtosecond pulse. The structure of this sample implies that the experiment is set up in reflexion geometry, which is sketched in Fig. 4. A Ti:Sapphire laser delivers 120 fs pulses with 76 MHz repetition rate. These are then split into two parts, the first

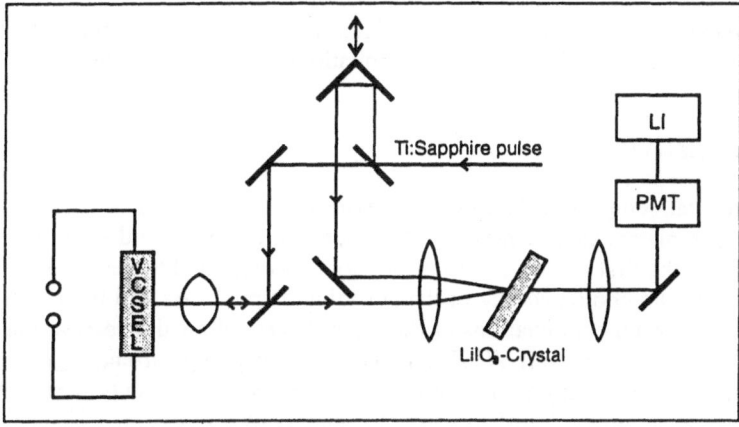

Figure 4 Setup for the experiments on the quantum well VCSEL.

of which perturbs the already operating VCSEL and the second of which is used as the reference pulse for the up-conversion. The point $t = 0$ in this case corresponds to the situation when the perturbation pulse arrives at the sample. The up-converted signal is detected by a photomultiplier (PMT) and measured by a lock-in amplifier (LI).

The second sample consists of 1 μm bulk $In_{0.53}Ga_{0.47}As$ material as the active medium, which has epitactically been grown lattice-matched to an InP substrate. After removing the substrate by selective etching, alternating $\lambda/4$ layers of SiO_2 and Ta_2O_5 have been evaporated on both surfaces ($\lambda = 1.55\,\mu$m, i.e. 0.73 eV), thus sandwiching the active medium between dielectric mirrors with a reflectivity of 99%. The bulk sample (Fig. 3(b)) serves as a model structure and has no electrical contacts, therefore it has to be both pumped and perturbed by an optical pulse. Because of its high threshold (large active volume, poor heat conduction), it requires the high pulse energies obtainable from a cavity-dumped Ti:Sapphire laser, which in our case was homebuilt in a similar way as the one presented in [9]. The experiment is set up analogously with Fig. 4, but it now consists of three beams: The first one serves to pump the sample, which is again held under ambient conditions, the second one perturbs it, and the third one is the reference pulse for the up-conversion. The point $t = 0$ now marks the situation when the pump pulse arrives at the sample, and the point when the microlaser is perturbed can independently be varied. Ref. [10] reports on former experiments that have been performed on this sample, characterizing the spectrally and temporally resolved response to one-pulse excitation. One feature that is described therein will be seen in the following as well: The emission due to the (first) pump pulse sets on with a certain delay, depending on the pump intensity and ranging from about 5 to 15 ps. This delay is due to the fact that (cf. section 3) the gain is proportional to the inversion at the spectral position of the laser mode, and hence stimulated emission will not start until the initial distributions have sufficiently both thermalized and cooled down. Within the rate equation picture it can be attributed to hot phonon effects [7,10], and it is also reproduced by other calculations such as [11] or the coupled Maxwell-Bloch equations presented in section 3.

2.3 Experimental Results

The experimental setup described in the previous paragraph (Fig. 4) gives a three-dimensional parameter space, consisting of the pump intensity (i.e. either the cw current or the energy of the first pulse), the pulse energy that is applied to the already operating VCSEL, and the photon energy of the perturbation pulse. As will be shown in the following, only certain combinations of these parameters allow the observation of the switch-off for the quantum well VCSEL, namely low cw pump current, high pulse energies, and high photon energies. The influence of the cw current can be seen in Fig. 5. For a given pulse energy of 0.9 nJ and a photon energy of 1.73 eV, i.e. 275 meV above the VCSEL mode, up-conversion scans are shown for various currents. The threshold of the VCSEL is 2.8 mA, so the lowest current corresponds to $1.2\,I_{thr}$, and the highest current is $1.8\,I_{thr}$. Within this range, the behaviour of the VCSEL changes very clearly: At low currents, the laser switches off completely after the perturbation. When the current is increased, the switch-off gradually becomes both slower and less pronounced. The recovery and subsequent increase in emission (due to the stimulated recombination of the additionally excited carriers) sets on earlier, and eventually no sign of switching is visible. At higher currents and hence at higher photon densities[2], the VCSEL is obviously more robust. The corresponding up-conversion scans on the bulk sample (Fig. 6) confirm that this behaviour is inherent to the switching process. In the case of the bulk microlaser the arrival time of the second pulse has been adjusted to the onset of laser emission after the first pulse, which is indicated by the arrows. Decreasing the photon

---

[2] The stationary carrier density is *clamped*, i.e. independent of the pump intensity.

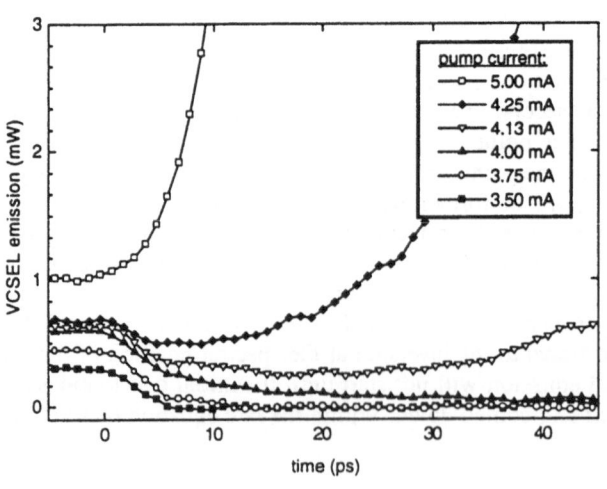

Figure 5
Quantum well VC-SEL: Variation of the pump intensity when perturbed by a 0.9 nJ pulse at 1.73 eV. The threshold corresponds to 2.8 mA.

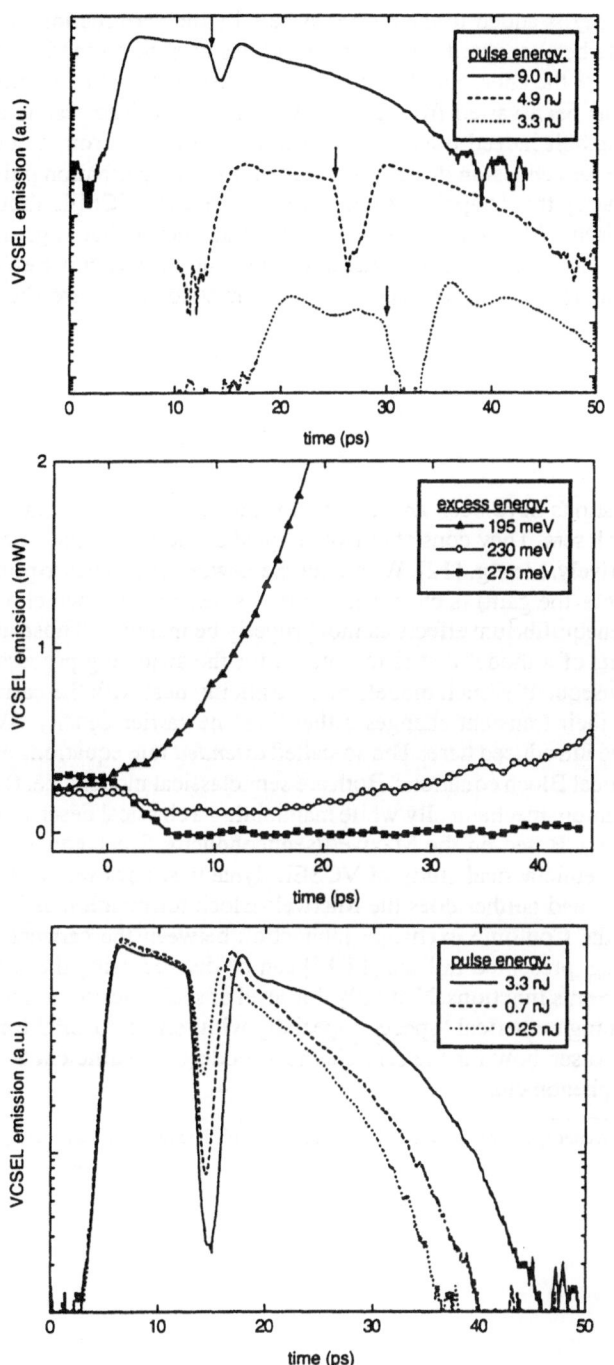

Figure 6
Bulk VCSEL: Variation of the pump intensity when perturbed by a 0.25 nJ pulse at 1.5 eV. The threshold corresponds to 2.0 nJ.

Figure 7
Quantum well VCSEL: For a fixed pump current (3.5 mA) and pulse energy (0.9 nJ) the photon energy of the perturbation pulse is varied.

Figure 8
Bulk VCSEL: For a fixed energy of the first pulse (9.0 nJ), the energy of the second pulse is varied.

energy means that less energy is added to the carrier system by the perturbation pulse so that the switching should be attenuated. This experiment has been performed on the quantum well sample[3], where the energetic distance between the perturbation photons and the absorption edge has been varied from 275 meV to 195 meV. The results are displayed in Fig. 7. They exhbit indeed the expected behaviour, ranging from a complete switch-off to an increased emission directly subsequent to the perturbation pulse. A variation of the pulse energy that is applied to the already operating VCSEL should produce a similar effect: Each additional carrier is excited at an energy that is greater than the mean energy, so if more carriers contribute to the heating, the switching effect will naturally be more pronounced. The corresponding up-conversion scans are shown in Fig. 8.

3   Theory

The conventional rate equations have been an important means for the theoretical description of semiconductor lasers. They consist of two coupled equations for the carrier and photon density, respectively, see e.g. [12]. Within the framework of this description, the carrier system (and hence the gain) is characterized by its density only, which implies that many-body or nonequilibrium effects cannot properly be included. Those are, however, crucial ingredients of a model that is to account for the switching processes reported in this article. Consequently, such models must explicitly deal with the carrier distribution functions and their transient changes rather than the carrier density. Two different approaches will be introduced here: The so-called *extended* rate equations and the combined Maxwell/optical Bloch equations. Both are semiclassical models, i.e. they treat the active medium quantum mechanically while maintaining a classical description of the light field in the cavity. Based on the Maxwell/semiconductor Bloch equations, Ref. [13] presents another semiclassical study of VCSEL dynamics, however without allowing for heating effects, and neither does the Maxwell-Bloch formulation of [14], which takes into account the Coulomb exchange interaction between the carriers. A fully quantized theory as e.g. presented in Refs.[11,15] can be derived using the technique of nonequilibrium Green's functions. Naturally this level of sophistication implies very high demands concerning numerical aspects, especially with respect to CPU time. Therefore it is interesting to see how far the semiclassical models are sufficient for a description of the relevant phenomena.

---

[3] For the bulk VCSEL the photon energy has been fixed at 1.50 eV, i.e. 0.77 eV above the InGaAs bandgap.

3.1 The extended rate equations

As can already be seen from the name, this model is still very closely related to the standard rate equations. For a three-dimensional[4] active medium the equations describing the coupled dynamics of the carrier density n and the photon density p are given by

$$\dot{n} = -cgp - \frac{n}{\tau_{eh}} + \Gamma_{\text{pump}} \tag{3.1}$$

$$\dot{p} = +cgp + \beta\frac{n}{\tau_{eh}} - \frac{p}{\tau_p}, \tag{3.2}$$

which looks the same as in the conventional description: cgp is the rate of stimulated emission, n/τ_{eh} describes spontaneous emission, of which a fraction β will couple into the laser mode, Γ_{pump} is the pump rate, and p/τ_p allows for cavity losses. Yet these equations implicitly contain the very important difference that the gain now becomes a function of the carrier distributions, i.e. not the carrier density as a whole but the occupation numbers of electrons and holes, respectively, at the energetic position of the laser mode ω_l determine the gain:

$$g^{3D} = g_0^{3D} \times \theta(E - \hbar\omega_l)\sqrt{E - \hbar\omega_l}\,(f_e + f_h - 1) \tag{3.3a}$$

$$g^{2D} = g_0^{2D} \times \theta(E - \hbar\omega_l)\,(f_e + f_h - 1). \tag{3.3b}$$

As in the standard rate equations, carrier-carrier scattering is assumed to be so fast that f_e and f_h can be taken as Fermi functions, uniquely characterized by temperature and chemical potential[5]. But instead of clamping the temperature at $T_0 = 300$ K, we now introduce another equation, which describes the dynamic behaviour of the carrier temperature. In order to obtain this equation, we consider the total energy density $\bar{E}_{\text{tot}}$ of the system:

$$\bar{E}_{\text{tot}} = \bar{E}_e + \bar{E}_h + E_{\text{res}}n_{\text{res}} + \hbar\omega_l p, \tag{3.4}$$

where $\bar{E}_e$ and $\bar{E}_h$ are the energy densities of the carriers. E_{res} and n_{res} are the energy and density of the carriers that are injected/excited from a reservoir in either of the pump processes, i.e. E_{res} corresponds to the difference between the energy at which carriers are excited (or injected) and the absorption edge of the gain medium, hence it can also be termed excess energy E_{exc}. Eq. (3.4) provides an expression $\bar{E}_{\text{tot}}(n,T_{eh},n_{\text{res}},p)$[6], and we now have to look at the total temporal derivative of $\bar{E}_{\text{tot}}$. Applying the chain rule of differentiation yields

---

[4] The modifications for the description of a laser containing m quantum wells of length L_z consist in replacing the term for stimulated emission in eq. (3.1) by $cgp \times \frac{L_z}{m}$ and the term for spontaneous emission in eq.(3.2) by $\beta\frac{m}{L_z}\frac{n_{eh}}{\tau_{eh}}$.

[5] The chemical potential is calculated from the carrier density in the usual way, necessitating either Boltzmann or degenerate approximation in the two-dimensional case and the Padé approximation [16] in the three-dimensional case.

[6] For simplicity we will assume a common temperature for electrons and holes. An exemplaric estimation with separate temperatures for the two types of carriers shows that this simplification is justified.

$$\frac{d\bar{E}_{tot}}{dt} = \frac{\partial \bar{E}_{tot}}{\partial n}\,\dot{n} + \frac{\partial \bar{E}_{tot}}{\partial T_{eh}}\,\dot{T}_{eh} + \frac{\partial \bar{E}_{tot}}{\partial n_{res}}\,\dot{n}_{res} + \frac{\partial \bar{E}_{tot}}{\partial p}\,\dot{p}. \tag{3.5}$$

Energy conservation implies that the total differential $d\bar{E}_{tot}$ and hence the derivative (eq. (3.5)) equals zero. Therefore we can solve for $\dot{T}_{eh}$:

$$\dot{T}_{eh} = -\frac{\left(\frac{\partial \bar{E}_{tot}}{\partial n}\right)}{\left(\frac{\partial \bar{E}_{tot}}{\partial T_{eh}}\right)}\,\dot{n} - \frac{\left(\frac{\partial \bar{E}_{tot}}{\partial n_{res}}\right)}{\left(\frac{\partial \bar{E}_{tot}}{\partial T_{eh}}\right)}\,\dot{n}_{res} - \frac{\left(\frac{\partial \bar{E}_{tot}}{\partial p}\right)}{\left(\frac{\partial \bar{E}_{tot}}{\partial T_{eh}}\right)}\,\dot{p} := \Gamma_{heat}. \tag{3.6}$$

In order to find a closed expression for the partial derivatives $\partial \bar{E}_{tot}/\partial n$ and $\partial \bar{E}_{tot}/\partial T_{eh}$ we use the following two relations:

$$n_i = \int D_i(E)f_i(E)dE \tag{3.7}$$

$$\bar{E}_i = \int E D_i(E)f_i(E)dE, \tag{3.8}$$

where $D_i(E)$ is the corresponding density of states function, $i = $ e,h, and $n_e = n_h = n$. The evaluation of eqns.(3.7) and (3.8) requires some approximations, namely the degenerate approximation for electrons and the Boltzmann approximation for holes. The partial derivatives $\partial \bar{E}_{tot}/\partial n_{res}$ and $\partial \bar{E}_{tot}/\partial p$ are obtained from eq.(3.4). Inserting those expressions into eq.(3.6) yields:

$$\dot{T}_{eh} = -\left(\frac{1}{3}\left(\frac{3}{8\pi}\right)^{2/3}\frac{h^2}{k_B m_e}n^{-1/3} + \frac{T_{eh}}{n}\right)\dot{n} - \frac{2}{3}\frac{E_{res}}{k_B n}\,\dot{n}_{res} - \frac{2}{3}\frac{\hbar\omega_l}{k_B n}\,\dot{p}. \tag{3.9}$$

So far we have only considered heating processes through both pumping and stimulated emission. Apart from this, there is also energy relaxation by which hot carriers lose their energy to the crystal lattice, or rather to the LO-phonon population, which will then decay into acoustic phonons. The rate of energy transfer from the carrier system to the LO-phonon system is proportional to the difference between the Bose factors $N(T_{eh})$ and $N(T_{LO})$, i.e. the difference between the LO-phonon number that would correspond to the carrier temperature and the actual LO-phonon number [17,18]. A large Bose factor, corresponding to a high LO-phonon temperature, prevents further energy removal from the carrier system. In other words, the LO-phonons form a "bottleneck" for the cooling, which becomes narrower with increasing LO-phonon lifetime. This behaviour is incorporated into our model by writing the cooling term for the carrier temperature as

$$\Gamma_{cool} = \frac{\hbar\omega_{LO}}{k_B}\frac{N(T_{eh}) - N(T_{LO})}{\tau_{eh-LO}}, \tag{3.10}$$

thus necessitating another rate equation, which describes the dynamics of the LO-phonon occupation number. This fourth equation can be derived by performing the summation

$$E_{LO} = \sum_{|q| \leq 2k_F} N(q)\hbar\omega_{LO} \tag{3.11}$$

and taking into account that both energy and carrier conservation apply to the carrier-phonon system. The limitation $|q| \leq 2k_F$ describes that a carrier, which at most has a momentum $k = k_F$, cannot be scattered farther than to $k = -k_F$, and that the phonon momentum equals the difference. Actually one would have to consider the energy loss of the carrier due to the phonon emission, thus limiting the phonon momentum to $|q| = k_F + \sqrt{k_F^2 - \frac{2m_e\omega_{LO}}{\hbar}}$. Within the degenerate and Boltzmann limit for electrons and holes, respectively, we obtain

$$\dot{N}(T_{LO}) = -\frac{3}{8}\frac{k_B}{\hbar\omega_{LO}}\dot{T}_{eh}. \tag{3.12}$$

After considering that the LO-phonon population itself will loose its energy to the lattice, we can finally write down the closed set of extended rate equations for a three-dimensional gain medium, where we have rearranged eq. (3.6) after inserting eqns.(3.1) and (3.2)[7]:

$$\frac{dn}{dt} = -cgp - \frac{n}{\tau_{eh}} + \Gamma_{pump} \tag{3.13a}$$

$$\frac{dp}{dt} = +cgp + \beta\frac{n}{\tau_{eh}} - \frac{p}{\tau_p} \tag{3.13b}$$

$$\frac{dT_{eh}}{dt} = \left(+\frac{T_{eh}}{n} + \frac{\hbar^2}{k_B m_e}\left(\frac{\pi^4}{3n}\right)^{1/3} - \frac{2}{3}\frac{(\hbar\omega_l - E_g)}{k_B n}\right)cgp$$

$$+ \left(-\frac{T_{eh}}{n} - \frac{\hbar^2}{k_B m_e}\left(\frac{\pi^4}{3n}\right)^{1/3} + \frac{2}{3}\frac{(\hbar\omega_{pump} - E_g)}{k_B n}\right)\Gamma_{pump} \tag{3.13c}$$

$$-\frac{\hbar\omega_{LO}}{k_B}\frac{N(T_{eh}) - N(T_{LO})}{\tau_{eh-LO}}$$

$$\frac{dN(T_{LO})}{dt} = \frac{3}{8}\frac{N(T_{eh}) - N(T_{LO})}{\tau_{eh-LO}} - \frac{N(T_{LO}) - N(T_0)}{\tau_{LO}}. \tag{3.13d}$$

The modifications for describing a two-dimensional gain medium in a three-dimensional cavity have already been mentioned above. Apart from the fact that we have to distinguish between two types of pumping

$$\Gamma_{pump} = \Gamma_{pump}^{cw} + \Gamma_{pump}^{pulse} \tag{3.14}$$

with corresponding excess energies E_{exc}^{cw} and E_{exc}^{pulse}, the equations are derived on the analogy to the case of a bulk microlaser.

---

[7] Spontaneous emission and cavity losses have been neglected at this point.

3.2   The Combined Maxwell/Optical Bloch Equations

As in the previous paragraph, the explanations will concentrate on the bulk VCSEL. However, we will no longer assume quasi-equilibrium but calculate the actual distributions. Modelling the laser dynamics, we will allow for the following aspects: The main purpose is to calculate the electrical field inside the cavity, or rather its squared amplitude, which is proportional to the laser output. In order to keep the notation consistent with the Bloch equations, we will actually deal with the Rabi frequency $\Omega_R = d_{cv}\mathcal{E}/\hbar$ instead of the electrical field. The corresponding equation of motion is the so-called self-consistency equation [19], which is derived by inserting the ansatz of a harmonic mode into the wave equation obtained from Maxwell's equations. Basically it reads

$$\frac{\partial \mathcal{E}}{\partial t} = -\frac{\omega}{2\varepsilon_0\varepsilon_b}\mathrm{Im}(\mathcal{P}) - \frac{\mathcal{E}}{2\tau_p}, \qquad (3.15)$$

where ε_b is the background dielectric constant and the term containing the photon lifetime τ_p considers cavity losses. The contribution from spontaneous emission still has to be added. The macroscopic polarization $\mathcal{P}$ is related to the interband transition amplitude p_{vc}, which in turn is coupled to the dynamics of the carrier system. The interplay between occupation probabilities and transition amplitudes is described by the optical Bloch equations, to which the rotating wave approximation has been applied, i.e. the ansatz $p_{vc} = \bar{p}_{vc}\exp(i\omega t)$ has been employed, and only resonant phenomena are considered. Furthermore the occupation functions $f_{e,h}$ are changed by the fact that carriers are added to the system through pumping and removed from the system by spontaneous emission. Finally we consider that the non-equilibrium distribution will relax towards a Fermi distribution $f^0_{e,h}$. Due to the energy transfer between the subsystems (carriers, phonons, photons) the Fermi distribution itself will not be constant but change according to the actual carrier and energy densities: The equation of motion for the carrier density is obtained from a k-summation over the rates of change in the distribution functions, considering that carrier-carrier scattering does not affect the carrier density. A separate equation for the dynamics of the energy density will then allow the calculation of the corresponding temperature, so that the chemical potential of the actual quasi-equilibrium distribution can be calculated after the Padé approximation [16]. Hot phonon effects, which slow down the energy transfer between carriers and phonons, are again taken into account. The complete set of equations is given by

$$\frac{\partial \Omega_R}{\partial t} = -\frac{\omega}{2\varepsilon_0 \varepsilon_b \hbar} d_{cv} \mathrm{Im}(\mathcal{P}) - \frac{\Omega_R}{2\tau_p} + \beta \frac{\partial \Omega_R}{\partial t}\Big|_{spont} \qquad (3.16a)$$

$$\mathcal{P} = \frac{2}{V} \sum_k d_{cv}^* \bar{p}_{vc} = \int dE_{cv} D(E_{cv}) \bar{p}_{vc}(E_{cv}) \qquad (3.16b)$$

$$\frac{\partial \bar{p}_{vc}(E_{cv})}{\partial t} = -\left(\frac{i}{\hbar}(E_{cv} + E_g - \hbar\omega) + \frac{1}{T_2(E_{cv})} \right) \bar{p}_{vc}(E_{cv})$$

$$+ i\frac{\Omega_R}{2}(1 - f_e(E_{cv}) - f_h(E_{cv})) \qquad (3.16c)$$

$$\frac{\partial f_{e,h}(E_{cv})}{\partial t} = -\frac{f_{e,h}(E_{cv}) - f_{e,h}^0(E_{cv})}{T_1(E_{cv})} - \Omega_R \mathrm{Im}(\bar{p}_{vc})$$

$$- \frac{f_e(E_{cv}) f_h(E_{cv})}{\tau_{eh}} + \frac{\partial f_{e,h}}{\partial t}\Big|_{pump}. \qquad (3.16d)$$

In eq. (3.16a) d_{cv} is the dipole matrix element, and β is again the spontaneous emission coupling coefficient. $D(E_{cv})$ is the joint density of states (eq. (3.16b)), where E_{cv} is the sum of kinetic energies of electrons and holes. T_1 and T_2 are the population and phase relaxation times, respectively. Their energy dependence is taken into account by a linear interpolation between the value at the band edge, which can be varied as a parameter, and

$$T_1\big|_{E_{cv} \geq 800 \text{ meV}} = T_2\big|_{E_{cv} \geq 800 \text{ meV}} = 10 \text{ fs}.$$

Γ_{pump} is again the pump rate, and τ_{eh} the spontaneous emission lifetime for electrons and holes.

3.3 Numerical Results

In order to visualize the dynamics of the switching, it is instructive to look at the temporal evolution of the three quantities n, p, T_{eh} as calculated by the extended rate equations for the 2D case, which is shown in Fig.9[8]. The simulation starts at $t = -6$ ns so that at $t = -5$ ps all quantities have reached their stationary value. At time $t = 0$ the perturbation pulse arrives, causing the carrier density to rise in a step-like way. Since the additional carriers are excited with an excess energy of 275 meV, the mean energy is clearly increased, resulting in a sharp rise of 200 K in the carrier temperature. Hence the gain (cf. eqns. (3.3a) and (3.3b)) is decreased, thus reducing the rate of stimulated emission (eq. (3.2)) so that the laser switches off. When the carrier system has sufficiently cooled down, gain is reestablished, and the laser emission starts again, exceeding the stationary value until all the additional carriers have recombined. The influence of carrier heating and of the "phonon bottleneck" can be seen in Fig. 10 where the results are shown for the simulation including hot LO-phonons with $\tau_{LO} = 2.45$ ps [20,21], $\tau_{LO} = 0.1$ ps, and for a fixed carrier temperature. Varying the excess energy in the simulations (Fig.11) delivers results that are consistent with the experiments and our intuitive explanation, and

---

[8] Hot phonons have been included, though their dynamics is not shown here.

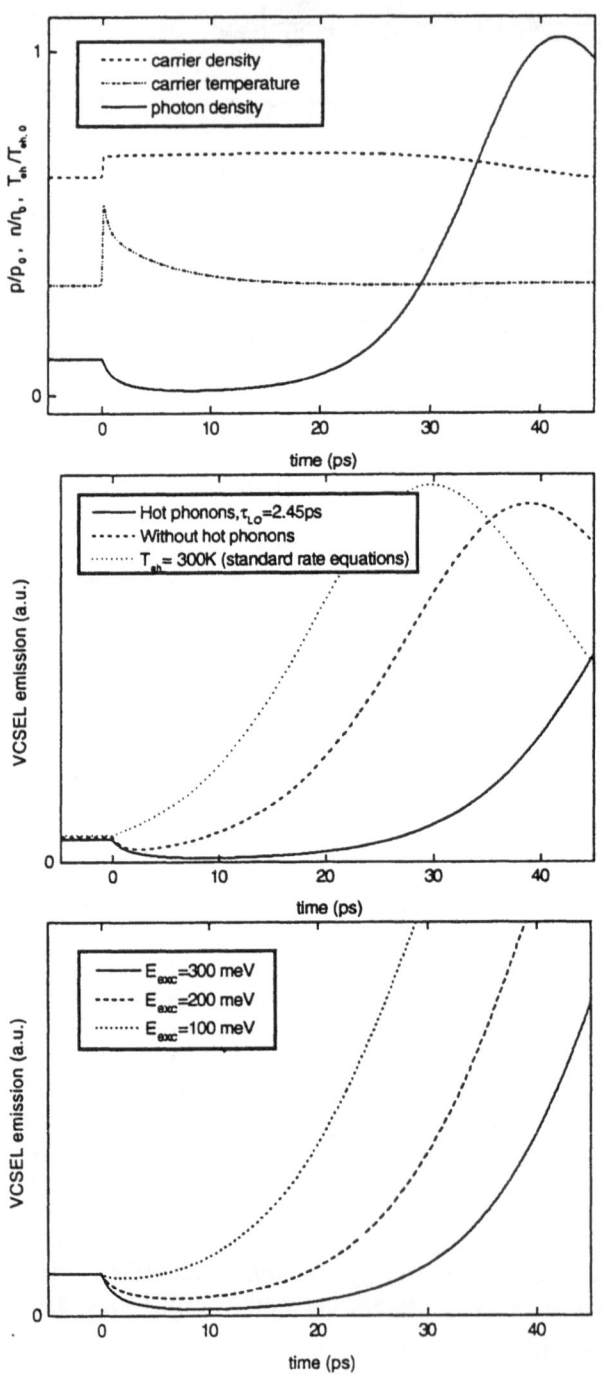

Figure 9
Quantum well VCSEL: Coupled evolution of photon density, carrier density and carrier temperature as calculated by the extended rate equations ($\Gamma_{pump}^{cw} = 2 \times 10^9$ ps$^{-1} \simeq$ $1.7\Gamma_{thr}$, $\Gamma_{pump}^{pulse} = 8.6 \times$ 10^{11} ps$^{-1} \times \exp(-(t/72 \text{ fs})^2)$. The curves were normalized by $p_0 = 1.5 \times 10^{15}$ cm^{-3}, $n_0 = 1.8 \times 10^{12}$ cm^{-2}, and $T_{eh,0} = 1000$ K.

Figure 10
Quantum well VCSEL: Influence of hot carrier and hot phonon effects in the extended rate equation model.

Figure 11
Quantum well VCSEL: Dependence of the switching on the excess energy of the additional carriers.

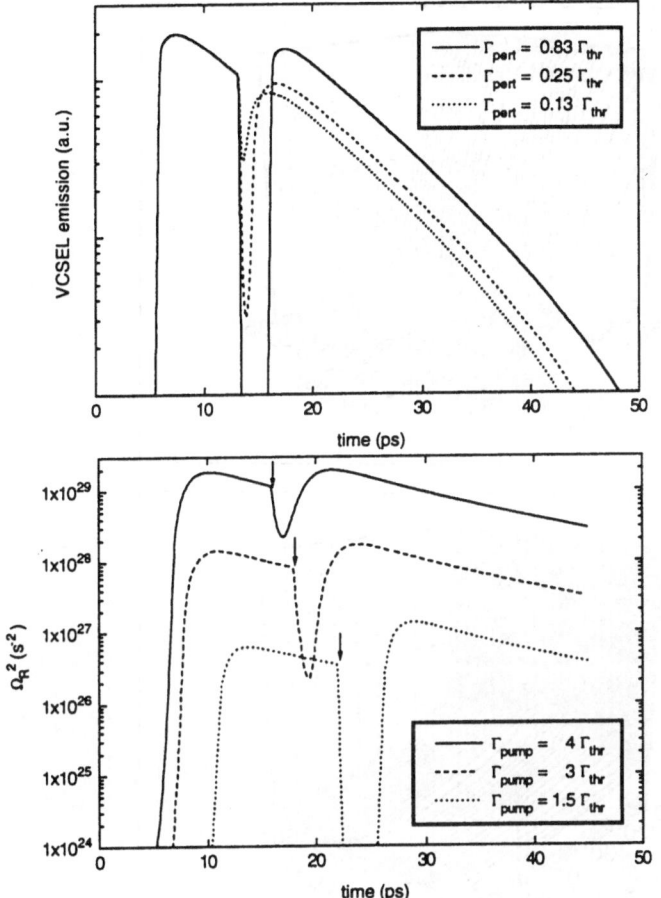

Figure 12
Bulk VCSEL: Influence of the perturbation intensity on the switching as calculated by the extended rate equations.

Figure 13
Bulk VCSEL: Attenuation of the switching with increasing pump rate as calculated by the coupled Maxwell/Bloch equations.

the same applies to the variation of the perturbation intensity (Fig. 12): If the amount of additional energy is increased via pulse energy and photon energy, respectively, the heating and hence the switch-off will be more pronounced. The third parameter that has been varied in the experiments is the pump intensity, i.e. the cw-current in the case of the quantum well VCSEL and the energy of the first pulse for the bulk sample. In both cases the switching gradually disappears with increasing pump rate. This effect cannot be reproduced completely by the extended rate equation model, whereas the integration of the combined self-consistency/optical Bloch equations still delivers results that are consistent with the experiments, which is shown in Fig. 13. The parameters for the simulations based on the extended rate equations are given in Refs. [6,7], those for the coupled Maxwell/Bloch equations in Ref. [22]. Apart from calculating the quantities that are accessible experimentally, one may also gain insight into the underlying physics by looking at those which can only be simulated numerically. This was already done for the extended rate equations in Fig. 9, but it is of course more interesting in a

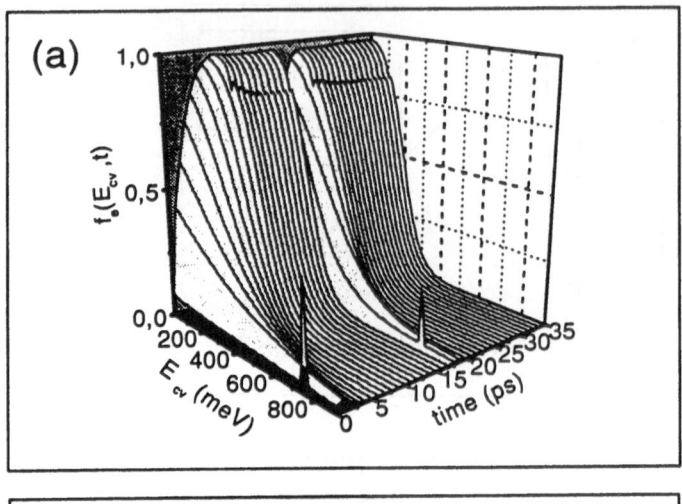

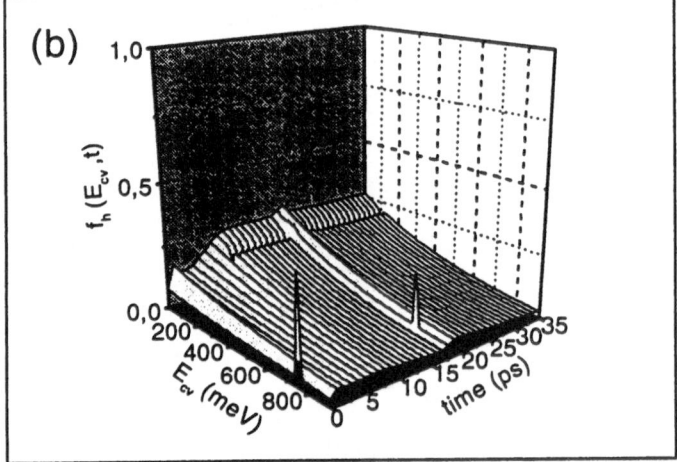

Figure 14 Bulk VCSEL: Spectrally resolved evolution of the carrier distributions as calculated by the coupled Maxwell/Bloch equations: (a) electrons, (b) holes.

more sophisticated model such as the combined Maxwell/optical Bloch equations. In Fig. 14 we have plotted the temporal evolution of the spectrally resolved carrier distributions for the case of a bulk microlaser excited by two pulses. Immediately after excitation the electron distribution follows the spectral shape of the pump pulse, but the electrons rapidly thermalize towards a hot Fermi distribution. Due to the subsequent cooling, the occupation at the spectral position of the cavity mode will increase. The change in the inversion factor $-(1 - f_e - f_h)$ makes the transition amplitude change accordingly, cf. eq.(3.16c), so that the imaginary part of the transition amplitude takes

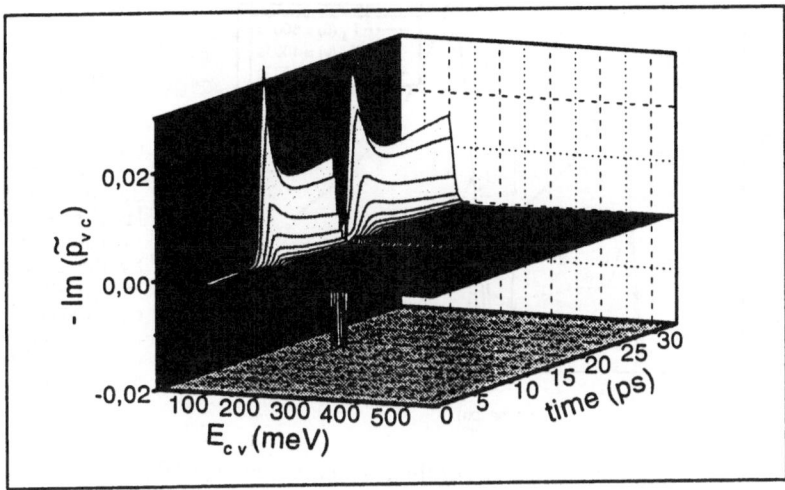

Figure 15 Bulk VCSEL: *Negative* of the imaginary part of the transition amplitude as calculated by the coupled Maxwell/Bloch equations.

on large negative values and stimulated emission will set on, which is shown in Fig. 15. Although all scattering processes and relaxation mechanisms basically apply to both kinds of carriers, the hole distribution evolves in a somewhat different way. Only a fraction $m_r/m_h \approx 0.1$[9] of the excess energy is directly added to the hole distribution, while most of it heats up the electron distribution. The latter will in turn lose its energy to the LO-phonon system and thence to the hole distribution, so the hole temperature does not rise instantaneously but with a certain delay. The rate of stimulated emission at the laser mode exceeds carrier-carrier scattering, or in other words, a spectral hole is burned into the distribution. It vanishes gradually as the photon density due to the first pulse starts to decay. The second pulse is then applied, and the distributions are heated, once again first the electrons and the the holes. Correspondingly the inversion factor changes again so that the laser switches into the absorptive regime, which can clearly be seen in Fig. 15. The occurence of a spectral hole can also explain why the switch-off disappears at larger pump rates and hence higher photon densities: At a very high pump rates, the spectral hole will be so deep that carrier-carrier scattering subsequent to the perturbation will at first fill up the hole, thus conserving inversion and maintaining stimulated emission. In order to understand the complex interplay of the underlying physical processes, which influence the characteristics of the transient laser output, it may be useful to vary the simulation parameters, especially the relaxation times T_1 and T_2. The variation of T_1 does not bring about any surprises: For higher scattering rates T_1^{-1} the laser output is higher since pumping is more effective due to reduced Pauli-blocking, and the switch-off is slightly more pronounced. On the other hand, varying T_2 in the range of 20

---

[9] $m_r = (m_e^{-1} + m_h^{-1})^{-1}$ is the reduced mass of electrons and holes.

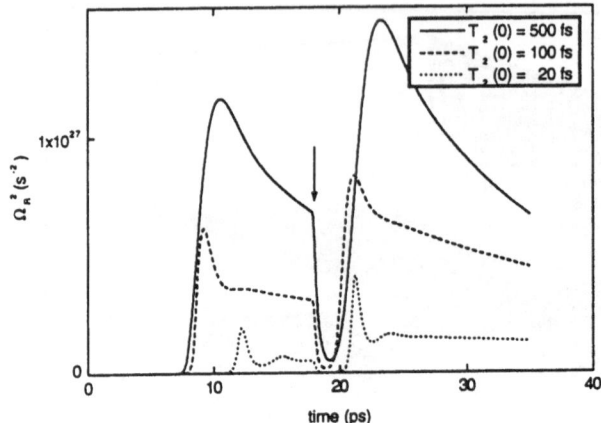

Figure 16
Bulk VCSEL: Influence of the phase relaxation time T_2 on the dynamics of the laser emission as calculated by the coupled Maxwell/Bloch equations under the adiabatic approximation for $\bar{p}_{vc}$. $T_1(0) = 100$ fs.

... 500 fs results in qualitative changes of the laser emission on a 10 ps scale, which is shown in Fig. 16. One additional approximation has been made here, the so-called adiabatic approximation, which consists in neglecting the explicit time dependence of the interband polarization, i.e. equating the RHS of eq.(3.16c) to zero. With higher phase relaxation rates, i.e. smaller values of T_2[10], clear structures appear in the laser emission. They consist of a fast decay followed by a shoulder, which even delevops into an oscillation for the smallest value of T_2. This behaviour seems similar to the one observed in the experimental curves (Figs. 6, 8), and it can possibly be attributed to an interplay between heating due to stimulated emission and subsequent cooling after the laser output has been reduced, i.e. the analogue of relaxation oscillations regarding inversion at the laser mode instead of the carrier density.

4   Summary and Conclusions

In conclusion, we have presented evidence for an ultrafast switching of semiconductor microlasers based on transient carrier heating. It is achieved by perturbing the already operating microlaser with an additional femtosecond optical pulse and can be observed both on model structures such as a bulk InGaAs microlaser and on actual devices with GaAs quantum wells as the active material. As could be expected, the switching vanishes if either the pulse energy of the perturbation or the photon energy, i.e. the excess kinetic energy of the additional carriers, is decreased. Surprisingly, the switching becomes also less pronounced if the lasers are pumped at a higher level. The first two dependences can be described by the extended rate equation model, which still assumes quasi-equilibrium but allows for the dynamics of the carrier temperature. On the other hand, the third one is attributed to spectral hole burning and therefore necessitates the

---

[10] The given values are those at the band edge. Assuming $T_1(800 \text{ meV}) = T_2(800 \text{ meV}) = 10$ fs, the values at the pump energy are obtained from linear interpolation.

more sophisticated description based on the coupled Maxwell/Bloch equations. Within the framework of this model we also obtain the surprising result that a variation of the phase relaxation time T_2 from 20 fs to 500 fs brings about a change in the emission behaviour on a timescale that exceeds T_2 by up to three orders of magnitude.

We acknowledge support by the Krupp-Stiftung and by the DFG through the Schwerpunktprogramm "Optische Signalverarbeitung".

Bibliography

[1] K. Iga, F. Koyama, and S. Kimoshita. *IEEE J. Quant. El.*, **24**, 1845 (1988).

[2] J. L. Jewell, J. P. Harbison, A. Scherer, Y. H. Lee, and L. T. Florez. *IEEE J. Quant. El.*, **27** 1346 (1991).

[3] S. L. McCall, A. F. J. Levi, R. E. Slusher, S. J. Parton, and R. A. Logan. *Appl. Phys. Lett.*, **60**, 289 (1992).

[4] M. A. Newkirk and K. J. Vahala. *Appl. Phys. Lett.*, **52**, 770 (1988).

[5] F. Jahnke and S. W. Koch. *Appl. Phys. Lett.*, **67**, 2278 (1995).

[6] S. G. Hense and M. Wegener. *Phys. Rev. B* **55**, 9255 (1997).

[7] M. Elsaesser, S. G. Hense, and M. Wegener. *Appl. Phys. Lett.*, **70**, 853 (1997).

[8] B. Tell, Y. H. Lee, K. F. Brown-Goebeler, J. L. Jewell, R. E. Leibenguth, M. T. Asom, L. Luther, and V. D. Mattera. *Appl. Phys. Lett.*, **57**, 1855 (1990).

[9] M. Ramaswamy, M. Ulman, J. Paye, and J. G. Fujimoto. *Optics Letters*, **18**, 1822 (1993).

[10] G. Pompe, T. Rappen, and M. Wegener. *Phys. Rev. B*, **51**, 7005 (1995).

[11] F. Jahnke and S. W. Koch. *Phys. Rev. A*, **52**, 1712 (1995).

[12] W. W. Chow, S. W. Koch, and M. Sargent III. *Semiconductor Laser Physics*. Springer, 1994.

[13] W. W. Chow. *Optics Letters*, **20**, 2318, (1995).

[14] C. M. Bowden and G. P. Agrawal. *Phys. Rev. A*, **51**, 4132 (1995).

[15] K. Henneberger and S. W. Koch. *Phys. Rev. Lett.*, **76**, 1820 (1996).

[16] H. Haug and S. W. Koch. *Quantum theory of the optical and electronic properties of semiconductors*. World Scientific, 1993.

[17] L. F. Lester and B. K. Ridley. *J. Appl. Phys.*, **72**, 2579 (1992).

[18] C. Y. Tsai, L. F. Eastman, and Y. H. Lo. *Appl. Phys. Lett.*, **63**, 3408 (1993).

[19] P. Meystre and M. Sargent III. *Elements of Quantum Optics*. Springer, second edition, 1991.

[20] M. Rieger, P. Kocevar, P. Lugli, P. Bordone, L. Reggiani, and S. M. Goodnick. *Phys. Rev. B*, **39**, 7866 (1989).

[21] J. Menéndez and M. Cardona. *Phys. Rev. B*, **29**, 2051 (1984).

[22] If not stated explicitly, the parameters in the simulations are as follows: $d_{cv} = 2.1661 \times 10^{-28}$ Asm for $E_{cv} \leq 480$ meV, $= 0$ elsewise, $\beta = 10^{-4}$, $\tau_p = 1.77$ ps, $\tau_{eh} = 100$ ps, $\tau_{e-LO} = 5$ ps, $\tau_{h-LO} = 1$ ps, $\tau_{LO} = 2.45$ ps, $T_0 = 300$ K (lattice temperature), $m_e = 0.041 \times m_0$, $m_h = 0.377 \times m_0$ (m_0 is the free electron mass), $E_g = 0.73$ eV, $\hbar\omega = 0.83$ eV (energy of the VC-SEL mode), $E_{exc} = 0.75$ eV, $\hbar\omega_{LO} = 35$ meV, $T_1(0) = 100$ fs, $T_2(0) = 500$ fs, $\Gamma_{pump} = 1.5 \times 10^{14}$ s$^{-1} \times \exp(-(t/96\,\text{fs})^2)$ ($=3 \times \Gamma_{pump}^{thr}$), and $\Gamma_{pert} = 3 \times 10^{13}$ s$^{-1} \times \exp(-(t/96\,\text{fs})^2)$.

Nonlinear Semiconductor Microcavity Reflectance and Photoluminescence from Normal-Mode Coupling to Lasing

H. M. Gibbs[1], D. V. Wick[1], G. Khitrova[1], J. D. Berger[1],
O. Lyngnes[1], T. R. Nelson[1] Jr., E. K. Lindmark[1], S. Park[1], and
J. Prineas[1],
M. Kira[2], F. Jahnke[2], S. W. Koch[2], and W. Rühle[2],
S. Hallstein[3], and K. Tai[4]

[1] Optical Sciences Center, University of Arizona, Tucson, AZ 85721
[2] Fachbereich Physik und Zentrum für Materialwissenschaften
Philipps-Universität, Renthof 5, D-35032 Marburg, Germany
[3] Max-Planck-Institu für Festkörperforschung, Heisenbergstraße 1, D-70569,
Stuttgart, Germany
[4] Institute of Electro-Optical Engineering, National Chiao Tung University,
Hsinchu, Taiwan, Republic of China

Abstract: The transition from the nonperturbative reversible emission regime of normal-mode coupling all the way to the perturbative regime of irreversible lasing emission is studied. The microcavity samples contain one or two InGaAs/GaAs quantum wells with very narrow absorption linewidths (1 meV) resulting in record normal-mode-coupling splitting-to-linewidth ratios. For zero exciton-cavity detuning, the transmission peaks and reflectance dips vanish with increased carrier density with little change in splitting. This new nonlinear behavior is observed because exciton broadening with little reduction in oscillator strength is the dominant nonlinearity at low densities for such narrow-linewidth excitons. A microscopic theory, where the effects of carrier and polarization scattering are included at the microscopic level, explains these experiments. The photoluminescence emitted perpendicular to the layers also shows curious density-dependent behavior. When the cavity mode is tuned energetically above the exciton resonance, the upper-polariton peak is weaker (stronger) than the lower-polariton peak at low (high) carrier densities. This crossover occurs at a carrier density less than a factor of two below that for lasing threshold. This behavior is in good agreement with a full quantum mechanic description for both light and carrier, where the mutual Coulomb interaction of carriers is included.

1 Introduction

In the 1980's nonlinear Fabry-Pérot etalons were studied extensively as potential optical switches, logic gates, and memory devices [1–3]. Most of those studies were performed at room temperature with phonon-broadened quantum-well linewidths of $\approx$ 10 meV. They were based on the bleaching and/or shift of an exciton or band edge resulting in a change in transmission and/or shift of the single peak of the cavity. Weisbuch *et al.* [4] showed that at low temperatures a modern-day quantum-well (QW) sample can have a narrow linewidth of only a few meV. In the vicinity of the exciton resonance, the reduced linewidth results in very pronounced index changes so that the Fabry-Pérot resonance condition requiring an integral number of wavelengths between the two mirrors can be satisfied at three different frequencies. This is analogous to the Casperson [5] single-mode instability for a hole-burned gain transition. In our absorption case, the high absorption at line center destroys the transmission at the central-frequency Fabry-Pérot solution. That leaves the two sideband solutions giving the characteristic normal-mode coupling (NMC) double-peaked transmission spectrum. The semiconductor NMC experiments are analogous to the many-atom quantum optics experiments [6] of the mid-eighties and are far [7] from the quantum statistical limit predicted in the early sixties [8] and just recently reached in the atomic experiments [9]. Although the quantum-statistical limit might result in curious semiconductor devices, it will require structures with much higher finesse and much smaller volumes. The results are found to differ interestingly from the atomic case because of the very different nature of exciton nonlinearities and electron-hole luminescence. The explanation for these experimental studies of nonlinear NMC reflectance and luminescence has required major theoretical improvements [7,10–12]. In Section 2 the samples are described and their linear transmission and reflectance are shown. Furthermore, a brief summary of the microscopic theory explaining the experiments is presented. In Section 3 the luminescence is studied using off-resonance cw or fs excitation and time-integrated as well as time-resolved detection. Excellent agreement is found with a recently developed quantum-electrodynamic theory [12] for electron-hole luminescence in a microcavity.

2 Nonlinear Transmittance and Reflectance

The microcavities studied consist of two 8 nm $In_{0.04}Ga_{0.96}As$ QWs between thick GaAs barriers within GaAs/AlAs Bragg mirrors. The In concentration is sufficiently large for the heavy-hole exciton peak to be around 834 nm at 4 K, so that the GaAs substrate does not have to be removed for transmission studies. Nonetheless, the strain shifts the light-hole exciton peak to 826 nm, so that it does not interfere with NMC studies with the heavy-hole exciton. The small exciton linewidth (1 meV = 0.6 nm at 4 K) leads to record splitting-to-linewidth ratios of 6.8 (17) for Bragg mirrors consisting of 14 and 16.5 (19 and 21.5) periods for the top and bottom mirrors, respectively, with 99.6% (99.94%) calculated reflectivity for samples labeled NMC22(NMC20); see Fig. 1.

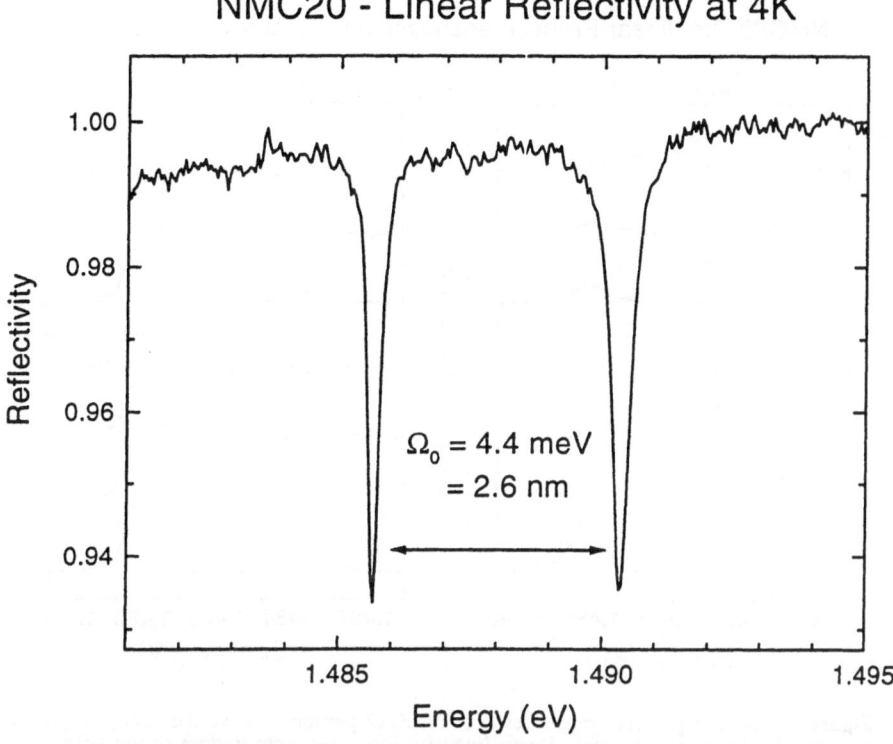

Figure 1 Linear reflectance at 4 K of NMC20, a microcavity with 2 quantum wells in a $3\lambda/2$ spacer between DBR mirrors with $R = 99.94\%$.

We have performed cw pump-probe measurements of the exciton saturation and the nonlinear transmission of a microcavity exhibiting NMC. A light-emitting diode with peak wavelength at 850 nm and a spectral FWHM of 50 nm is used as a broadband probe; it is square-wave modulated at 6 kHz and detected with a Hamamatsu R636 photomultiplier tube and lock-in amplifier. The pump beam from a Ti:sapphire laser is focused on the microcavity to a 46 μm diameter, sufficiently larger than the 30 μm probe diameter. Figure 2 shows the nonlinear transmission for NMC22 for pumping above the cavity stopband and for pumping into the lower-energy peak. The striking feature of the data is the loss of transmission with negligible change in splitting. At high enough carrier density for excitation above the stopband, NMC is destroyed and single-peak transmission is restored; at still higher densities, lasing occurs at the single-peak wavelength.

As explained in the Introduction, only two of the three Fabry-Pérot resonances can be detected due to the strong absorption at the middle resonance. The unusual carrier density dependence of the remaining resonances can be explained by investigating the absorption at their position. Figure 3a shows the exciton absorption as measured from

NMC22- Nonlinear Probe Transmission with Increasing Pump

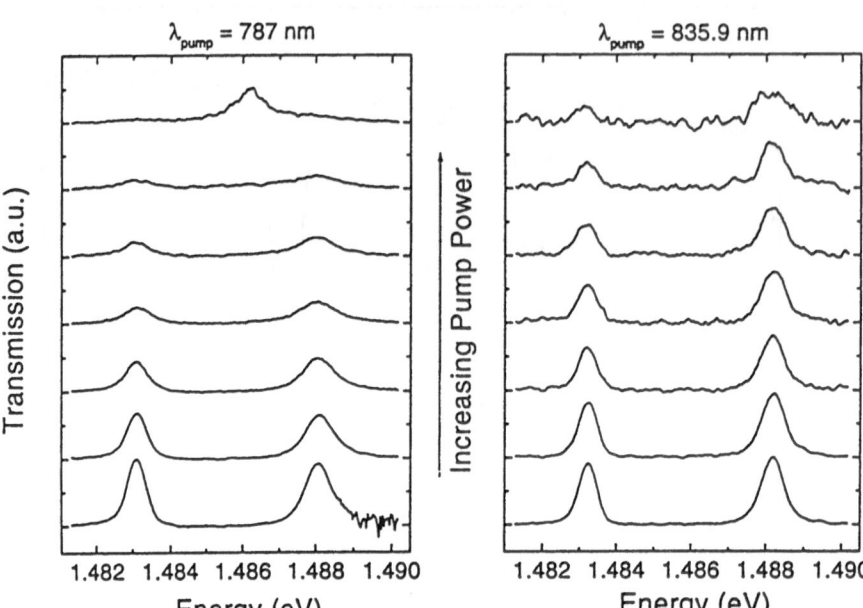

Figure 2 Cw pump-probe transmission of NMC22 pumping above the cavity stopband and into the lower-energy peak. Pump intensity increases from bottom to top with each trace offset for clarity.

the transmission of $\approx$ 100 fs pulses $[\ln(I_T/I_I)]$. Figure 3b shows that the linewidth increases markedly with the light intensity, whereas the oscillator strength computed as the area under the curve changes very little. The curious nonlinear behavior of the NMC saturation is now easily understood: at low carrier densities the absorption is low at the two off-resonance cavity peaks, so their transmissions are high; as the carrier density increases the dominant change in the exciton absorption is broadening which *increases* the absorption at the two peaks thereby lowering their transmissions. The constant NMC splitting can be directly attributed to the constant oscillator strength.

The explanation for the nonlinear behavior of the exciton absorption is found from the microscopic theory which identifies the physical origin of the broadening to be dephasing by carrier-carrier and polarization scattering [7]. Figure 4 shows the results of the computation. To obtain agreement with the experiments, it is essential to include off-diagonal dephasing (in-scattering) which largely compensates for diagonal dephasing due to carrier-carrier scattering (out-scattering).

It is the narrow exciton linewidth of the QWs which makes the broadening so apparent, before the loss of oscillator strength due to phase space filling and screening bleaches the exciton. Loss of oscillator strength, which dominated previous nonlinear

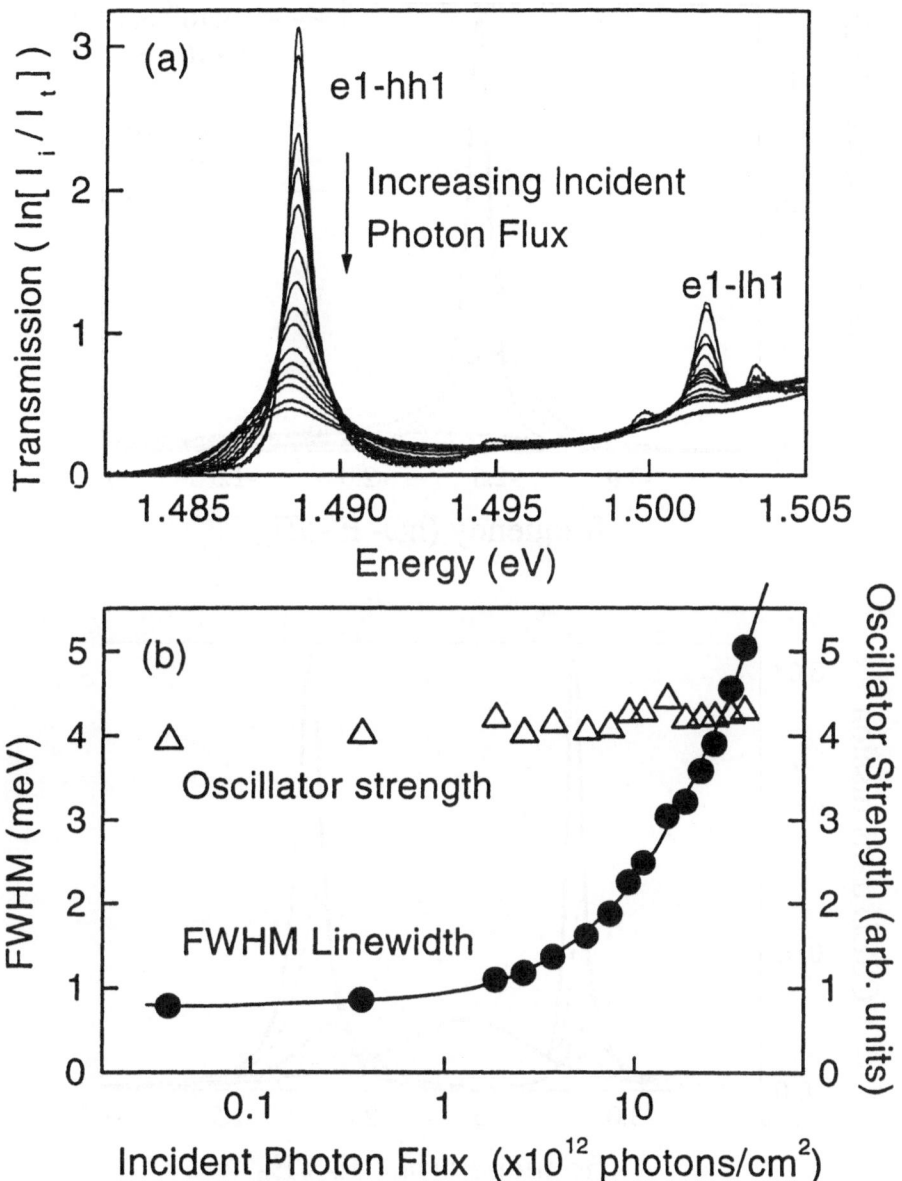

Figure 3 a) Exciton absorption of a 20-QW sample measured by the transmission of 100 fs pulses. b) FWHM linewidth and relative oscillator strength as a function of incident photon flux estimated from (a).

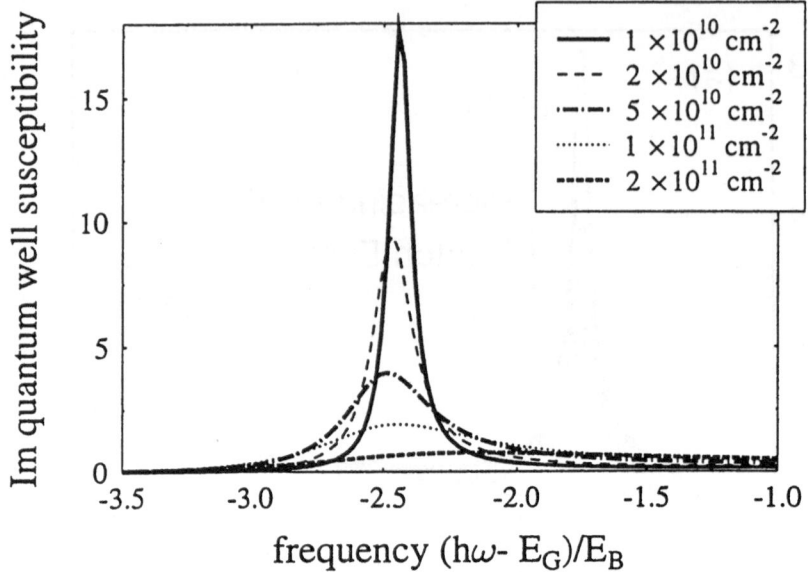

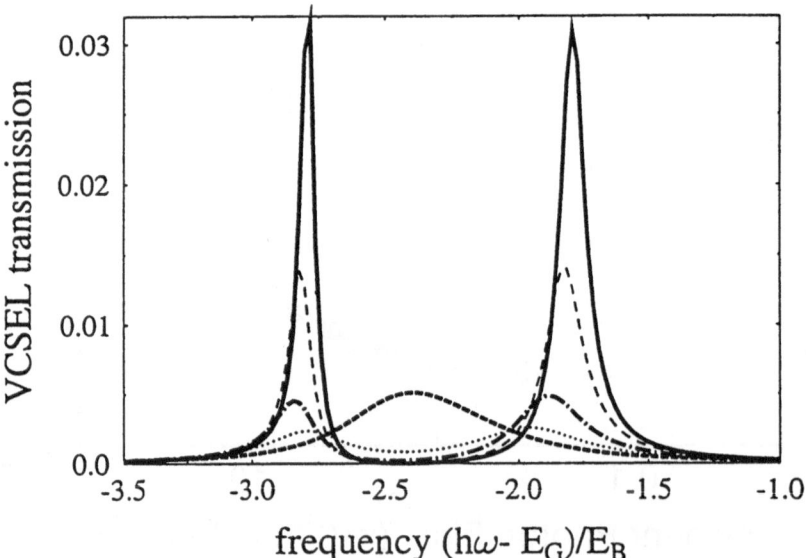

Figure 4 a) Imaginary part of the optical susceptibility calculated for an 8 nm QW and plasma excitation with various densities at carrier temperature 77 K. b) Calculated transmission of the QW microcavity for increasing plasma density and bleaching of the exciton according to a). The cavity resonance has been tuned from -2.05 (full line) to -2.14 (short dashed line) to compensate for the small numerical exciton shift.

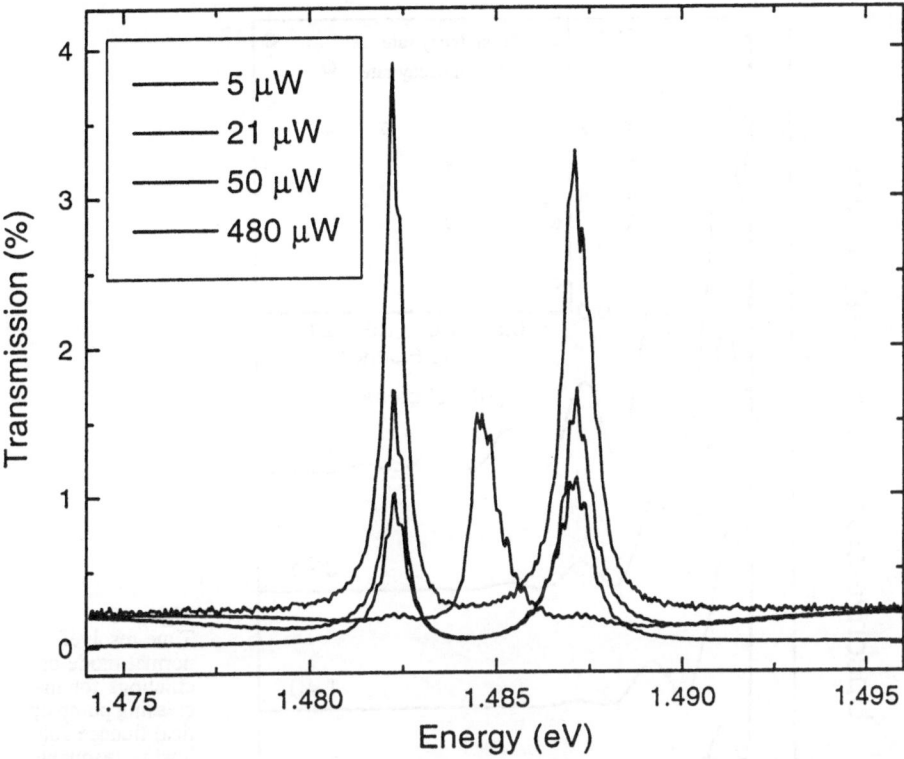

Figure 5 Time-integrated nonlinear transmission of NMC22 using 100 fs pulses centered at 1.4847 eV.

NMC experiments [14–16], is important here only in the transition from NMC to lasing at densities above the regime where loss of transmission occurs with little change in splitting.

Nonlinear transmission was also studied by time-integrated detection of the through-put of 100 fs laser pulses as shown in Fig. 5. Again loss of transmission with little reduction in splitting is seen just as in the cw case.

The time-resolved measurements were performed using a standard upconversion setup; the beam of fs pulses from a Tsunami mode-locked Ti:Sapphire laser was split into a reference and a pump beam. The spectral FWHM of the fs pulses was found to be 27 meV. The pump beam was focused at normal incidence onto the sample with a 58 μm spot size. The spot could be moved on the sample so that the detuning between the cavity Fabry-Pérot resonance and the quantum well exciton energy could be tuned due to the thickness variation of the sample as one scans across the wafer. The sample was mounted in a helium cryostat and held at 6 K temperature. The reflected signal from the sample was collected at normal incidence and imaged onto a pinhole, so that only the central portion of the sample spot with uniform intensity was collected. It was then

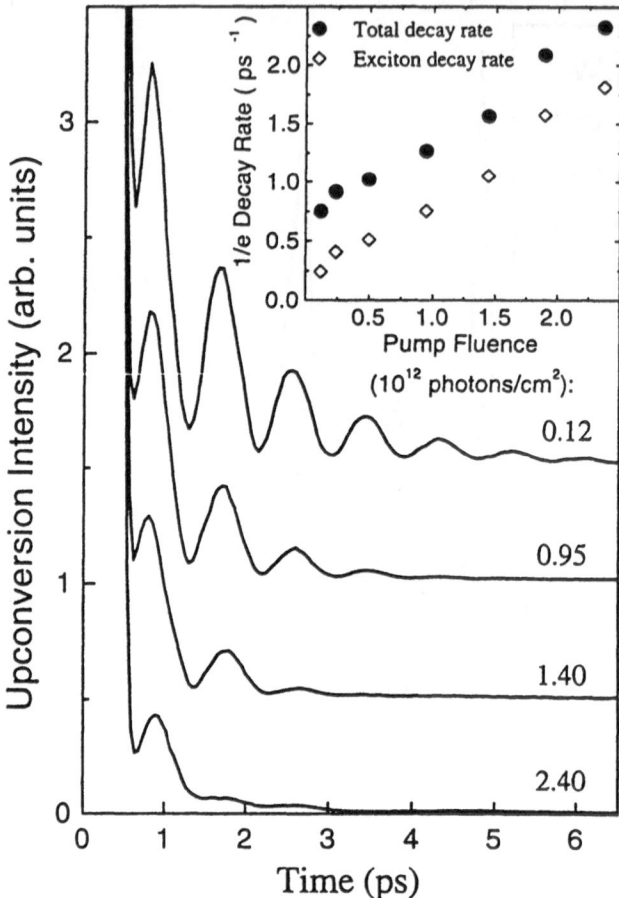

Figure 6
Time-resolved
normal mode os-
cillations for in-
creasing pump op-
tical fluence fol-
lowing resonant
100 fs pulsed ex-
citation. The inset
shows the oscil-
lation decay rate
versus pump opti-
cal fluence.

focused together with the reference beam onto a BBO crystal. The sum frequency signal generated in the BBO crystal was collected with an RCA8575 PMT after angular
selection, and the signal was fed to a lock-in amplifier. The delay of the reference beam
was controlled with a stepper motor delay track with 0.1 μm resolution. Both the lock-
in amplifier and the delay track were interfaced to a PC. The spectra of the reflected
excitation pulses were measured using a spectrometer and were not time-resolved.

Time-resolved measurements of the reflection of $\approx$ 100 fs laser pulses from NMC22
were made by the upconversion technique and are shown in Fig. 6 [17]. The NMC
modes observed in the static investigations show up as an oscillating signal in the
dynamic measurements. The beating signal can be interpreted as oscillation between
cavity-like (when emission from the cavity is at a maximum) and exciton-like polariton. The time-integrated reflectivity, shows two dips at the NMC peaks, the only energies at which light enters efficiently into the microcavity to excite polarization which

re-radiates and beats to give the oscillations.

With increasing excitation, see Fig. 6, the oscillations damp out more rapidly, but the oscillation frequency hardly changes. This is the time-resolved equivalent of the cw reduced transmission without a reduction in splitting. The decay rate plotted in the inset of Fig. 6 increases linearly with carrier density (proportional to pump fluence) corresponding to the exciton broadening.

Figure 7 compares time-resolved data for weak, intermediate, and strong pulse excitation with the corresponding microscopic theory calculations. As in Fig. 6, the intermediate-excitation decay shows fast damping with little change in oscillation period. The high-excitation decay shows very rapid damping of NMC oscillations due to rapid excitation induced dephasing followed by a slower decay (longer cavity storage time and narrower reflectance dips).

In the linear regime, the pulse propagation can be described by Fourier transforming the incoming pulse and then applying the transfer matrix method for each of its frequency components. The exciton only enters through its linear time-independent susceptibility computed from the microscopic theory. For a QW microcavity, radiative broadening and radiative coupling between the QWs can be included in the transfer matrix description [10]. However, this linear model is not applicable for the intense pulse propagation studied here. The strong excitation generates nonlinear excitonic effects that have to be treated self-consistently with the nonlinear pulse propagation problem [10].

In the theory used to analyze the experimental observation, we directly solve the time-dependent Maxwell's equations where the source term is determined by the macroscopic QW polarization $P_{QW}(t)$. In a Bloch basis with the in-plane carrier momentum k, we use $P_{QW}(t) = \frac{1}{L^2} \sum_k d_{cv}^* \Psi_k(t) + c.c.$ where d_{cv} is the dipole matrix element. The coherently driven excitonic polarization $\Psi_k(t)$ together with the occupation probability $f_k^{e,h}(t)$ for electrons and holes follow from generalized semiconductor Bloch equations

$$\left[\hbar \frac{\partial}{\partial t} - \varepsilon_k^e(t) - \varepsilon_k^h(t) \right] \Psi_k(t) - \left[1 - f_k^e(t) - f_k^h(t) \right] \Omega_k(t)$$
$$= i \left[\Gamma_k(t) \Psi_k(t) + \frac{1}{L^2} \sum_{k'} \Gamma_{k,k'}(t) \Psi_{k'}(t) \right], \quad (2.1)$$

$$\hbar \frac{\partial}{\partial t} f_k^a(t) + \Omega_k(t) \Psi_k^*(t) - \Omega_k^*(t) \Psi_k(t)$$
$$= i \left\{ \Sigma_k^{in,a}(t) [1 - f_k^a(t)] - \Sigma_k^{out,a}(t) f_k^a(t) + \Sigma_k^{pol,a}(t) \right\}. \quad (2.2)$$

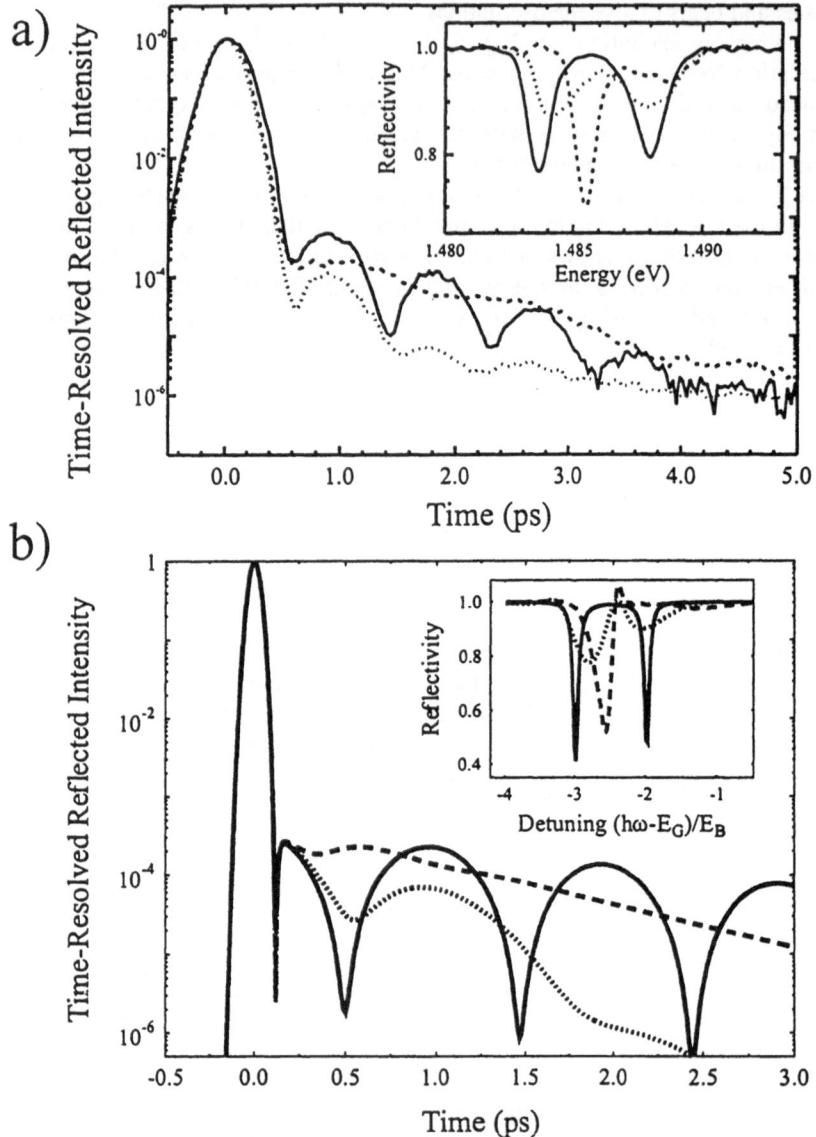

Figure 7 a) Time-resolved reflected signal (normalized to 1 at $t = 0$ ps) from sample NMC22 close to zero detuning after 100 fs pulse excitation with a photon flux of ($\pm 20\%$ uncertainty in absolute scale) 4.8×10^{11} (solid line), 3.8×10^{13} (dotted line), and 7.6×10^{13} (dashed line) photons/cm^2. Corresponding spectra of the reflected signal are shown in the inset. b) Calculated emission spectra from a structure similar to sample NMC22 after excitation by 100 fs pulses with $\Omega_R = 0.01 E_B$ (dotted line), and $\Omega_R = 2 E_B$ (dashed line). The inset shows the corresponding calculated reflection spectra.

The LHS of Eqs. (2.1) and (2.2) correspond to the well-known Hartree-Fock equations with phase-space filling nonlinearities, the renormalized Rabi-energy Ω_k (which contains the cavity field at the QW position), and the renormalized free-particle energies $\varepsilon_k^{e,h}$. The RHS of Eqs. (2.1) and (2.2) account for correlation contributions. Within second Born approximation, correlations lead to carrier and polarization scattering as well as screening contributions to the Hartree-Fock self-energies. The complex rates Γ_k and $\Gamma_{k,k'}$ describe dephasing of the excitonic polarization and the corresponding energy renormalization whereas Σ_k contains the influence of carrier and polarization interaction on the occupation probabilities. The theory microscopically describes nonlinear saturation of coherently driven excitons due to bandgap renormalization, dephasing, and reduction of oscillator strength. In the nonlinear regime, the field dynamics has to be solved directly from time-dependent Maxwell's equations together with Eqs. (2.1) and (2.2). The full light-propagation problem in the microcavity is treated using Maxwell's boundary conditions for the multi-layer system. The details of the theory are reported in Ref. [10].

The calculated reflected emission intensity for a semiconductor microcavity with parameters corresponding to the experiment is shown in Fig. 7b. The results fully reproduce the dynamic behavior of the measured results shown in Fig. 7a. The $t = 0$ peak is broader in the measurements due to the time resolution of our upconversion setup. This reduces the modulation depth and hides the feature at about 0.1 ps. The reflectance spectra shown in the inset of Fig. 7b are obtained by taking the Fourier transform of the calculated emission.

As the carrier density increases, carrier and polarization scattering increases the effective dephasing leading to a broadening of the excitonic transition without decreasing the oscillator strength. The broadening increases the absorption at the wavelengths of the reflectance dips thus reducing the depth of and broadening the reflectance dips. This translates in the time domain to a reduction of the modulation depth of the NMC oscillations. However, since the oscillator strength is conserved by scattering processes, the NMC splitting and oscillation period stay constant. Phase space filling and Coulomb screening do eventually reduce the oscillator strength, but not until the NMC reflectance dips and oscillations are almost completely gone. Then the bare-cavity reflectance dip appears, signaling the transition to the perturbative regime. The decay of the corresponding time-resolved signal (dashed line in Figs. 7a and 7b) is determined by the cavity lifetime.

3   Nonlinear Semiconductor Microcavity Luminescence

Photoluminescence (PL) from NMC microcavities has been used to determine the linear dispersion curve by angularly resolved detection [18] and the nonlinear response as a function of excitation density [14]. As pointed out above, the broad inhomogeneous linewidth prevented studies of the exciton broadening regime emphasized here.

a)   b)

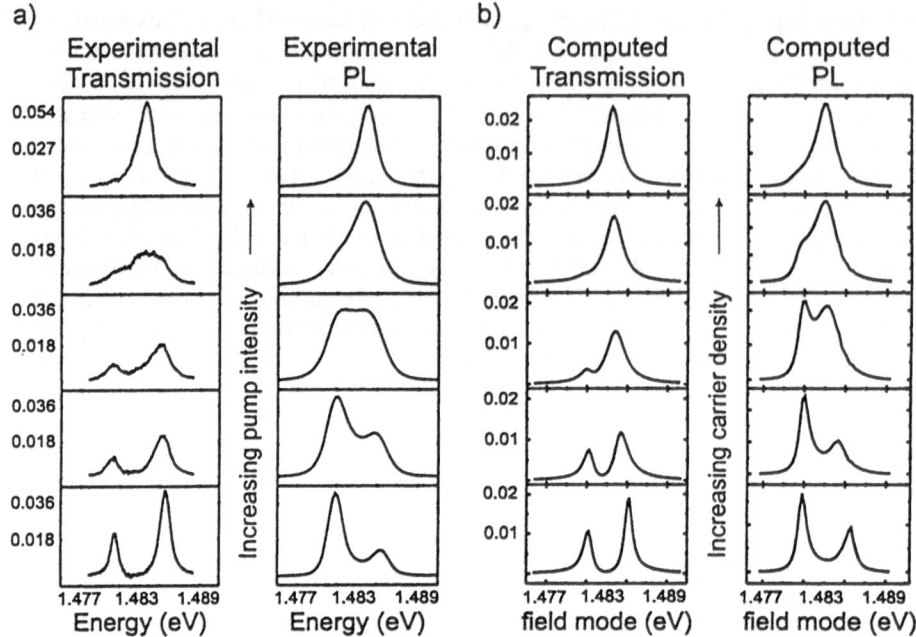

Figure 8   a) Nonlinear cw pump-probe transmission and corresponding photoluminescence of NMC22 for a cavity detuning from the exciton of ≈ +1.9 meV. b) Calculated
transmission and photoluminescence based on microscopic theory for a cavity detuning of
+1.6 meV.

For equal reflectance dips, i.e. for the cavity detuned to slightly higher energy than
the exciton, the PL spectrum of NMC22 at 4 K consists of two narrow peaks with the
lower-energy peak more than an order of magnitude more intense than the upper-energy
peak. At 110 K, for which kT is approximately 2 times larger than the NMC splitting,
the lower energy peak is only 50% higher and the upper peak is substantially broadened.

PL as a function of above-stopband excitation intensity has been measured. The pump
Ti:sapphire laser is incident at a 15° angle from normal and focused to a 100 μm diameter spot on the microcavity. The laser is chopped using an acousto-optic modulator
into 0.5 μs pulses with a 10% duty cycle to avoid heating the sample. The sample is
mounted in a helium cryostat and kept at about 4 K. Because the microcavity is grown
with a taper, the cavity resonance can be tuned by moving the sample within the focal
plane. PL is collected in the normal direction from a small solid angle $\sim 4\pi \times 10^{-3}$ sr,
and imaged onto an aperture. The area of the aperture is an order of magnitude smaller
than the image of the PL spot, such that only PL coming from the central, most uni-
formly excited area of the sample is collected. The spectra are then measured using an
optical multichannel analyzer with a linear array detector.

Figure 8a shows the simultaneous measurement of the normal incidence transmission
and normally emitted PL as a function of excitation intensity. At low carrier densities,

the PL-spectrum shows two NMC peaks. We perform a comparison between transmission and PL-spectra for various carrier densities. The relative peak heights of these two cases behave differently but the peak positions behave similarly. For positive cavity-exciton tunings, the low energy peaks in transmission and PL become smaller with increased excitation, whereas the high energy peak shifts to lower energy. The transmission and PL peaks broaden with increased excitation as a result of exciton broadening as discussed above for transmission spectra of the centrally tuned case.

A microscopic calculation shows exactly the same features as the experimental observations. Figure 8b is a steady-state computation, assuming electron and hole distributions in equilibrium at 80 K. Hot carriers excited by a short laser pulse relax due to carrier-carrier scattering and LO-phonon emission. We perform a full quantum analysis of the luminescence properties by studying both the field dynamics and the many-body system of carriers microscopically.

In the following, we shall give a brief summary of the theory. The field is quantized by introducing boson operators b_q for each field mode having a wave vector q. We treat the carrier system at a microscopic level to describe properly the many-body Coulomb effects [3]. The luminescence properties are determined from the equations of motion for expectation values $\langle b_q^\dagger b_{q'} \rangle$ and $\langle b_q^\dagger \hat{P}_k \rangle$ where $\hat{P}_k$ is the microscopic polarization of the matter.

Our microscopic theory shows that $\langle b_q^\dagger \hat{P}_k \rangle$ is directly driven by spontaneous emission involving recombination of an electron hole pair

$$\Omega_{SE}(k,q) = \mu_{cv}(k) f_k^e f_k^h E_q I_q. \tag{3.3}$$

Besides the product of electron and hole occupations $f_k^e f_k^h$, the rate of spontaneous emission depends on the dipole matrix element $\mu_{cv}(k)$ and on the field strength at the QW position ($E_q I_q$-term).

The resulting equations are evaluated numerically for a microcavity containing two 8 nm QWs. The theory curves in Fig. 8b are in very good agreement with the data in Fig. 8a.

The microcavity PL can also be studied by time-integrated detection of the PL spectra following 100 fs excitation above the stopband as shown in Fig. 9 for an even larger positive detuning of the cavity peak above the exciton peak than in Fig. 8a. At low excitations the lower-polariton intensity is much stronger than the upper-polariton intensity. With an increase in pump power the upper-polariton emission increases rapidly, while the lower-polariton emission tends to saturate. This curious increase in the upper-polariton emission has led to the speculation that electrons and holes form Bosons and Bose-condensate into the upper-polariton eigenstate [19–21]; this supposed bosonic behavior has been related to so called Boser. However, the electron-hole (e-h) pair does not have bosonic properties since the commutator of e-h Boson operator [3] has typical value of 0.5, far from the ideal bosonic 1, when the high energy PL-peak starts to grow rapidly. The explanation for the experiment is found from the QED theory developed without introducing any Bosons. The comparison of theory and experiment presented in Fig. 8 shows an excellent agreement for various experimental conditions. The measurements and theory both show that the high and low energy peak heights equalize at

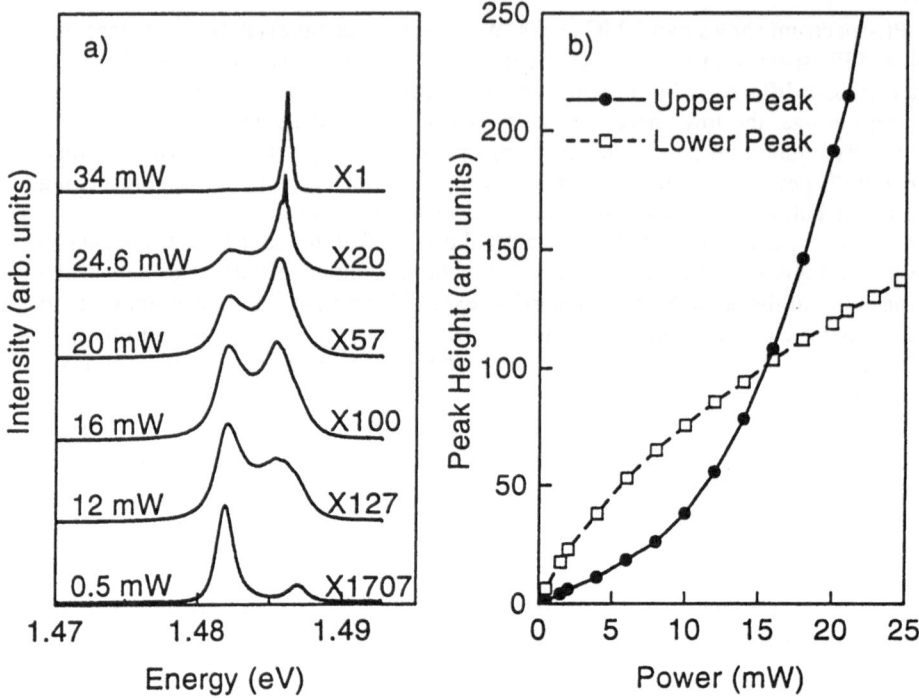

Figure 9 a) Microcavity luminescence spectra for increasing pump power following 100 fs pulsed excitation above the cavity stopband. The cavity mode is tuned to 2.83 meV above the exciton resonance. b) Peak luminescence intensity versus pump power from the spectra of a). High energy peak: circles, solid line. Low energy peak: squares, dashed line.

high densities (1 to 2×10^{11} cm^{-2}) and less than a factor of two below lasing threshold density. Part of the Boser argumentation is that the crossover occurs when there are still two peaks in PL, so there must still be NMC. However, in the detuned case the generalized Rabi splitting at low densities is $\Omega = (\Omega_0^2 + \Delta^2)^{1/2}$, but at high densities Ω goes to Δ, not zero, if the on-resonance NMC splitting Ω_0 goes to zero. In fact, our data and theory show that at the crossover, splitting is determined by detuning ($\Omega \approx \Delta$) and consequently the system is already in the perturbative irreversible decay regime. Furthermore, the upper-energy peak hardly shifts from crossover to lasing.

A streak camera has been used to time resolve the PL spectra following above-stopband excitation, as shown in Fig. 10. The crossover scenario occurs twice in time for a single excitation density because at earlier times (< 20 ps in Fig. 10e) the carriers have not yet cooled, so the lower-polariton emission dominates. Just as an increase in cw pumping caused the upper polariton to overtake the lower one, in the time-resolved spectra cooling results in the upper one overtaking the lower at about 50 ps. As emission occurs the carrier density dies away, so the crossover plays out in reverse as shown in Fig. 10e and b. Good agreement is obtained with a microscopic theory except that

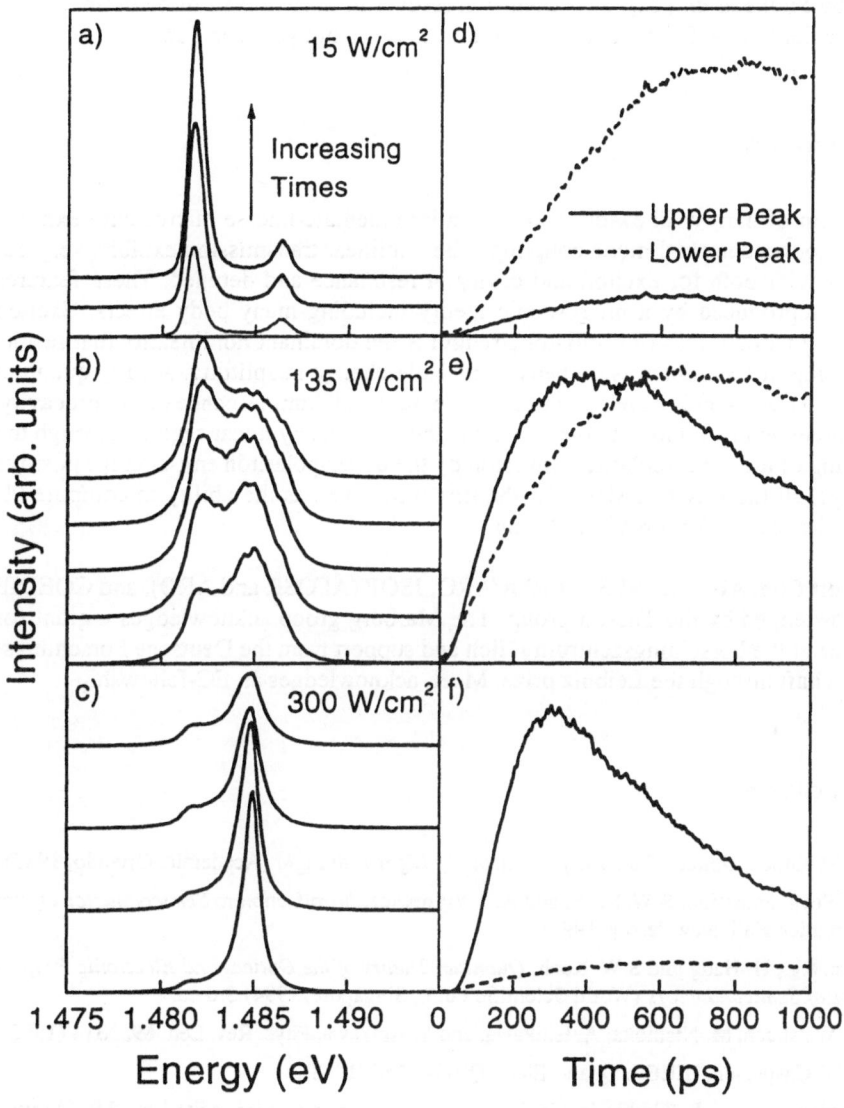

Figure 10 Temporal evolution of microcavity luminescence following 2 ps excitation above the mirror stopband for pump intensities of (a,d) 15 W/cm^2, (b,e) 135 W/cm^2, and (c,f) 300 W/cm^2. (a) shows spectral cuts at times (bottom to top) 219, 443, and 662 ps, (b) at times 219, 443, 518, 662, and 886 ps, and (c) at times 219, 443, 662, and, 886 ps. (d-f) show the high energy (solid line) and low energy (dashed line) peak intensities versus time.

the cooling times are longer than calculated because of hot-phonon effects as explained earlier by Snoke *et al.* [22]. This double crossover in time is even more dramatic in a magnetic field of 14 T, but the theory for that case has not yet been done.

4 Summary

The system of narrow-linewidth QWs in an intermediate-finesse microcavity exhibits very pronounced normal mode coupling. The nonlinear transmission exhibits very curious behavior both for exciton and cavity in resonance and detuned. These features are fully reproduced by a microscopic theory including many-body effects. Exciton broadening without loss of oscillator strength is the dominant nonlinearity that results in vanishing of the transmission peaks with little change in splitting. A fully quantum mechanical treatment is developed to treat quantum well luminescence in a microcavity. Good agreement is found with the data from the low-density linear regime, through the overtaking of the lower-polariton emission by the upper-polariton emission for positive detunings, all the way to lasing. For the first time, one has the ability to compute PL from a microcavity from NMC to lasing.

Support from AFOSR, NSF, DARPA/ARO, JSOP (AFOSR and ARO), and COEDIP is acknowledged by the Tucson group. The Marburg group acknowledges a grant for CPU time at the Forschungszentrum Jülich and support from the Deutsche Forschungsgemeinschaft through the Leibniz prize. M. K. acknowledges an EC-fellowship.

Bibliography

[1] H.M. Gibbs, *Optical Bistability: Controlling Light with Light* (Academic, Orlando, 1985).

[2] N. Peyghambarian, S.W. Koch, and A. Mysyrowicz, *Introduction to Semiconductor Optics* (Prentice Hall, New Jersey, 1993)

[3] See, e.g., H. Haug and S.W. Koch, *Quantum Theory of the Optical and Electronic Properties of Semiconductors* (World Scientific Publ., Singapore, 1994) 3rd ed.

[4] C. Weisbuch, M. Nishioka, A. Ishikawa, and Y. Arakawa, Phys. Rev. Lett. **69**, 3314 (1992).

[5] L.W. Casperson, IEEE J. Quant. Elect. **QE-14**, 756 (1978).

[6] Reviewed by H.J. Kimble in *Cavity Quantum Electrodynamics*, edited by P.R. Berman (Academic Press, San Diego, 1994).

[7] F. Jahnke, M. Kira, S.W. Koch, G. Khitrova, E.K. Lindmark, T.R. Nelson, Jr., D.V. Wick, J.D. Berger, O. Lyngnes, H.M. Gibbs, and K. Tai, Phys. Rev. Lett. **77**, 5257 (1996).

[8] E.T. Jaynes and F.W. Cummings, Proc. IEEE **51**, 89 (1963).

[9] R.J. Thompson, G. Rempe, and H.J. Kimble, Phys. Rev. Lett. **68**, 1132 (1992).

[10] F. Jahnke, M. Kira, and S.W. Koch, *Linear and Nonlinear Optical Properties of Quantum Confined Excitons in Semiconductor Microcavities*, Z. Phys. B, to be published.

[11] F. Jahnke in this volume.

[12] M. Kira, F. Jahnke, S.W. Koch, J.D. Berger, D.V. Wick, T.R. Nelson, Jr., O. Lyngnes, G. Khitrova, and H.M. Gibbs, *Quantum Theory of Nonliear Semiconductor Microcavity Luminescense Explaining "'Boser"' Experiments*, Phys. Rev. Lett., to be published.

[13] Y. Zhu, D.J. Gauthier, S.E. Morin, Q. Wu, H.J. Carmichael, and T.W. Mossberg, Phys. Rev. Lett. **64**, 2499 (1990).

[14] R. Houdré, J.L. Gibernon, P. Pellandini, R.P. Stanley, U. Oesterle, C. Weisbuch, J. O'Gorman, and B. Roycroft, Phys. Rev. B **52**, 7810 (1995).

[15] J.-K.Rhee, D.S. Citrin, T.B. Norris, Y. Arakawa, and M. Nishioka, Solid State Commun. **97**, 941 (1996).

[16] The nonlinear behavior here is different from Refs. [14,15] because the inhomogeneous broadening here is much smaller than the NMC splitting so that the homogeneous broadening can have such a dramatic effect on the two transmission peaks before loss of oscillator strength.

[17] O. Lyngnes, J.D. Berger, J.P. Prineas, S. Park, G. Khitrova, H.M. Gibbs, F. Jahnke, M. Kira, and S.W. Koch, *Nonlinear Emission Dynamics from Semiconductor Microcavities in the Nonperturbative Regime*, Solid State Commun., to be published.

[18] R. Houdré, C. Weisbuch, R.P. Stanley, U. Oesterle, P. Pellandini, and M. Ilegems, Phys. Rev. Lett. **73**, 2043 (1994).

[19] J.D. Berger, S. Hallstein, W.W. Rühle, O. Lyngnes, G. Khitrova, H.M. Gibbs, M. Kira, F. Jahnke, and S.W. Koch, Quantum Optoelectronics Conference, Incline, Nevada (March 1997).

[20] A. Imamoglu, R.J. Ram, S. Pau, and Y. Yamamoto, Phys. Rev. A **53**, 4250 (1996).

[21] S. Pau, H. Cao, J. Jacobson, G. Björk, Y. Yamamoto, and A. Imamoglu, Phys. Rev. A **54**, R1789 (1996).

[22] D.W. Snoke, W.W. Rühle, Y.-C. Lu, and E. Bauser, Phys. Rev. Lett. **68**, 990 (1992).

Hanle-Oscillations in the Stimulated Emission of Microcavity Laser

M. Oestreich[1], S. Hallstein[2], J. D. Berger[3], M. Hilpert[2],
F. Jahnke[1], G. Khitrova[3], W. W. Rühle[1] S. W. Koch[1],
H. M. Gibbs[3], and H. C. Schneider[1]

[1] Fachbereich Physik der Philipps-Universität, Renthof 5, D-35032 Marburg, Germany
[2] Max-Planck-Institut für Festkörperforschung, Heisenbergstraße 1, D-70569 Stuttgart, Germany
[3] Optical Science Center, University of Arizona, Tuscon, AZ 85721, USA

Abstract: We demonstrate a new phenomenon in the emission dynamics of semiconductor microcavities. Hanle-oscillations of the electron spins modulate the gain of microcavities up to extremely high frequencies. So far, we achieved modulation frequencies of 22 GHz with a modulation depth of 96% but we expect much higher pulse repetition rates for higher magnetic fields and for semiconductors with larger electron g factors. In principle, the maximum frequency is only limited by the photon life-time and the maximum gain of the microcavity. The pulse repetition rate is extremely stable since it only depends on the internal clock of the Larmor precession of the electron spins and, therefore, is insensitive to scattering and energy relaxation of the electrons. A microscopic theory is developed to analyze the non-equilibrium carrier and laser dynamics in the framework of Hartree-Fock equations. Measurements with coupled microcavity lasers show a density dependent switching between two distinct oscillation frequencies. The switching turns out to be an interesting method to characterize the active gain medium in coupled microcavities.

1 Introduction

The high speed dynamics of semiconductor microcavity lasers is attracting a lot of interest since the lasers are possible candidates for optical communication devices. However, to our knowledge, so far only two kinds of optical experiments are reported which study the ultrafast switching dynamics of these lasers, and both depend on the energy relaxation of carriers. The first one studies the laser dynamics after excitation of the microcavity with a short laser pulse above the stop band. [1,2] The relaxation of the excited carriers due to electron-phonon scattering leads to inversoin at the laser resonance with a typical delay. The second kind of experiment rapidly increases the energy of the electrons in the conduction bands in the already lasing microcavity by a short external laser puls far above the band gap. Despite the higher total carrier density in the microcavity, the gain at the emission wavelength decreases due to the higher carrier energy. Thereby, the laser can be switched off on a timescale less than a picosecond. [3,4] Here, we will demonstrate a new gain switching method which does not depend on energy relaxation processes but on Hanle-oscillations [5–9] – the optical gain modulation is caused by the Larmor precession of the electron spin.

2 Experimental Realization

Circularly polarized excitation of a microcavity above the stop band creates a preferential electron spin orientation in the conduction band according to the different transition probabilities for heavy and light hole excitation. As a consequence, two different gain curves for right (σ^+) and left (σ^-) circularly polarized light exist found for sufficiently high carrier densities at time $t = 0$, as depicted schematically in Fig. 1(a). Just after excitation only σ^+ polarized light exceeds the losses at the emission energy of the microcavity and is therefore strongly amplified. Emission of σ^- light is, at this time, only of spontaneous nature. The electron spins, oriented perpendicular to a magnetic field, start to precess around the field direction with the Larmor frequency $\omega_L = g_e^* \mu_B B / \hbar$, where g_e^* is the electron Landé g factor, B is the magnetic field, and μ_B is Bohr's magneton. After a quarter precession period ($t = \frac{1}{4}T$), both spin states in emission direction of the microcavity are populated equally and the corresponding gain curves drop below threshold (Fig. 1(b)). Lasing is switched off and only spontaneous emission – which is half σ^+ and half σ^- – is present. After half a Larmor period ($t = T/2$), the laser exceeds threshold for σ^- light, while σ^+ light is emitted spontaneously (Fig. 1(c)).

For our experiment we use a semiconductor microcavity laser consisting of a $3/2\lambda$-cavity with two 8 nm $In_{0.04}Ga_{0.96}As$ quantum wells (QWs) separated by GaAs barriers. [10,11] The wells are placed in the antinodes of the intracavity electric field, which is formed by two GaAs/AlAs distributed Bragg reflectors with 99.6% reflectivities. The sample is held at a temperature of 15 K in a 16 Tesla superconducting magnet. The microcavity is excited with circularly polarized pulses of a Ti:Sapphire laser with 2 ps pulse length, 80 MHz repetition rate, a spot diameter of 100 μm, and 780 nm wavelength. The pump wavelength is above the microcavity stopband, so that both the quantum wells and the barriers are excited. Choosing the z-axis of our coordinate system in the growth direction, we align the magnetic field in the plane of the wells, i.e. in the x-direction (Voigt geometry, see Fig. 2). The microcavity emission at 835 nm is detected in reflection geometry, analyzed with a circular polarizer, spectrally dispersed in a 0.32 m spectrometer, and temporally resolved in a synchroscan streak-camera with a time resolution of 7 ps (FWHM of the exciting laser pulse).

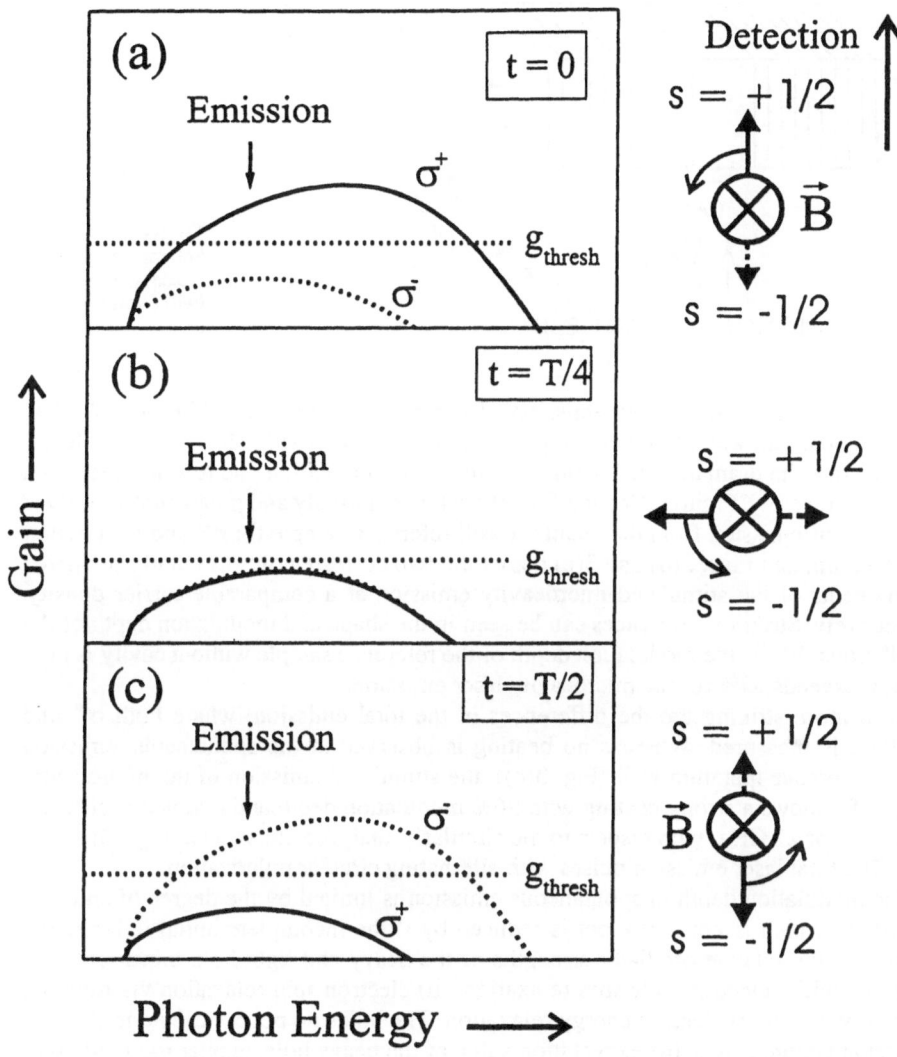

Figure 1 Schematic gain spectra (a) at excitation ($t = 0$), (b) after a quarter Larmor precession period ($t = T/4$), and (c) after a half Larmor precession period ($t = T/2$). The right side demonstrates the spin precession about the magnetic field.

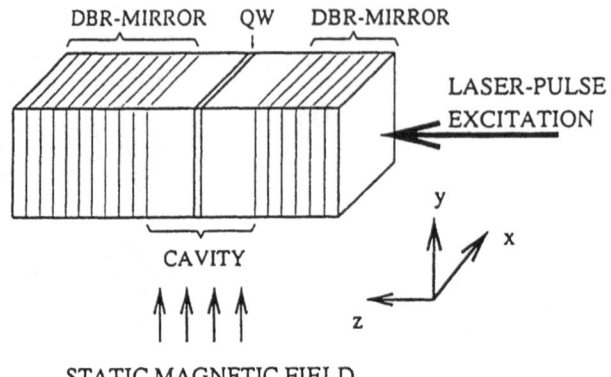

DBR-MIRROR   QW   DBR-MIRROR

LASER-PULSE
EXCITATION

y
x

CAVITY
z
STATIC MAGNETIC FIELD

Figure 2
Scheme of a QW-
Laser with magnetic
field (Voigt geome-
try).

We compare the stimulated emission of our micrcavity system (Fig. 3(d)–3(f)) to that of a quantum well reference sample without microcaviy (Fig. 3(a)–3(c)), to clearly demonstrate the dramatic enhancement of the light modulation due to spin precession in the microcavity. Figures 3(a) and 3(b) show the temporally and polarization resolved spontaneous emission from the quantum well reference sample for σ^+ and σ^- circular polarization, and Figs. 3(d) and 3(e) depict the two corresponding circularly polarized components of the stimulated microcavity emission at a comparable carrier density, respectively. Striking differences can be seen in the shape and modulation depth of the oscillations. While the modulation depth of the reference sample without cavity is only 27%, it exceeds 99% for the microcavity laser emission.

Even more striking are the differences in the total emission, where both σ^+ and σ^- light is measured. Whereas no beating is observed in the spontaneous emission of the reference quantum well (Fig. 3(c)), the stimulated emission of the microcavity (Fig. 3(f)) shows a strong beating with 96% modulation depth and doubled oscillation frequency of 22 GHz, with respect to the circularly analyzed emission in Figs. 3(d) and 3(e). The total laser emission pulses with alternating circular polarization.

The modulation depth in spontaneous emission is limited by the degree of spin polarization, which is not 100% but is reduced by i) the incomplete initial polarization caused by the relative oscillator strengths of the heavy and light hole transition, ii) a very fast initial electron-hole spin relaxation, iii) electron spin relaxation via trapping in the QWs, and iv) electron energy relaxation. The holes do not contribute to this polarization beating, since the expectation value of the heavy hole angular momentum in the observation direction vanishes in this geometry. [12]

The modulation depth in the total emission depends strongly on the excitation power, since the oscillations arise from a modulation of the optical gain. Figure 4 demonstrates the influence of excitation power on the total microcavity laser emission. We see that in the experiment with low excitation power, the laser remains below threshold at all times and no modulation is observed. Higher excitation leads to an increased maximum gain for σ^+ and σ^- light and to a larger stimulated part in the total emission. Oscillations begin to appear on top of a spontaneous emission background. The modulation depth

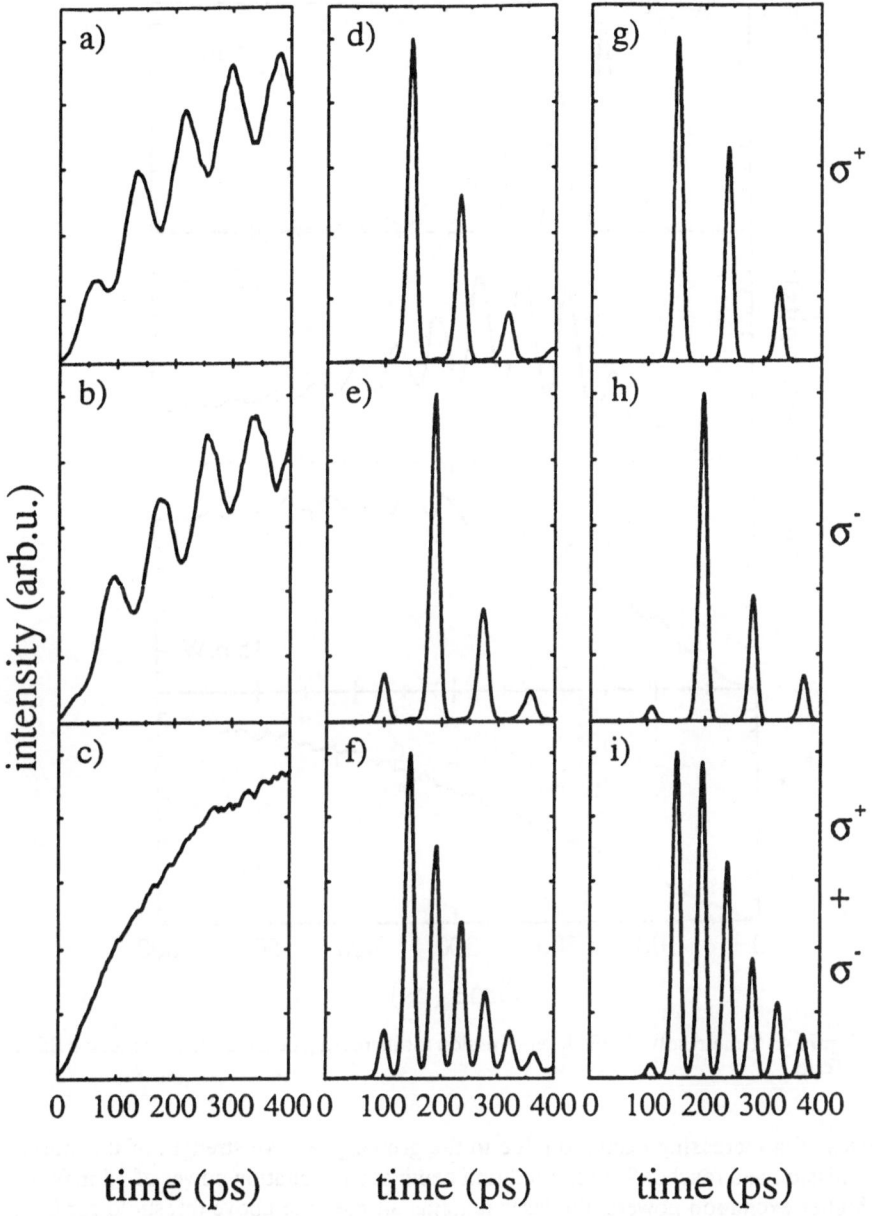

Figure 3 (a,b,c) Time-resolved spontaneous emission of the reference sample and (d,e,f) time-resolved stimulated emission of the microcavity laser in a 2 Tesla magnetic field. The rightmost column (g,h,i) shows numerical results with parameters corresponding to the conditions of d, e, f. The top row of figures shows the σ^+ and the middle row the σ^- components. The bottom row gives the total emission.

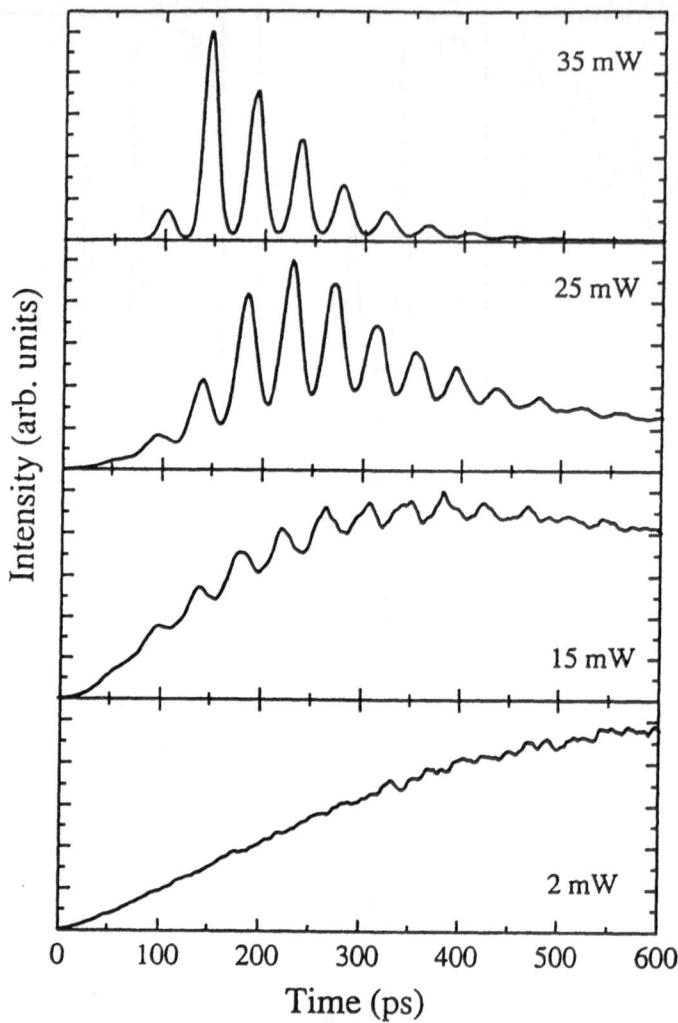

Figure 4 Time-resolved total laser emission for increasing densities at a magnetic field of 2 Tesla.

increases with increasing excitation due to the growing relative strength of the stimulated emission and reaches 96% modulation depth at an excitation power of 35 mW. At even higher excitation powers, the laser remains all the time above threshold for both polarizations and, therefore, the modulation depth decreases again. [13]

3 Theoretical Description

For the theoretical description we choose the static magnetic field in x-direction $B = Be_x$. The growth direction (z-direction) of the microcavity laser is orthogonal to the direction of the magnetic field, see Fig. 2. This geometry determines the selection rules in the two-band model for the quantum-well active material: Because of symmetry we use for the quantum numbers the z-component and the square of the total angular momentum. The electron states are labeled by $s_z = \pm 1/2$, the hole states by $j_z = \pm 3/2$. The light-hole band in the sample used for the experiments is energetically separated from the heavy-hole band due to strain effects.

3.1 Coupling of Electron Bands

For the magnetic dipole interaction we have the Hamiltonian

$$H_{\mathrm{mag}} = -\frac{\hbar \omega_L}{2} \sum_{s_z = \pm 1/2} \sum_{k} \alpha^{\dagger}(k, s_z)\, \alpha(k, -s_z), \qquad (3.1)$$

where $\alpha^{\dagger}(k, s_z)$ denotes the creation operator for an electron with crystal momentum $\vec{k}$ and spin component in z-direction s_z. The Larmor energy $\hbar \omega_L$ is defined by

$$\hbar \omega_L = g_e^* \mu_B B. \qquad (3.2)$$

The interaction Hamiltonian couples electrons with different spins. In the geometry chosen, the heavy holes are not influenced by the magnetic field since the matrix element.

$$\langle j_z | \mu_x | j_z' \rangle = 0 \qquad (3.3)$$

vanishes.

In a fashion analogous to the derivation of the semiconductor Bloch-equations (SBE) [14] we obtain screened Hartree-Fock equations for the distribution functions in the bands, $f_{\pm}^e$ and for the polarization between the electron bands, ψ:

$$\frac{\partial}{\partial t} f_{\pm}^e(k) \Big|_B = 2 \operatorname{Im} \left[\Omega_L^*(k) \psi(k) \right] \qquad (3.4)$$

with the renormalized Larmor frequency

$$\hbar \Omega_L(k) = \frac{\hbar \omega_L}{2} - \sum_{k'} V_{k,k'} \psi(k'). \qquad (3.5)$$

For the Larmor polarization

$$\psi(k) = \left\langle a^\dagger(k, -1/2) a(k, +1/2) \right\rangle \tag{3.6}$$

we obtain

$$i\frac{\partial}{\partial t}\psi(k) = \omega^e(k)\psi(k) - \Omega_L(k)\left[f^e_-(k) - f^e_+(k)\right] \tag{3.7}$$

where the free polarization rotation is due to the Coulomb potential:

$$\hbar\omega^e(k) = -\sum_{k'} V_{k,k'}\left[f^e_-(k') - f^e_+(k')\right]. \tag{3.8}$$

Note that the Larmor polarization $\psi(k)$ is an intraband correlation function as opposed to the interband polarization $P(k)$ known from the SBE.

In our analysis, the fast relaxation of hole spins and the thermalization of electrons and holes are taken into account in a k-dependent rate-equation approximation. The plasma screening at high densities is modeled by a single plasmon-pole approximation for the dielectric function.

3.2 Coupling of Electron and Hole Bands

Optical excitation of the quantum-well with σ^+-polarized light drives the interband polarization P_+ which depends on the in-plane momentum k and couples the $s_z = -1/2$ and $j_z = -3/2$ bands, whereas σ^--polarized light couples $s_z = +1/2$ and $j_z = +3/2$. Hence, only the magnetic field couples the σ^+ with the σ^- subsystems.

The microcavity laser is modelled by rate equations for the intensities of the σ^+ and σ^--emission, respectively. Here, the interband polarizations created by the laser are adiabatically eliminated in favor of the susceptibilities for σ^+ and σ^- polarizations. The susceptibilities are computed from the non-equilibrium distribution functions by using a matrix inversion technique [14]. The resulting computed laser emission curves, which correspond to Fig. 3(d)–(f), are plotted in Fig. 3(g)–(i). Theory and experiment are in excellent agreement.

4 Coupled Microcavities

In the following, we extent the experiments from a single microcavity to a coupled microcavity system with two different quantum well widths, see Fig. 5(a) and Ref. [16]. The coupled microcavity laser is comprised of two λ-sized $Al_{0.2}Ga_{0.8}As/Al_{0.5}Ga_{0.5}As$ cavities, separated by a common mirror. The top mirror consists of 17 pairs, the middle mirror of 9 pairs, and the bottom mirror of 23 pairs of $Al_{0.2}Ga_{0.8}As/AlAs$. The bottom cavity contains three 10 nm thick GaAs QWs whereas the top cavity contains three 16 nm thick GaAs QWs. The Larmor frequencies of the two QWs differ by about a factor of two due to the strong dependence of g^*_e on QW thickness.[9]

The experimental arrangement is identical to the experiment described in chapter 2 but with an excitation wavelength of 734 nm. The emission wavelength of the coupled microcavities is fixed at 811 nm between the emission wavelength of the ground states of the wide and the narrow QWs at a sample temperature of 15 K. The microcavity emits at a single wavelength at our experimental conditions. Figure 5(b) shows the temporal evolution of the σ^+ polarized microcavity laser emission for different excitation densities at a magnetic field of 9 Tesla. The experiment clearly shows a switching between fast oscillations at excitation powers below 20 mW and slow oscillations at excitation powers above 40 mW. The fast oscillations originate from the Larmor precession of the electron spins in the wide quantum well while the slow oscillations originate from the Larmor precession of the electron spins in the narrow quantum well. The assignment is unequivocal since the electron g factors of the QWs are well known. We observe fast oscillations at low excitation densities because only the wide QW gain spectra overlaps with the cavity resonance. The oscillation amplitude is small in this case since, firstly, the spontaneous emission is not negligible in comparision to the stimulated emission (compare with Fig. 4 at 15 mW). Secondly, due to the different band filling of electrons with spin up and spin down, photons emitted by the narrow QW are absorbed in the wide QW most strongly, if their circular polarization is opposite to the emitted circular polarization of the wide QW. This process destroys the initial spin polarization and, thereby, suppresses especially at medium densities Hanle oscillations shortly after excitation of the microcavity, as seen in Fig. 5 at an excitation density of 30 mW. At high excitation densities the gain spectra broaden and shift to lower energies due to the large electron-hole densities and band gap renormalization. Thereby, the gain spectra of the narrow QW dominates the stimulated emission above a certain carrier density and we observe a slow beating instead of the fast beating at low densities.

In conclusion, Hanle oscillations in coupled microcavity lasers can be used (a) to modulate the lasers with two different frequencies and with a density dependent switching between the two frequencies and (b) as a tool to characterize the active gain medium.

5 Summary

In summary, we have demonstrated a new method for ultrafast modulation of the stimulated emission of a microcavity laser. The stimulated emission is synchronized to the stable electron spin precession about the magnetic field axis. The transfer of electron spin coherence to the optical field yields a pulsed total emission with twice the Larmor frequency and alternating circular polarization. The data is described by a microscopic model for the spin-split states in the framework of a Hartree-Fock theory. Experiments with coupled microcavities exhibit a switching between two distinct oscillation frequencies with increasing excitation density.

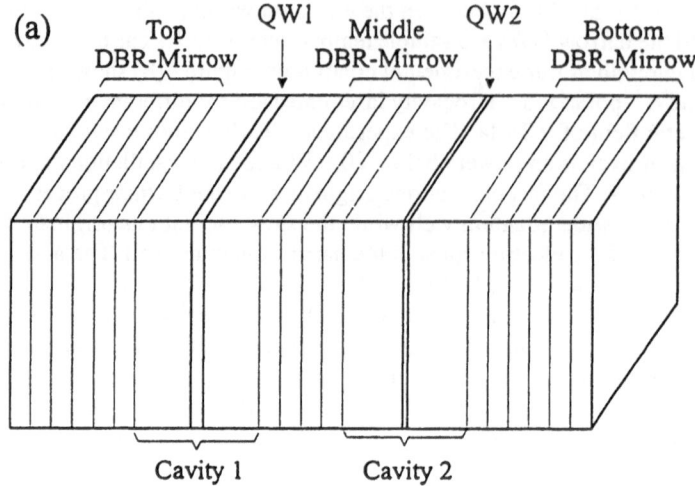

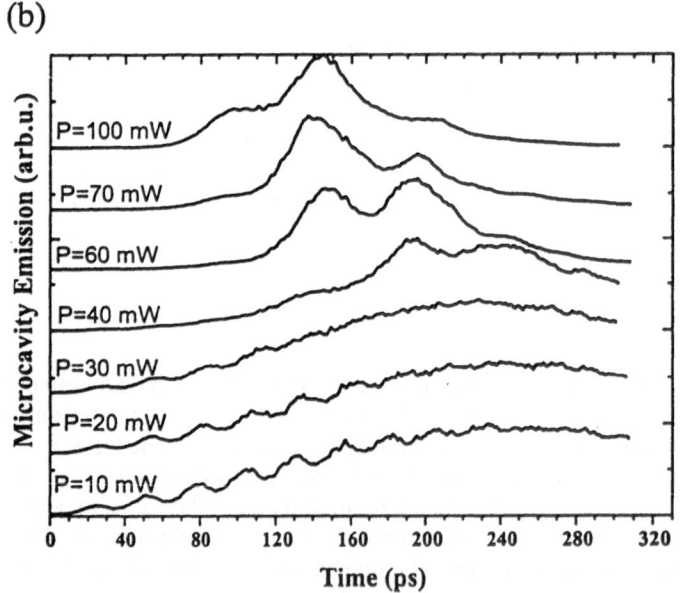

Figure 5 (a) Coupled microcavity laser, schematically. (b) Time-resolved σ^+ polarized laser emission for increasing excitation densities at a magnetic field of 9 Tesla.

6 Acknowlededgement

The authors thank H. J. Queisser and H. Gießen for helpful discussions and G. Reiner, K. J. Ebeling, and P. Michler for the coupled microcavity sample. The technical assistance of K. Rother and H. Klann and a grant for CPU time at the Forschungszentrum Jülich are gratefully acknowledged. The work has been partially supported in the framework of the Schwerpunkt "Quantenkohärenz in Halbleitern" of the Deutsche Forschungsgemeinschaft and by NSF Atomic, Molecular, and Optical Physics and Lightwave Technology and DARPA/ARO Ultraphotonics.

Bibliography

[1] P. Michler, A. Lohner, W. W. Rühle, and K. Ploog, Appl. Phys. Lett. 66, 1599 (1995).

[2] P. Michler, W. W. Rühle, G. Reiner, K. J. Ebeling, and A. Moritz, Appl. Phys. Lett. 67 1363, (1995)

[3] F. Jahnke and S. W. Koch, Appl. Phys. Lett. 67, 2278 (1995).

[4] S. G. Hense, M. Elsässer, and M. Wegener, published in this volume.

[5] A. P. Heberle, W. W. Rühle, and K. Ploog, Phys. Rev. Lett. 72, 3887 (1994).

[6] M. Oestreich, S. Hallstein, A. P. Heberle, K. H. Schmidt, K. Eberl, E. Bauser, and W. W. Rühle, Phys. Rev. B 53, 7911 (1996).

[7] M. Oestreich, A. P. Heberle, W. W. Rühle, R. Nötzel, and K. Ploog, Europhys. Lett. 31, 399 (1995).

[8] M. Oestreich and W. W. Rühle, Phys. Rev. Lett. 74, 2315 (1995).

[9] R. M. Hannak, M. Oestreich, A. P. Heberle, and W. W. Rühle, Solid State Comm. 93, 313 (1995)

[10] A. Tredicucci, Y. Chen, V. Pellegrini, M. Börger, L. Sorba, F. Beltram, and F. Bassani, Phys. Rev. Lett. 75, 3907 (1995).

[11] C. Weisbuch, M. Nishioka, A. Ishikawa, and Y. Arakawa, Phys. Rev. Lett. 69, 3314 (1992)

[12] S. Bar-Ad and I. Bar-Joseph, Phys. Rev. Lett. 66, 2491 (1991).

[13] The modulation depth also decreases with increasing magnetic field as soon as the Larmor frequency becomes comparable to the photon life-time in the microcavity. We observe this effect in principle but the measurements are obscured by the limited time resolution of our experimental setup.

[14] For a textbook discussion, see H. Haug and S. W. Koch, Quantum Theory of the Optical and Electronic Properties of Semiconductors, 3rd ed. (World Scientific, Singapore, 1994).

[15] The material parameters used in the calculations are: cavity-damping time 2 ps, $m_e = 0.0665m_0$, $m_h = 0.234m_0$, carrier-carrier scattering rate 100 fs, thermalization rate 30 ps, hole-spin relaxation rate 5 ps, spontaneous emission rate 800 ps and coupling 10^{-3} and transverse mode overlap 0.2. The pump field polarization and Rabi-energy are taken to be 0.6 and $1.6E_B$, respectively .

[16] P. Michler, M. Hilpert, and G. Reiner, Appl. Phys. Lett. 70, 21st April (1997).

Ultrafast Dynamics of the Surface Plasmonin Gold Nanoparticles

Gero von Plessen

Sektion Physik, Ludwig-Maximilians-Universität München,
Amalienstrasse 54, D-80799 München, Germany

Abstract: Noble-metal nanoparticles show pronounced optical resonances that are caused by coupling of the light to the surface plasmon excitation. Here we study the damping of the surface plasmon resonance in gold nanoparticles as a consequence of high-intensity optical excitation of the electron gas. Femtosecond pump-probe experiments are performed on gold nanoparticles embedded in a dielectric sol-gel matrix. Optical excitation of single-electron interband transitions leads to a pronounced broadening of the surface plasmon line. A similar behavior is observed for resonant excitation of the surface plasmon. This broadening is the dominant optical nonlinearity of the system, and reflects the excitation-induced damping of the surface-plasmon resonance. The time evolution of the damping rate follows that of the electron temperature, showing that the damping rate is strongly influenced by transient variations in the electron scattering rate. The dependence of the damping rate on the excitation energy shows evidence for fast relaxation of the optically generated d-band holes.

1 Introduction

The brilliant colors of noble-metal nanoparticles embedded in glass have been known for many hundreds of years. They have been used for the staining of glass products since Roman times, and are still employed in decorative coloring today. Further applications of the optical properties of metal nanoparticles include their use in passive optical elements such as polarizers [1], and in biosensors [2]. Their potential use in nonlinear all-optical devices is currently under discussion [3,4].

Noble-metal nanoparticles in dielectric matrices can be produced by a variety of techniques, including ion-implantation [5] and sol-gel processing [6], and in a wide range of different sizes, shapes, and concentrations [7]. The remarkable optical properties of metal nanoparticles have their origin in the surface plasmon, i.e. the collective oscillation of the conduction band electrons in the metal nanoparticle. The principle of this oscillation is shown schematically in Fig. 1. The alternating electric field of an incident light wave forces the conduction electrons of a metal nanoparticle, which is embedded ·

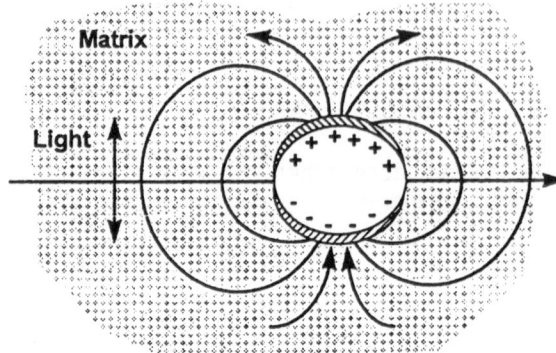

Figure 1
Schematic representation of the surface-plasmon oscillation in a metal nanoparticle embedded in a dielectric matrix material. The alternating electric field of an incident light wave forces the conduction electrons of the nanoparticle into an oscillatory collective motion. The eigenfrequency of the oscillation is determined by the restoring Coulomb force between the electrons and the positively charged ion cores.

in a dielectric matrix, into an oscillatory collective motion. Quantum mechanically, this oscillation can be described in terms of a boson-like quasiparticle, the surface plasmon (SP). The eigenfrequency of the oscillation, i.e. the energy of the SP, is determined by the attractive force between the electrons and the positively charged ion cores, and can be calculated from the dielectric functions of the metal nanoparticle itself and the surrounding dielectric matrix. Experimentally, the SP oscillation is observed as a pronounced optical resonance, whose spectral position can be varied over a wide range by suitable choice of the metal, the dielectric, and the geometry of the nanoparticle [7]. For instance, the SP resonance of noble-metal nanoparticles embedded in glass lies in the visible, giving rise to the brilliant colors exploited technically.

Thin films containing metal nanoparticles possess large optical nonlinearities that originate from local-field enhancements at the SP frequency, which are due to the enormous polarizability of the metal nanoparticle. This resonant field enhancement tends to enlarge the intrinsic optical nonlinearities of the nanoparticle material. Among the latter is the third-order nonlinearity related to optical interband transitions from the d-bands to empty conduction-band states, and the nonlinearity due to phase-space filling effects in the vicinity of the Fermi energy ("hot-electron effect") [8]. For very small clusters, a strong intraband nonlinearity due to quantum-confinements effects is expected [7].

The linear and nonlinear optical properties of the SP are crucially influenced by its limited lifetime. Decay, or damping, mechanisms of the SP have been extensively studied in the literature. Quantum mechanically, the SP can decay into electron-hole pair excitations (Landau damping) [9–11]. In a semiclassical picture, the decay is often described in terms of a loss of phase coherence of the collective electron oscillation due to scattering of single electrons participating in the oscillation [7]. Both pictures predict a strong correlation between the SP linewidth and the calculated rate of electronic surface scattering, which has indeed been observed experimentally [7]. All experimental studies of the SP decay in metallic structures have so far been performed in the low-excitation regime, where the energy optically transferred to the system is too small to strongly perturb the single-particle distribution [12]. In contrast, for high-intensity optical excitation the electron gas can reach very high temperatures, and conditions for SP decay may be very different. Evidently this physical situation is of high interest, both for fundamental reasons and with a view to nonlinear device applications where large energies are optically transferred to the system.

In the present paper, we study the damping of the SP in gold nanoparticles induced by high-intensity laser excitation of the electron gas [13]. We perform femtosecond pump-probe experiments on gold nanoparticles embedded in a dielectric sol-gel matrix. The excitation of the electron gas is provided by pulsed optical pumping of single particle interband transitions from the d band into s-p states far above the Fermi level. We observe a SP line broadening which builds up within ~ 1 ps after excitation. A similar behavior is observed for resonant excitation of the SP. In both cases, the broadening is the absolutely dominant optical nonlinearity in this intensity regime and reflects the excitation-induced damping of the SP resonance. The time evolution of the damping rate follows that of the electron temperature, showing that the damping rate is strongly influenced by transient variations in the electron scattering rate. The dependence of the damping rate on the excitation energy shows evidence for the relaxation of the optically generated d-band holes.

2   Linear Optical Properties

For the present work, gold nanoparticles have been prepared in a dielectric nanocomposite matrix by sol-gel processing [6]. One set of samples studied here consists of 2.6 μm thick sol-gel composite films which contain near-spherical gold nanoparticles with a mean diameter of 30 nm and a volume fill factor of 0.2%, embedded in a dielectric inorganic-organic sol-gel matrix of refractive index $n = 1.5$. The transmission electron microscopy (TEM) photograph in Fig. 2 shows an area of such a film. The optical density of this sample as a function of photon energy E is shown in Fig. 3 (solid line). The resonance of the SP lies at 2.29 eV, below the interband absorption edge at $E_{d-E_F} = 2.38$ eV, giving the films a reddish appearance. The gradual increase of the optical density for higher energies (dashed line in Fig. 3) is caused by single-particle interband transitions from the 5d to the 6sp band (Fig. 4) [14]. A second set of samples consists of gold nanoparticles embedded in a dielectric matrix of refractive index $n = 2.2$. The optical density of this sample as a function of photon energy is shown in Fig. 5 (solid line). The resonance of the SP lies at 1.96 eV, giving the films a bluish appearance. The comparison between the two kinds of samples illustrates very nicely how the spectral position of the SP in gold nanoparticles can be shifted by varying the dielectric matrix of the film.

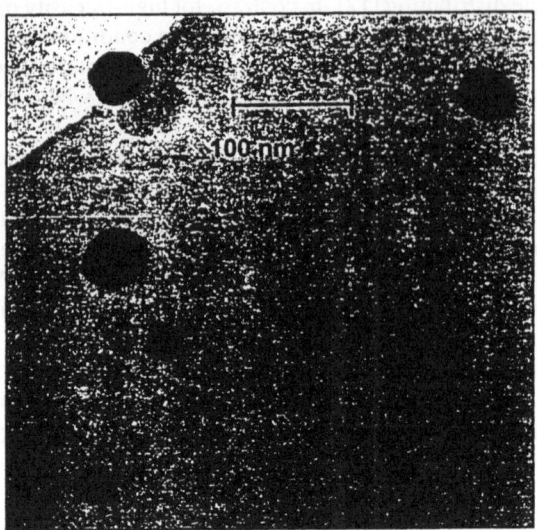

Figure 2
Transmission electron microscopy (TEM) photograph of an area of a nanocomposite film. Gold nanoparticles are shown as dark areas.

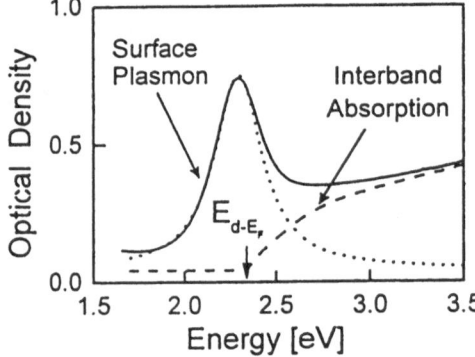

Figure 3
Optical density vs. photon energy (solid line) of gold nanoparticles embedded in a dielectric matrix material of refractive index 1.5. Dotted line: Lorentzian lineshape fit of the SP resonance. Dashed line: single-particle interband absorption obtained from subtracting the dotted line from the solid line; E_{d-E_F} marks the onset of the interband transitions.

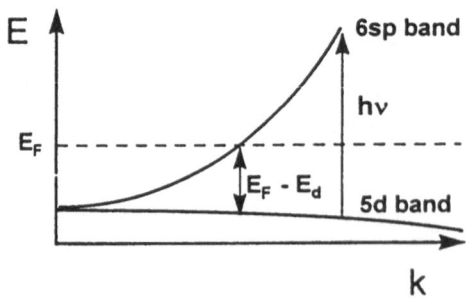

Figure 4
Schematic representation of interband transition due to an optical excitation. $E_F - E_d$ marks the energy difference between the Fermi energy and the 5d band.

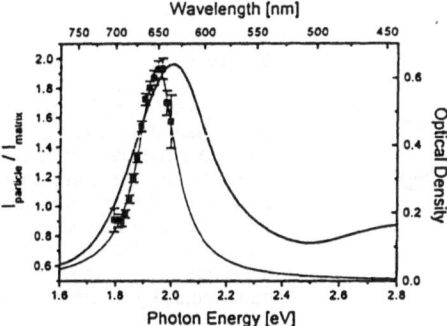

Figure 5
Optical density vs. photon energy (solid line) of gold nanoparticles embedded in a dielectric matrix material of refractive index 2.2. Squares: scattered light intensity from a single nanoparticle, normalized to the light intensity through the bare matrix. Thin line: Lorentzian fit to the data.

In both cases, the SP has a large spectral width, which is much larger than, e.g., excitonic linewidths in semiconductors. Essentially this a consequence of the very efficient damping of the SP resonance in the nanoparticle. Another contribution to the total linewidth is given by inhomogeneous broadening effects due to fluctuations of the nanoparticle size and local dielectric constant of the sol-gel matrix. This is illustrated in Fig. 5, which shows transmission data of single gold nanoparticles taken with a scanning near-field optical microscope (SNOM) as the light source [15]. The high spatial resolution of the SNOM allows to take light-scattering spectra of individual nanoparticles, and thus avoids the usual averaging over an inhomogeneous distribution that is characteristic of conventional far-field spectra. The thin line in Fig. 5, which is a Lorentzian fit to the SNOM data, shows an individual SP line. The comparison with the far-field absorption spectrum shows that the ratio of homogeneous linewidth to inhomogeneous broadening is about 1:1. This agrees with conclusions drawn by Aussenegg and collaborators on the ratio of homogeneous linewidth to inhomogenous broadening in silver nanoparticles [16]. A Lorentzian fit of the individual nanoparticle spectrum in Fig. 5 gives a homogeneous linewidth of $\Gamma_{\text{hom}} = 0.16$ eV.

This translates into a SP damping rate of $\gamma_{\text{SP}} = \Gamma_{\text{hom}}/2\hbar = (8 \text{ fs})^{-1}$ This very large value illustrates the highly efficient damping that is effective in the nanoparticle even at room temperature.

3 Pump-Probe Experiments: The Case of Interband Excitation

In this section we will show that high-intensity laser excitation of the nanoparticle leads to an enhanced damping of the SP. For this purpose, we perform time-resolved pump-probe experiments on the samples described in Section 2. The principle of the experiment is illustrated in Fig. 6. A short pump laser pulse excites the sample, and a spectrally broad second pulse probes the pump-induced changes in the optical transmission spectrum of the sample. Time resolution is obtained by delaying the probe pulse with respect to the pump pulse by a variable time delay.

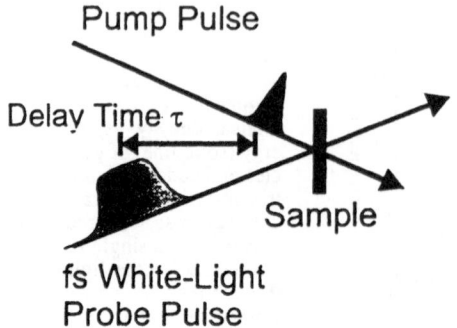

Pump Pulse

Delay Time τ

Sample

fs White-Light
Probe Pulse

Figure 6
Principle of the time-resolved pump-probe experiment: A short pump laser pulse excites the sample, and a spectrally broad second pulse probes the pump-induced changes in the optical transmission spectrum of the sample. Time resolution is obtained by delaying the probe pulse with respect to the pump pulse by a variable time delay τ.

We use the frequency-doubled output of a regenerative Ti:sapphire amplifier (1 kHz repetition rate) as the pump pulse and a white-light continuum as the probe pulse. The pump beam is focussed down to a spot diameter of 300 μm; the pump pulse has a duration of 200 fs, and is centered at 3.07 eV. Alternatively, pump pulses with a duration of < 200 fs and central photon energies between 2.08 eV and 2.6 eV are generated by a tunable optical parametric amplifier. The white-light continuum is generated by focussing part of the laser power into a water flow cell, and has a temporal duration of 200 fs. The probe pulse can be delayed with respect to the pump pulse by a variable time delay τ, and is spectrally analyzed in a monochromator after transmission through the sample. The differential transmission ΔT , i.e. the difference in transmission with and without pump pulse, is recorded as a function of probe photon energy $\hbar\omega$ at a fixed delay τ, or alternatively as a function of τ at a fixed probe photon energy. ΔT is thus sensitive to changes in the optical properties of the sample generated by the pump pulse.

A first set of experiments are performed with the pump pulse being situated at a central photon energy of 3.07 eV (vertical arrow in Fig. 7). The pump pulse thus excites electrons from the d band into s-p band states high above the Fermi level (see Fig. 4). This arrangement thus avoids any direct interaction of the pump pulse with the SP resonance. Normalized differential transmission spectra $\Delta T/T(\hbar\omega,\tau)$ are taken at a sample temperature of 35 K. A set of such spectra taken for a series of time delays τ is plotted in Fig. 7 versus the energy of probe photons $\hbar\omega$. We observe a pronounced nonlinearity at the spectral position of the SP, reaching peak values of up to $\Delta T/T \approx 15\%$. This nonlinearity consists of regions of induced transmission (center of the SP line) and induced absorption (wings), respectively. The symmetry of the nonlinearity with respect to the SP energy, and the observation that the spectrally-integrated induced transmission is approximately equal to the spectrally-integrated induced absorption both indicate that the nonlinearity is mainly caused by spectral broadening of the resonance. The differential transmission spectra are fitted to determine the magnitude of the line broadening. For simplicity a purely Lorentzian lineshape of the resonance is assumed [7]. This approach amounts to including in the Lorentzian lineshape any linewidth contributions from inhomogeneous broadening effects. This approximation is expected to work reasonably well for all our present purposes, for two reasons: the inhomogeneous broadening in our samples is relatively small in comparison with the homogeneous linewidth (see

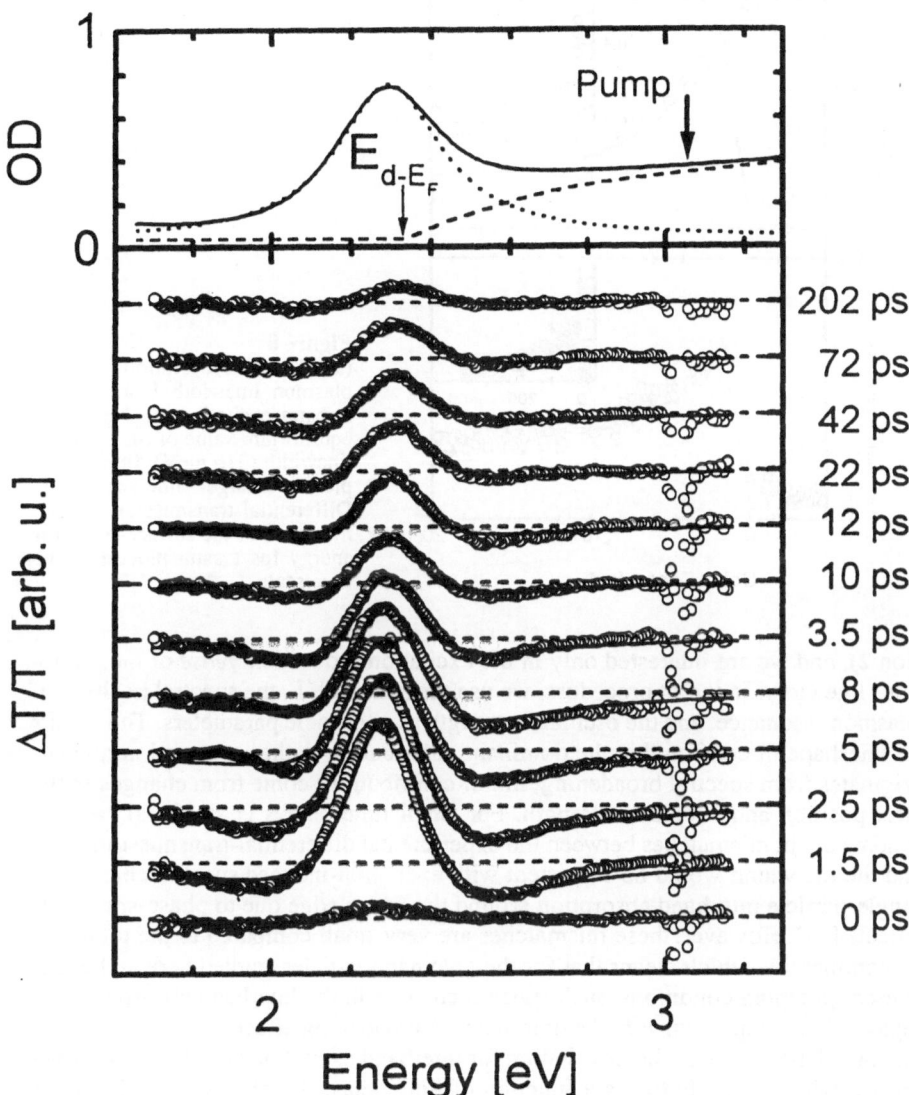

Figure 7 Differential transmission $\Delta T/T$ versus probe-photon energy at different time delays after interband excitation. Upper part: same as Fig. 3. The arrow indicates the central photon energy of the pump pulse.

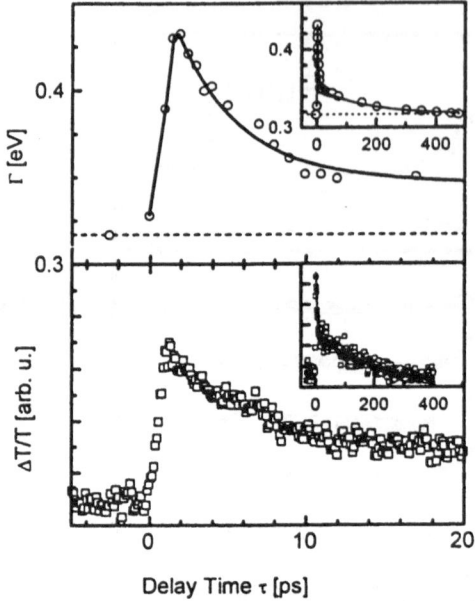

Figure 8
(a) Temporal evolution of the plasmon linewidth Γ after interband excitation; dashed line: equilibrium value of the plasmon linewidth (316 meV). Inset: same plot on a larger time scale. (b) Differential transmission $\Delta T/T$ measured at the surface-plasmon energy. Inset: same plot on a larger time scale.

Section 2), and we are interested only in the excitation-induced *increase* of the homogeneous linewidth. In the fit procedure, we use the linewidth Γ, the spectral position of the plasmon resonance, and the oscillator strength as adjustable parameters. The results of the lineshape fit confirm that the dominant contribution to the measured nonlinearity originates from spectral broadening; minor contributions come from changes in the spectral position and oscillator strength. For small time delays ($\tau < 10$ ps), we observe some slight mismatches between the experimental differential-transmission spectra and the fits which would be consistent with excitation-induced spectral changes in the single-particle interband absorption around the Fermi edge due to phase-space filling effects [17]. However, these mismatches are very small compared to the total size of the nonlinearity, which means that for the gold nanoparticles studied here and under the present pumping conditions, such spectral changes in the interband absorption can be neglected in comparison with the dominant SP broadening effect.

The time dependence of the broadening for interband excitation at 3.07 eV is shown in Fig. 8a, where Γ is plotted as a function of time delay. We observe a rapid initial rise of the linewidth from the equilibrium value of 316 meV to a maximum value of 432 meV. The duration of this time regime can be seen more clearly from Fig. 8b, where the induced transmission $\Delta T/T(\hbar\omega_{SP},\tau)$ measured at the SP energy is plotted versus time delay; the rise of $\Delta T/T(\hbar\omega_{SP},\tau)$, which is directly linked to that of the linewidth Γ, has a duration of ≈ 1 ps. The subsequent return of Γ to the equilibrium value is characterized by a two-component decay with time constants of 4 ps and ~ 200 ps, respectively.

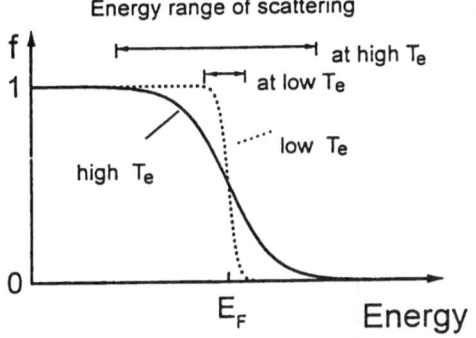

Figure 9
Schematic Fermi distribution functions at low (dashed line) and high (solid line) electron temperatures.

For an explanation of the time dependence of Γ, we note that a similar sequence of rapid initial rise and subsequent two-component decay has been observed in the electron temperature in metal films after femtosecond laser excitation [18]. Such temperature changes are accompagnied by variations in the electronic scattering rate [19]. It is expected that enhanced electron scattering will lead an accelerated loss of the phase coherence of the collective electron oscillation. We therefore suggest that *the increased line broadening shown in Fig. 8a is a result of SP damping due to an excitation-induced increase in the electronic scattering rate.* In more detail, our interpretation of the time dependence of the SP linewidth for interband excitation is as follows:

(i) The pump pulse induces single-particle transitions to 6sp-band states far above the Fermi level, and thus creates a nonthermal distribution of optically excited electrons. The electron gas thermalizes within ≈ 1 ps via electron-electron and electron-phonon scattering [18,20–22]. During the thermalization process, the excess energy of the optically excited particles is redistributed among the conduction band electrons. The Fermi edge smearing associated with this heating process (solid line in Fig. 9) results in an enhanced probability for electron scattering between states in the vicinity of the Fermi energy, due to the high density of both populated initial and unpopulated final states. The enhancement of the total electronic scattering rate leads to an increase of the SP decay rate that manifests itself as a line broadening in the differential transmission spectra in Fig. 7.

(ii) During and after thermalization, the highly excited electron gas loses energy to the gold lattice by phonon emission until the temperatures of the electron distribution and the lattice have equalized. Concomitant with this process is an increasing limitation of the electron scattering rate in accordance with the Pauli exclusion principle, due to the recovery of the energetic sharpness of the Fermi edge (dotted line in Fig. 9). The decay of the scattering rate leads to a decay of Γ with a time constant of 4 ps; this value is typical of equilibration processes between electron and lattice temperatures in metal colloids [22,23,25].

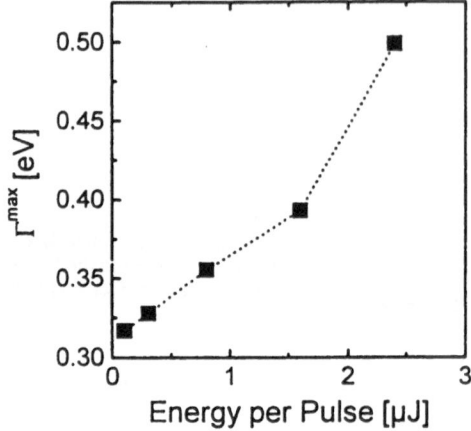

Figure 10
Maximum linewidth of the surface
plasmon resonance versus energy
per pump pulse for excitation at
3.07 eV pump-photon energy.

(iii) The temperature equilibration between the electron gas and the gold lattice leaves the joint system at a higher temperature than before the pump pulse; this is reflected in a non-vanishing line broadening at time delays $\tau > 10$ ps. The gradual decay of Γ with a time constant of 200 ps is explained by heat transfer from the gold colloids to the embedding matrix [23,25].

We can thus explain the time dependence of the SP linewidth with an excitation-induced increase in the SP damping rate. The linewidth Γ in Fig. 8a is observed to increase by as much as 120 meV during step (i), which corresponds to an increase of $\Delta\gamma_{SP} = (11 \text{ fs})^{-1}$ in the damping rate. This means that the excitation-induced increase in the SP damping rate is almost of the same size as the room-temperature damping rate expected from the SNOM spectra in Section 2.

In steps (i)–(iii), the observed time evolution of the SP damping rate follows that of the expected electron temperature. The magnitude of the electron heating is expected to depend on the total energy transferred onto the electron gas by the pump pulse. This expectation is indeed confirmed experimentally, as is shown in Fig. 10, where the line width reached at the end of step (i), Γ_{max}, is plotted versus the energy per pump pulse. Γ_{max} is clearly observed to increase with increased pump power, reflecting the fact that the electron gas reaches higher peak temperatures as the total energy optically transferred onto it increases. In Section 5, we will give an estimate of the peak electron temperature; however, before doing this, we need to clarify how large the transferred energy *per pump photon* is. This is the purpose of Section 4.

4   Electron Heating due to Hole Relaxation and Plasmon Decay

So far we have focussed on the heating of the electron gas by the optically excited electrons, ignoring in the discussion the photogenerated holes in the d-bands. It should be

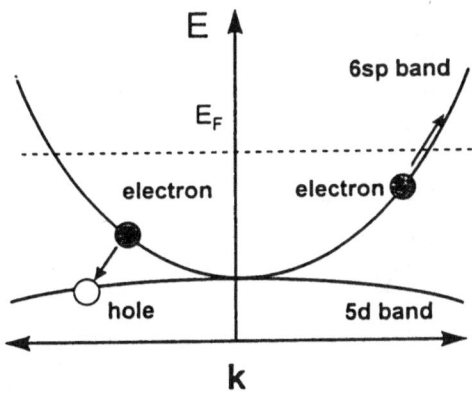

Figure 11
Schematic representation of the energy relaxation of d-band holes. The holes recombine with conduction band electrons via an Auger process in which the energy and momentum difference between electron and hole is transferred onto another conduction-band electron.

expected that these holes will relax in energy due to recombination with conduction band electrons via an Auger process in which the energy and momentum difference between electron and hole is transferred onto another conduction-band electron taking part in the process (Fig. 11). Indeed, very recently time-resolved two-photon photoelectron spectroscopy experiments have given indications of such an Auger process in silver and copper [24]. This raises the question on what time scales the hole relaxation in the present case occurs; in other words, how fast is the hole excess energy with respect to the Fermi energy released to the electron gas or the lattice? The answer to this question will ultimately allow us to estimate the peak electron temperature reached in the course of the heating process (see Section 5).

We have measured differential transmission transients at the spectral position of the SP resonance, $\Delta T / T(\hbar\omega_p, \tau)$. and varied the pump photon energy. At all pump photon energies, the transients exhibit a temporal profile that is essentially the same as that in Fig. 8b, reaching their maximum at the end of the thermalization process (not shown here). This maximum value, which we denote by DT_{max}, is directly linked to the SP linewidth, and thus gives a good measure of the peak electron temperature T_e^{max} reached at the end of the thermalization process. DT_{max} is plotted in Fig. 12 as dots versus pump photon energy, for the sample of Fig. 3; the linear optical density of the sample is shown for comparison. Let us focus for the moment on the part of the plot taken for interband excitation, i.e. at pump photon energies greater than $E_{d-E_F} = 2.38$ eV. For excitation at 2.60 eV, electrons are excited into conduction-band states that have relatively little excess energy with respect to the Fermi level, while for excitation at 3.07 eV an electron distribution with large excess energy is created by the pump pulse. If the electron excess energy with respect to the Fermi level was the only one to contribute to the electron heating, then T_e^{max}, and hence DT_{max}, should be much higher for excitation at 3.07 eV than at 2.60 eV. As the experimental values of DT_{max} show, this is not the case; the value found for excitation at 2.60 eV lies close to that at 3.07 eV. This experimental finding can only be explained if we assume that also the excess energy of the photoexcited d-band holes gives a substantial contribution to the electron heating during the first picosecond after excitation. In other words, *a large part of the excess energy of the*

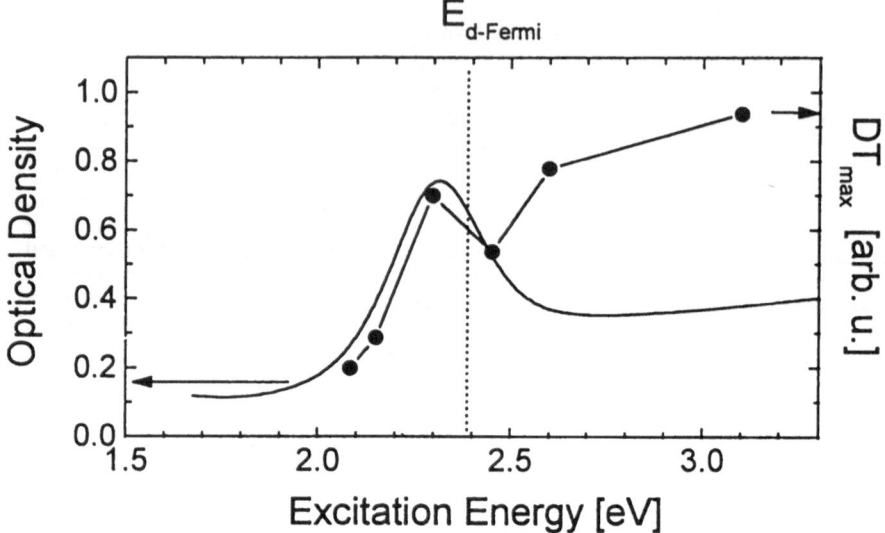

Figure 12 Dots show maxima of induced transmission DT_{max} for different pump-photon energies. For comparison, the optical density of the sample (refractive index of the matrix material: 1.5) is also shown.

photogenerated holes is transferred to the conduction-band electrons (and the lattice) via fast relaxation during the first picosecond after excitation.

We note that an essentially similar temporal dependence as in the case of interband excitation is observed for optical excitation at the spectral position of the SP (not shown here). This leads to the conclusion that also in this latter case, the SP damping is strongly influenced by electron scattering due to heating of the electron gas. However, an important difference is that here the electron gas is heated not by the energy of photogenerated electron-hole pairs (as in the interband excitation case) but by the energy released when the SP's excited by the pump pulse decay [26]. It is expected that the SP decays mainly into electron-hole pair excitations [11].

Another interesting difference in comparison with the interband excitation case becomes apparent when focussing on DT_{max} in Figs. 12 and 13. DT_{max} closely follows the measured optical density in the spectral region of the SP line. This is understandable since the total energy optically transferred to the nanoparticle is expected to be proportional to the number of absorbed photons. It is remarkable, however, that DT_{max} strongly deviates from the optical density when pumping takes place in the spectral range of the interband transitions. This means that pumping at interband transition frequencies results in a more efficient SP damping than at the SP resonance. This difference can be explained when we assume that part of the energy stored in the SP mode by the resonant pump pulse is not directly released to the electron gas but is relaxed through different dissipative channels (e.g. via SP-phonon coupling). An additional explanation may be

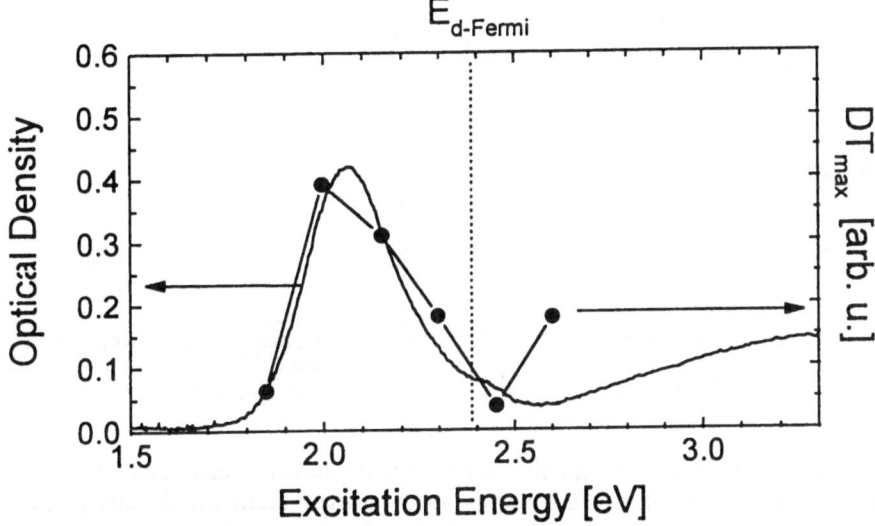

Figure 13 Dots show maxima of induced transmission DT_{max} for different pump-photon energies. For comparison, the optical density of the sample (refractive index of the matrix material: 2.4) is also shown. Here the optical density is from a different sample than in Fig. 5, and hence exhibits slight deviations from the spectrum shown there.

that part of the incident pump power is scattered off the nanoparticle rather than being absorbed by the SP resonance and so cannot contribute to the electron heating. However, Mie scattering calculations show that for the present nanoparticle size and matrix material, the scattered power should be only 30% of the absorbed power, and therefore be negligible.

Our explanation of the SP line broadening also casts a new light on observations made by other groups. In Refs. [22] and [25], transient broadenings of the SP line have been observed after optical excitation at the low-energy tail of the SP resonance in pump-and-probe experiments on noble-metal nanoparticles. While these studies include no quantitative analysis of the SP linewidth on which to base an unequivocal judgement, the published spectra point towards a possible interpretation of the line broadenings in terms of an excitation-induced SP damping, in analogy to the case discussed in the present paper.

5 Estimate of the Electron Temperature and the Electron-Electron Scattering Rate

In Section 3, we related the observed time dependence of the SP damping rate to that of the electron temperature. This understanding has been of a rather qualitative nature;

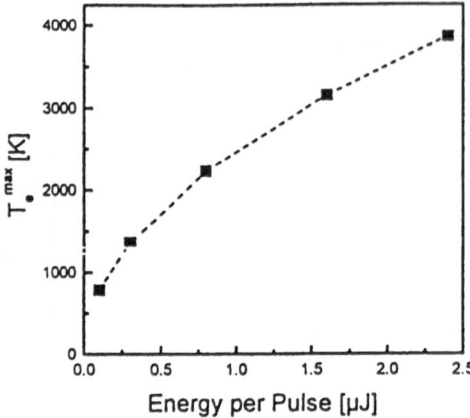

Figure 14
Estimated peak electron temperature T_e^{max} as a function of the energy per pulse.

we have so far provided no quantitative information on the actual electron temperatures reached in the experiment. It is the purpose of the present section to fill this gap, and draw a comparison between the measured SP damping rate and the estimated electron scattering rate. We restrict ourselves here to the case of interband excitation at 3.07 eV pump-photon energy; this is the situation treated in Section 3.

We start our discussion by considering the case of the highest pump power used in the experiment, i.e. 2.4 μJ per pulse. Taking into account the pump fluence, the measured optical density at the pump wavelength, the film thickness, and the known fill factor of the nanoparticles in the film, we calculate that each pump pulse excites $\sim 3 \cdot 10^4$ electrons per nanoparticle from the d bands into the sp band, leaving behind an equal number of holes in the d bands. In Section 4, we had concluded that a substantial part of the excess energy of the photogenerated holes is transferred to the conduction-band electrons (and the lattice) via fast relaxation processes during the first ps after excitation. For simplicity we will assume in the present estimate that this actually applies to all the hole excess energy. This means that the entire energy of each photogenerated electron-hole pair is transferred to the electron gas (and the lattice) during the first picosecond after excitation. Using the heat capacity of the electron gas as taken from Ref. [26], we then estimate that the thermalized electron gas has a peak temperature of $T_e^{max} \approx 3900$ K at the end of step (i). This value is obtained by neglecting the heat transfer to the gold lattice during step (i), and thus represents an upper limit on the electron temperature; a lower limit of ≈ 3500 K is obtained by assuming instantaneous thermalization and thus a maximum heat transfer rate during the first picosecond after excitation. The true value of T_e^{max} will lie somewhere between these two limits, and will depend on the details of the thermalization process [26]. In the same way, T_e^{max} is estimated also for the other pump powers used in the experiment. The pump-power dependence of T_e^{max} is shown in Fig. 14; it is not strictly linear, due to the temperature dependence of the electron heat capacity. Finally, we can also estimate the temperature of the nanoparticle after thermal equilibration between the electron gas and the gold lattice, i.e. at the end of step (ii). We obtain a temperature of ≈ 300 K at the highest pump power; this is much lower than the corresponding value of T_e^{max}, due to the much greater heat capacity of the gold lattice in comparison with that of the electron gas.

In Section 3, we explained the time dependence of the SP damping rate with an increase in the electron scattering rate induced by the heating of the electron gas. In the following, we will discuss how large the total increase in the electron scattering rate is expected to be. The Fermi edge smearing at high electron temperatures results in an enhanced probability for electron scattering between states around the Fermi energy, due to the high density of both populated initial and unpopulated final states (see Fig. 9). The scattering processes that are expected to be strongly enhanced at high electron temperatures are electron-electron (e–e), electron-phonon (e–ph), and electron-surface (e–surf) scattering. According to Fermi liquid theory, the electron-electron scattering rate τ_{ee}^{-1} for an electron at excess energy $\varepsilon = E - E_F$ above the Fermi level is given by

$$\frac{1}{\tau_{ee}} = K \frac{(\pi k_B T_e)^2 + \varepsilon^2}{1 + \exp(-\varepsilon/k_B T_e)} \qquad (5.1)$$

where $K = 0.1\ \mathrm{eV^{-2}\ fs^{-1}}$ is the e–e scattering-rate constant according to Ref. [26], and T_e is the temperature of the conduction band electrons. To gain some simple quantitative measure of the e–e scattering rate, we define an average e–e scattering rate as $\langle \tau_{ee}^{-1} \rangle = \int_0^\infty f(\varepsilon) \tau_{ee}^{-1}\, d\varepsilon / \int_0^\infty f(\varepsilon)\, d\varepsilon$, where $f(\varepsilon) = 1/(1 + \exp(\varepsilon/k_B T_e))$ is the Fermi function. From Eq. 5.1, we calculate an average e–e scattering rate of $\langle \tau_{ee}^{-1} \rangle = (10\ \mathrm{fs})^{-1}$ for the hot electron distribution at $T_e^{max} = 3900$ K. Conversely, the electron-phonon (e–ph) scattering rate is on the order of $\sim (100\ \mathrm{fs})^{-1}$ [27]. In contrast, the number of e–e and e–ph scattering events under equilibrium conditions at 35 K is negligibly small. The third electronic scattering process which can be strongly enhanced at high electron temperatures is electron-surface scattering, due to the high density of both populated initial and unpopulated final states at the same energy (see Fig. 9). However, the scattering rate associated with this process is presently difficult to quantify at high electron temperatures. We are thus left with the conclusion that the excitation-induced increase in the total electronic scattering rate is $\sim (10\ \mathrm{fs})^{-1}$ or even larger, and thus lies in the same range as the observed increase of $\Delta \gamma_{SP} = (11\ \mathrm{fs})^{-1}$ in the SP damping rate at the end of step (i) (see Section 3). This agreement is significant in that it emphasizes the strong correlation between the scattering of (single) electrons and the damping of the (collective) SP excitation at high electron temperatures. From a theoretical point of view, such a strong correlation is expected since high electron scattering rates facilitate the decay of the SP into an electron-hole pair of different energy, by providing the required energy difference via a scattering event.

6 Summary

We have studied the surface-plasmon decay induced by high-intensity optical excitation of the electron gas in gold nanoparticles. A pronounced broadening of the surface plasmon line in gold nanoparticles is observed both in the case of excitation of single-particle interband transitions, and for direct excitation of the surface plasmon. This line broadening is by far the dominant optical nonlinearity of the system, and reflects the excitation-induced damping of the surface plasmon resonance. The time evolution of the damping rate follows that of the electron temperature, reflecting that the damping rate is strongly enhanced by a transient increase in the electron scattering rate. The dependence of the damping rate on the excitation energy shows evidence for the relaxation of the optically generated d-band holes. We expect these results to be of importace for the understanding of the surface-plasmon decay in all situations where large energies are optically transferred to the electron gas, e.g. in nonlinear optical devices.

Acknowledgements

It is a pleasure to thank M. Perner, P. Bost, and T. Klar for their invaluable experimental contributions and many exciting discussions. Special thanks are due to J. Feldmann for initiating this work, and for many helpful discussions and suggestions. I also wish to thank U. Lemmer for his help with setting up the femtosecond pump-probe experiment, and S. Grosse for setting up the scanning near-field optical microscope. Useful discussions with W. Spirkl, and expert technical assistance by W. Stadler and M. Preis, are gratefully acknowledged. The time-resolved pump-probe experiments have been performed in the laboratories of E. O. Göbel at the University of Marburg. Finally I would like to thank U. Becker, M. Mennig, M. Schmia, and H. Schmidt of the Institut für neue Materialien in Saarbrücken for providing the nanoparticle samples.

Bibliography

[1] W. Gotschy, K. Vonmetz, A. Leitner, and F.R. Aussenegg, Opt. Lett. 21, 1099 (1996).

[2] R. Elghanian, J.J. Storhoff, R.C. Mucic, R.L. Letsinger, C.A. Mirkin, Science 277, 1078 (1997).

[3] C. Flytzanis, F. Hache, M.C. Klein, D. Ricard, and P. Roussignol, Progress in Optics 29, 321 (1991).

[4] C. Flytzanis, F. Hache, M.C. Klein, D. Ricard, and P. Roussignol, Progress in Optics 29, 321 (1991).

[5] R.F. Haglund, R.H. Magruder, L. Yang, K. Becker, J.B. Wittig, and R.A. Zuhr, Opt. Lett. 18, 373 (1993).

[6] M. Mennig, U. Becker, M. Schmitt, and H. Schmidt, in 8th CIMTEC, Florence, 1994, Adv. Materials in Optics, Electro-Optics and Communication Technologies (Techna).

[7] U. Kreibig and M. Vollmer, Optical Properties of Metal Clusters (Springer-Verlag, Berlin, 1995)

[8] F. Hache, D. Ricard, C. Flytzanis, and U. Kreibig, Applied Physics a – Solids and Surfaces 47, 347 (1988).

[9] D. Pines, Rev. Mod. Phys. 28, 184 (1956).

[10] D.B.T. Thoai and W. Ekardt, Sol. St. Comm. 41, 687 (1982).

[11] T. Inagaki, K. Kagami, and E.T. Arakawa, Appl. Opt. 21, 949 (1982).

[12] see, e.g., D. Steinmüller-Nethl, R. A. Höpfel, E. Gornik, A. Leitner, and F. R. Aussenegg, Phys. Rev. Lett. 68, 389 (1992)

[13] Some of the results reported here have been published in: M. Perner, P. Bost, U. Lemmer, G. von Plessen, J. Feldmann, U. Becker, M. Mennig, M. Schmitt, and H. Schmidt, Phys. Rev. Lett. 78, 2192 (1997).

[14] R. Rosei and D.W. Lynch, Phys. Rev. B 5, 3883 (1972).

[15] T. Klar, M. Perner, G. von Plessen, S. Grosse, W. Spirkl, and J. Feldmann, (to be submitted).

[16] J.R. Krenn, W. Gotschy, D. Somitsch, A. Leitner, and F.R. Aussenegg, Appl. Phys. A 61, 541 (1995).

[17] R.W. Schoenlein, W.Z. Lin, J.G. Fujimoto, and G.L. Eesley, Phys. Rev. Lett. 58, 1680 (1987).

[18] C.K. Sun, F. Vallee, L.H. Acioli, E.P. Ippen, and J.G. Fujimoto, Phys. Rev. B 50, 15337 (1994).

[19] D.-S. Kim, J. Shah, J.E. Cunningham, and T.C. Damen, Phys. Rev. Lett. 68, 2838 (1992).

[20] W.S. Fann, R. Storz, and H.W.K. Tom, Phys. Rev. B 46, 13592 (1992).

[21] J.Y. Bigot, J.C. Merle, O. Cregut, and A. Daunois, Phys. Rev. Lett. 75, 4702 (1995).

[22] T.S. Ahmadi, S.L. Logunov, and M.A. Elsayed, J. Phys. Chem. 100, 8053 (1996).

[23] T.W. Roberti, B.A. Smith, and J.Z. Zhang, J. Chem. Phys. 102, 3860 (1995).

[24] E. Knoesel, A. Hotzel, T. Hertel, G. Neuhold, M. Wolf, and G. Ertl, poster presented at the 'Workshop on Ultrafast Surface Dynamics' in Ascona, March 1997.

[25] T. Tokizaki, A. Nakamura, S. Kaneko, K. Uchida, S. Omi, H. Tanji, and Y. Asahara, Appl. Phys. Lett. 65, 941 (1994).

[26] R.H.M. Groeneveld, R. Sprik, and A. Lagendijk, Phys. Rev. B 51, 11433 (1995).

[27] C. Suarez, W.E. Bron, and T. Juhasz, Phys. Rev. Lett. 75, 4536 (1995).

Indirect Exchange Coupled Magnetic Multilayers

Horst Hoffmann and Rüdiger Scherschlicht

Institut für Experimentelle und Angewandte Physik der Universität Regensburg

Abstract: The indirect exchange coupling between ferromagnetic layers, separated by non magnetic metal layers of a few monolayer thickness was first found by Grünberg et al. [1] in Fe/Cr multilayers. In addition Baibich et al. [2] reported a Giant Magneto Resistance (GMR) in these systems. Since then a new field in the research of very thin magnetic films has been started because of the great interest in the coupling phenomena for basic research as well as for application. Indirect exchange coupling was also found in multilayer systems of ferromagnetic layers (Fe) and Rare Earth Metals (Tb) separated by non magnetic metal interlayers. As in the GMR systems the coupling strength oscillates with increasing thickness of the interlayers. The oscillation period depends on the material of the interlayers.

1 Introduction

Quantum well states in semiconductor multilayers are well known for a long time. Quantum well states exist, as soon as the thicknesses of the single layers are comparable to the Fermi wavelength. Such states should be expected also in metal layers as soon as their thickness comes down to the metal Fermi wavelength, which is much smaller than that of semiconductors. Modern vacuum technology enabled to produce continuous metal films of only a few monolayer thickness. This opened a new field of research in magnetic films and multilayers. It was not surprising that unexpected new phenomena were observed. One of these is the indirect exchange coupling of ferromagnetic layers separated by non magnetic metal interlayers.

Figure 1 gives schematically an example of a periodic multilayer system with ferromagnetic layers separated by non-magnetic metal interlayers. Depending on the thickness of the interlayers the magnetic moments of the ferromagnetic layers are parallel (Fig. 1b) or antiparallel (Fig. 1a) aligned.

Similar effects of indirect exchange coupling were found in periodic multilayers of Fe and Tb layers separated by non magnetic metal layers. In all cases the strength of the indirect exchange coupling oscillates with increasing thickness of the interlayers. The period length of the oscillation depends on the interlayer material and is in the range of 1 nm. In the case of multilayer systems with ferromagnetic layers the sign of the coupling strength oscillates also, while in the case of the Fe/Tb systems such change of the sign was not observed.

Figure 1 Coupling between ferromagnetic layers separated by non magnetic metal layer.
a) antiparallel coupling, b) parallel coupling.

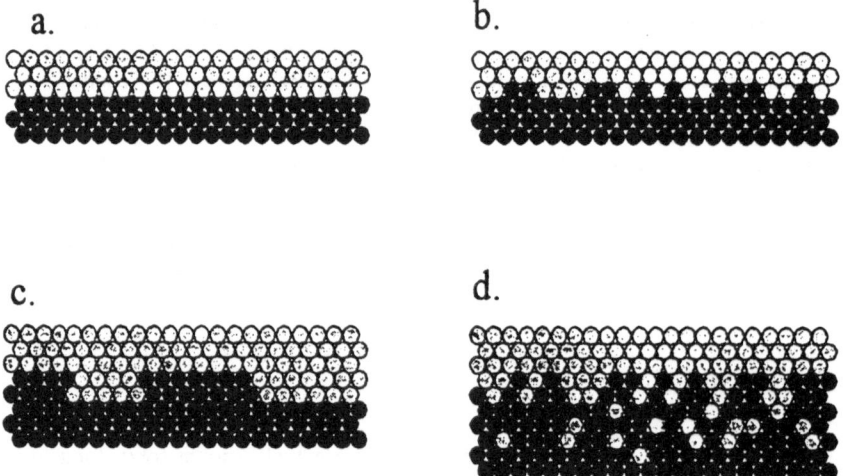

Figure 2 Characterization of interfaces. a) ideally smooth, b) microscopic roughness
(atomic scale), c) mesoscopic roughness (lateral coherence given by the crystallite size), d)
intermixing.

The thickness of each layer, especially of the interlayers is in the range of 0.5–5 nm,
i.e. this thickness can be counted in terms of monolayers. In such systems the roughness
of the interfaces cannot be neglected. The ideal interface is smooth on atomic scale
(Fig. 2a). Three typical roughnesses should be distinguished

1. microscopic roughness with lateral coherence on atomic scale (Fig. 2b),

2. mesoscopic roughness with lateral coherence length equals to the crystallite size
   (10nm) in polycrystalline layers (Fig. 2c),

3. intermixing at the interface (Fig. 2d).

Each special type of interface roughness influences the properties of the multilayer

system in a different way. The volume fraction of the system, including interface roughness, cannot be neglected. Therefore, special care has to be taken to investigate quantitatively the interface roughness by independent experiments.

2   The Indirect Exchange Coupling between Ferromagnetic Layers Separated by Non Magnetic Interlayers

The antiparallel coupling of the magnetic moments of Fe layers separated by Cr layers
(Fig. 1a) was first observed by Grünberg in investigations by Brillouin light scattering [1]. About simultaneously, Baibich et al. [2] reported a giant magnetoresistance
(GMR), observed in Fe/Cr/Fe multilayers. GMR is always observed for antiparallel
coupled multilayer systems. Brilloun light scattering (BLS) and GMR are the favorite
methods for investigating the indirect exchange coupling. By these methods, it was
found that the amplitude of the coupling strength (exchange constant) depends on the
material of the interlayers and that the sign of the coupling oscillates with increasing
thickness of the non magnetic interlayers [3–6].

2.1   Models of Indirect Exchange

The RKKY (Ruderman-Kittel-Kasuda-Yosida) indirect exchange: Magnetic ions polar-
ize electrons in their neighbourhood. The polarization oscillates in strength and sign
with increasing distance r from the magnetic ion (Fig. 3). The oscillation amplitude decreases with r^3. In the case of a ferromagnetic layer the decrease is with r^2. The oscillation period is half the Fermi wavelength of the free electrons. This short period needs
experiments, in which the interlayer thickness varies in steps of a monolayer. Especially
in Fe/Cr/Fe multilayer the short period could be detected by Brillouin light scattering

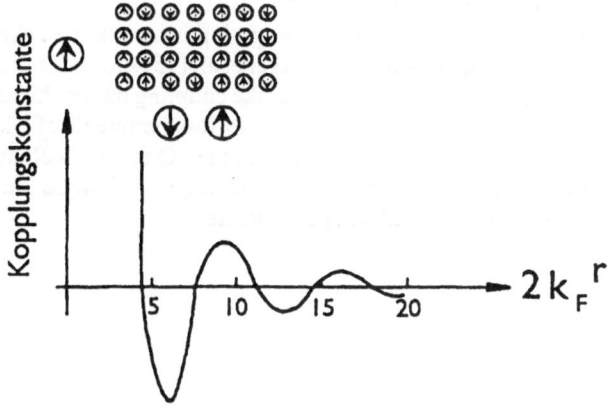

Figure 3
RKKY Coupling

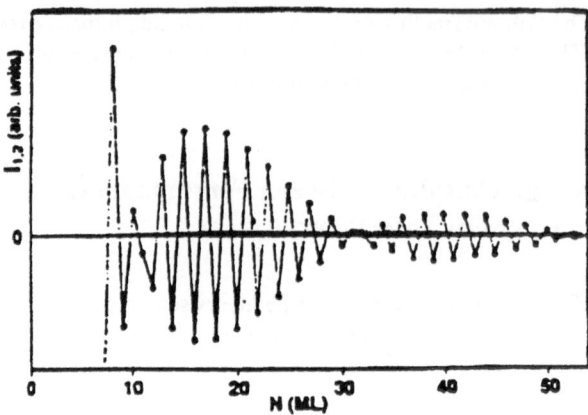

Figure 4
Modulation of
RKKY Cou-
pling [10].

[7] and spin polarized photoemission [8,9]. The usually investigated oscillation period by GMR is in the range of 1 nm. Bruno and Chappert [10] introduced irregularities at the interfaces and found that the interface roughness modulates the RKKY oscillations, leading to the short RKKY oscillations and an additional oscillation with periods of about 1 nm as observed in many experiments (Fig. 4). Similar theoretical results are reported by other authors [11]. According to this model the coupling depends only on the thickness of the interlayers and not on the thickness of the ferromagnetic layers.

In a later paper Bruno [12] proposed to discuss the influence of the interlayer by the quantum size effect. The interlayer is a quantum well, imbedded in the potential of the neighboured ferromagnetic layers. Spin up and spin down electrons are differentially reflected at the interface. Depending on the thickness of the interlayer the density of spin up and spin down electrons at the Fermi energy oscillates. This gives the oscillation of the sign of the coupling.

Spin polarized quantum well states have been found in spin polarized photo emission experiments of Cu layers on Co layers by Ortega et al. [13]. Similar results of Cu layers on Fe, Ni, Co layers were reported by Kläsges et al. [14].

The thickness of the ferromagnetic layers under investigation is also in the range to provide for quantum well states. Calculations by Wildberger et al. [15] showed that such states should exist leading to an oscillatory dependence of the coupling on the thickness of the ferromagnetic layer. This result is in agreement with GMR experiments of Okuno et al. [16] (Fig. 5), but not included in the original RKKY model. Quantum well states in the ferromagnetic layers as well as in the non-magnetic interlayers might be the key to explain indirect exchange coupling in the multilayer systems.

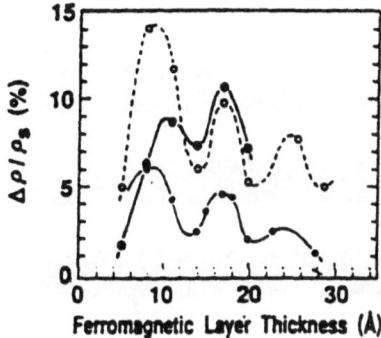

Figure 5
Oscillation of GMR with Fe thickness [16].

3 The Giant Magneto Resistance (GMR) of Exchange Coupled Films

The resistance of antiparallel coupled (Fig. 1a) multilayers is larger than that of parallel coupled layers (Fig. 1b). By applying a strong magnetic field in the plane of multilayers the magnetic moments of the ferromagnetic layers can be aligned along the magnetic field as soon as the strength of the applied field overcomes the indirect coupling field. In such strong fields all magnetic moments of the layers are orientated parallel to each other (Fig. 1b). The resistivity decreases with increasing applied field and saturates at a minimum when the moments are parallel aligned. The GMR ratio is defined by

$$\frac{\Delta R}{R_{\min}} = \frac{R_{\text{antiparallel}} - R_{\text{parallel}}}{R_{\text{parallel}}}$$

The field, necessary to overcome the antiparallel indirect exchange field is very large and not available in all laboratories. Therefore, in the literature the GMR ratio usually is defined by

$$\frac{\Delta R}{R} = \frac{R_{\max} - R_{\min}}{R_{\min}}$$

where $R_{\max}$ is the resistance at antiparallel orientation and $R_{\min}$ is the resistance at maximum applied field.

The GMR is due to spin dependent scattering of electrons. A very simple explanation is given in Fig. 6. Electrons with spin directions opposite to the direction of the magnetic moment of the ferromagnetic layer are scattered at the surface or inside the layer. Electrons with spins parallel to the magnetic moments are not scattered. The overall mean free path depends on the orientation of the magnetic moments in the multilayer system. In layers with antiparallel oriented magnetic moments (Fig. 6a) the mean free path of both spin directions is limited by the period length of the multilayer system. Parallel alignment of the magnetic moments by external fields (Fig. 6b) reduces the spin scattering of electrons with spins parallel to the magnetization of the layers. Their

a)   b)

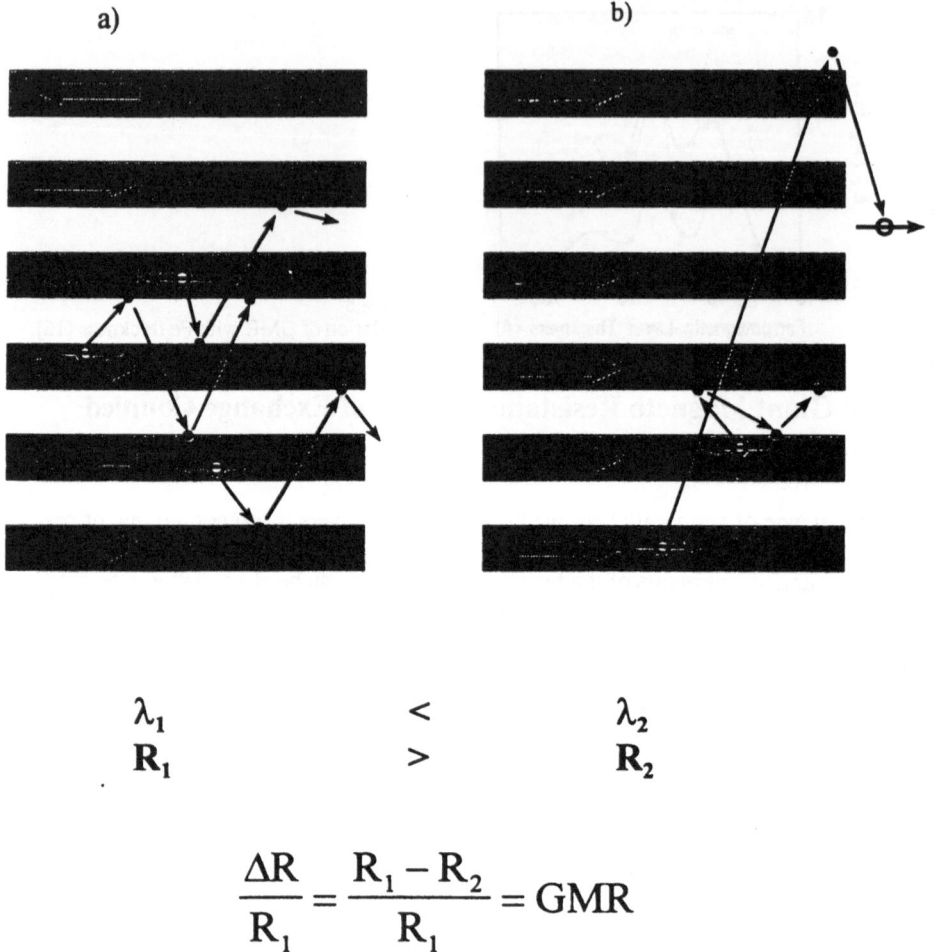

$$\lambda_1 \qquad\qquad < \qquad\qquad \lambda_2$$

$$R_1 \qquad\qquad > \qquad\qquad R_2$$

$$\frac{\Delta R}{R_1} = \frac{R_1 - R_2}{R_1} = GMR$$

Figure 6 Model of spin dependent scattering. a) antiparallel magnetic moments, b) parallel magnetic moments.

mean free path is now given by defect and phonon scattering which is much larger then the period length of the multilayer films. The resistivity is reduced. GMR cannot be observed in parallel coupled films. Therefore GMR is a measure for the type and strength of antiparallel coupling.

Parkin [3,5,6] investigated a great number of multilayer systems with various ferromagnetic transition metals and various non magnetic metal interlayers. He found that the GMR, i.e. the coupling, oscillated with decreasing amplitude at increasing thickness of the interlayers. The length of the oscillation period (about 1 nm) depends on the interlayer material. His experimental results were supported by many other investigators.

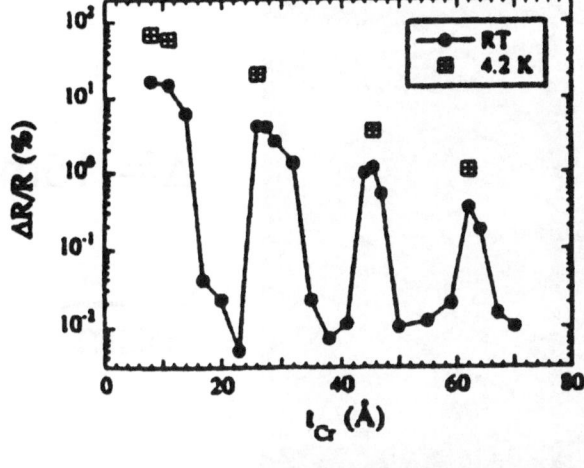

Figure 7
Oscillation of GMR with increasing thickness of the Cr interlayer [4].

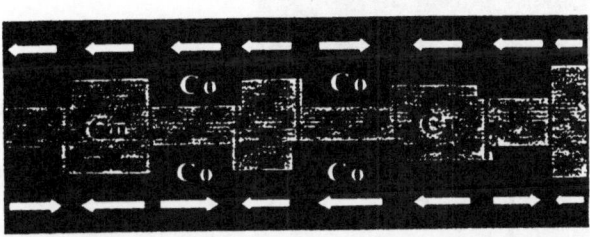

Figure 8
Thickness fluctuation due to mesoscopic interface roughness. The multilayer includes parallel and antiparallel coupled magnetic moments.

Figure 7 [4] gives an example of the oscillation of GMR in Fe/Cr multilayer films. The large amount of the GMR (50 to 100 %) is of interest for application. Unfortunately, large fields have to be applied, the sensitivity is low for very sensitive magnetic sensors. Much research is devoted to increase the sensitivity.

The reported values of the amount of GMR in polycrystalline multilayers differ strongly. This is thought to originate from the preparation conditions, leading to different interface roughness. Microscopic interface roughness acts on the electron scattering. We found that the mesoscopic roughness influences drastically the amount of GMR [17]. Due to the fluctuation of the interlayer thickness the multilayer film includes parallel and antiparallel coupled areas (Fig. 8). The parallel coupled fraction decreases the maximum possible value of GMR. Lorentz microscopy of such films (Fig. 9) show the different coupling areas. The related magnetoresistance curve shows hysteresis (Fig. 10).

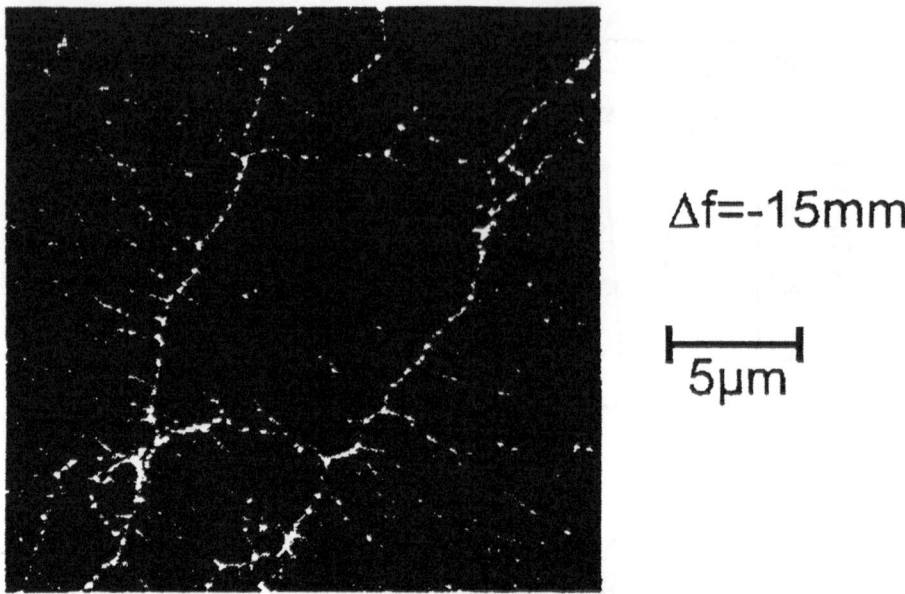

Figure 9 Lorentz microscopy of a polycrystalline Co/Cu sandwich. The part in the center is antiparallel coupled, the other parts are parallel coupled.

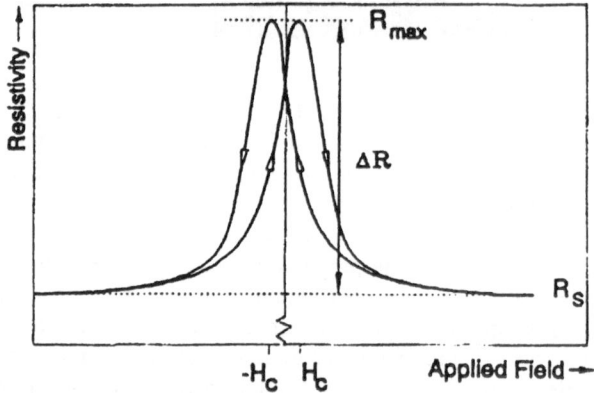

Figure 10
Magnetoresistance
curve with hystere-
sis [Co(1.1 mm)/
Cu(2.0 mm)],
$\frac{\Delta R}{R} = 20\%$.

4 Indirect Exchange Coupling between Ferromagnetic (Fe) Layers and Paramagnetic (Tb) Layers, Separated by Non Magnetic Metal Layers (Pt, Ta, Cu, Au)

4.1 Homogeneous Amorphous FeTb Layers

The exchange coupling between Fe and neighbouring Tb atoms is known to be antiparallel. Homogeneous amorphous $Fe_{1-x}Tb_x$ layers with $x > 0.23$, therefore, are fer-

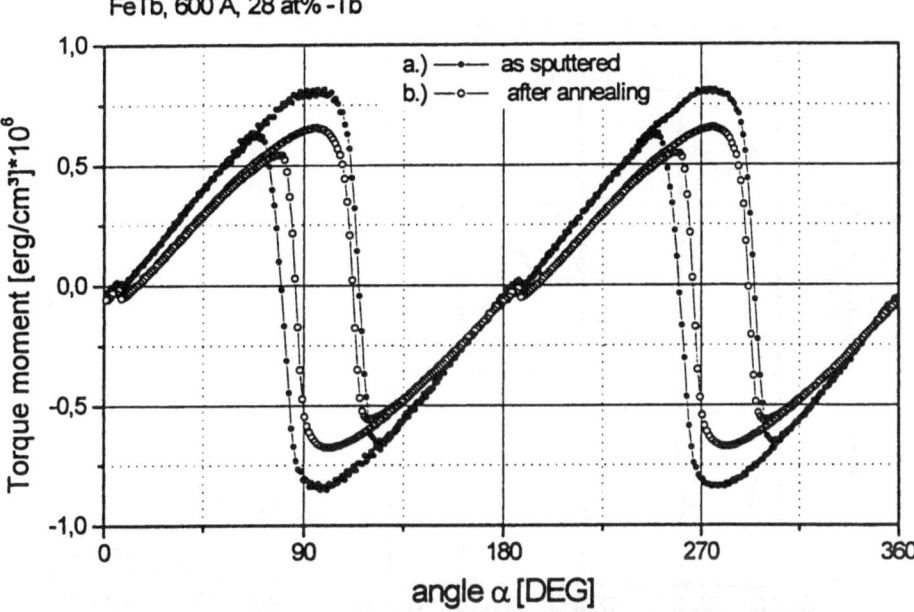

Figure 11 Torque curves of an amorphous $Fe_{72}Tb_{28}$ film, a) before annealing b) after 2 h annealing at 140 °C in an in-plane field of 1 040 kA/m [18].

rimagnetic materials. Sputter deposited $Fe_{1-x}Tb_x$ films show a perpendicular magnetic anisotropy, i.e. the easy axis of magnetization is oriented perpendicular to the plane of the film. This effect makes $Fe_{1-x}Tb_x$ films desirable for magneto-optical recording. A number of hypotheses are offered in the literature to explain the origin of the perpendicular anisotropy. One favorite model is the (Neel-Taniguchi) pair ordering model in which the anisotropy of the directional distribution of Fe-Tb pair axis is responsible for the magnetic anisotropy. The density of neighbouring Fe and Tb atoms along a selected direction can be seen from the pair distribution function, evaluated from the intensities of electron diffraction.

Annealing of amorphous $Fe_{1-x}Tb_x$ films in fields parallel to the film plane decreased the perpendicular anisotropy, simultaneously the density of Fe-Tb pairs parallel to the annealing field increased [18], while the Fe-Tb pair density perpendicular to the annealing field decreased.

Figure 11 shows the torque curve of a $Tb_{28}Fe_{72}$ film in the as sputtered case (a) and after 2 h annealing at 140 °C in fields parallel to the film plane (b). The maxima of the curves give the perpendicular anisotropy constant. It has been decreased by 20% during annealing. The decrease is caused by reorientation of the pair axes along the in-plane annealing field, decreasing the number of pairs perpendicular to the film. The in-plane distribution was determined from the pair distribution functions $g_x(r)$, $g_y(r)$ in two directions x, y, perpendicular to each other in the film plane. The difference $g_x(r) - g_y(r)$

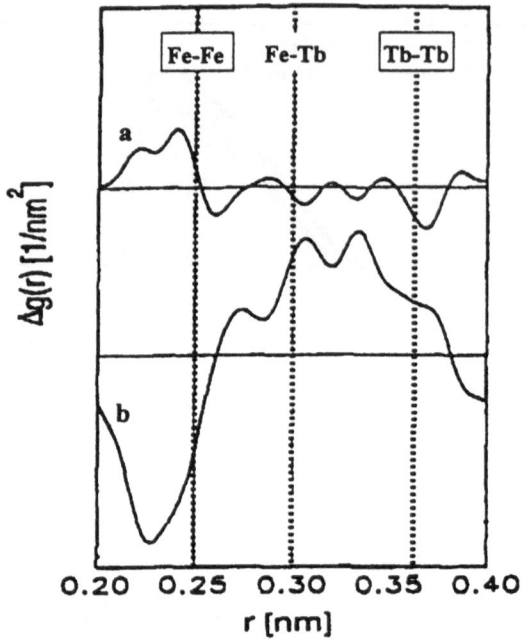

Figure 12
Pair distribution function $g_x(r) - g_y(r)$, a) before annealing, b) after annealing at 140 °C in a field of 1 040 kA/m parallel to the x-axis [18].

gives the inhomogeneity of the distribution. As shown in Fig. 12, before annealing the in-plane distribution is isotropic $(g_x(r) - g_y(r) = 0)$ (The structure $g_x(r) - g_y(r)$ to be seen in the Fig. 12a gives the uncertainty of the experiment). After annealing more Fe-Tb pairs are found along the x-axis, which is the direction of the field during annealing, than along the y-axis, as can be seen from $g_x(r) - g_y(r)$ in Fig. 12b. Because of the symmetry we assume that the number of pairs perpendicular to the film plane (z-axis) has been decreased in the same way like in the y-direction. This reorientation of pairs causes the decrease of the perpendicular anisotropy.

4.2 Fe/Tb Multilayers

The disk like multilayers with 17 mm in diameter were sputtered in a static mode onto glass substrates. The sputter deposition was done by sputtering from separated Fe- and Tb-targets. Special screening systems provided for deposition of only one element at the same time. The transmission electron microscopy of cross sections (Fig. 13) showed a clear layered structure. But, unfortunately, this method, applied to amorphous or polycrystalline layers cannot resolve interface roughness with atomic resolution. A method of two dimensional evaluation of the pair distribution function from the intensities of electron diffraction has been developed. Application of this method showed very clearly that the multilayers under investigation had sharply separated interfaces at short

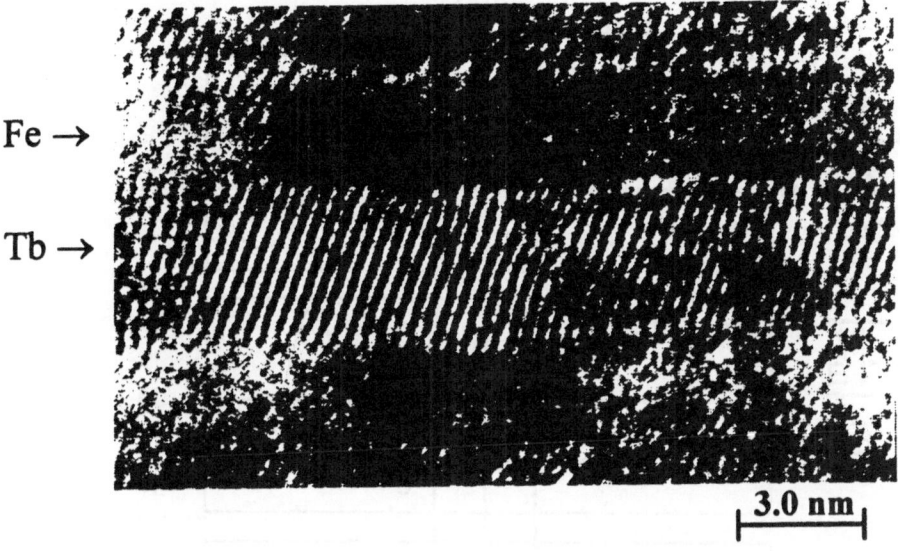

Fe →

Tb →

3.0 nm

Figure 13 TEM image of the cross section of a Fe/Tb multilayer film.

range order. Figure 14 compares the distribution of Fe-Tb pairs parallel to the interface of Fe/Tb multilayers (Fig. 14a,c) and the distribution in homogeneous amorphous $Fe_{68}Tb_{32}$ films with comparable composition (Fig. 14b). Within the resolution limit of the method, Fe-Tb pairs parallel to the interface could not be observed in the multilayer films. We are allowed to discuss the properties of Fe/Tb multilayers as resulting from clearly separated layers, separated at least in the short range order.

These multilayers showed a strong perpendicular anisotropy as long as the residual pressure of the sputter system was below $3 \cdot 10^{-8}$ mbar. At larger residual pressure between two sputtering steps (Fe or Tb) the surface of the grown layer is always contaminated. At contaminated interfaces the Fe and Tb atoms are no longer nearest neighbours. A perpendicular anisotropy was not observed in such multilayer films.

The perpendicular magnetic moment, due to the perpendicular anisotropy was determined from the magnetization loop with fields perpendicular to the film plane. Extrapolation of the magnetization loop to zero field gives the zero field moment m. Figure 15 shows the temperature dependence of the perpendicular moment $m(T)$ for various combinations of layer thicknesses. All combinations show the same Curie temperature 155 °C. This means the layered Fe/Tb films show the same Curie temperature as homogeneous amorphous $Fe_{67}Tb_{23}$ films. The Tb-layers are ferrimagnetically coupled to the Fe-layers. From the combination of various thicknesses of the Fe-layers, d_{Fe}, and of the Tb-layers, d_{Tb}, it was found that at the interface always three monolayers of Fe ($d_{Fe} = 0.75$ nm) and of Tb ($d_{Tb} = 1.1$ nm) couple ferrimagnetically. This interface area is ferrimagnetic with perpendicular anisotropy. In multilayer films Fe/Tb each layer is bounded by two interfaces. A $(Fe/Tb)_n$ multilayer with $d_{Fe} = 1.5$ nm and $d_{Tb} = 2.2$ nm

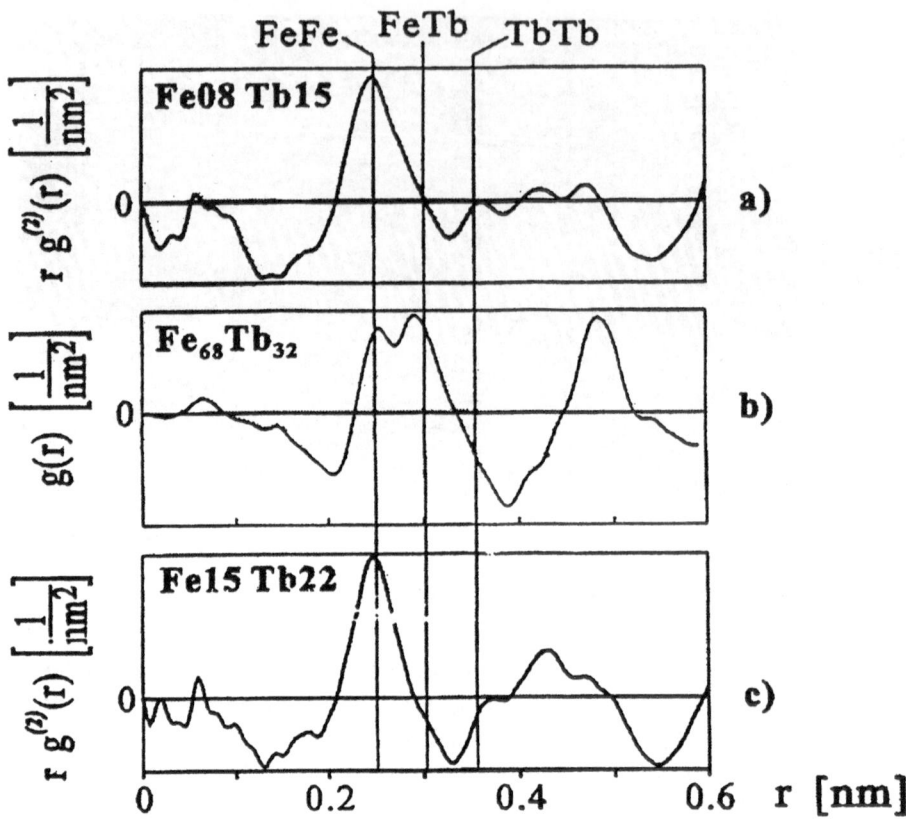

Figure 14 Pair distribution functions parallel to the surface, respectively to the interface, a) [Fe(0.8 nm)/Tb(1.5 nm)]$_{12}$ multilayer, b) Homogeneous amorphous Fe$_{68}$Tb$_{32}$ film, c) [Fe(1.5 nm)/Tb(2.2 nm)] multilayer.

therefore, is a ferrimagnetic material with perpendicular magnetic anisotropy. The coupling area is determined by ($d_{Fe\ coupl} + d_{Tb\ coupl} = 1.85$ nm) (Fig. 16). In a very simple model the multilayer system with the original material period $d_{Fe} + d_{Tb}$ shows now the periodicity: ferromagnetic Fe (in-plane anisotropy) – ferrimagnetic Fe/Tb (perpendicular anisotropy) – paramagnetic Tb – ferrimagnetic Fe/Tb (perpendicular anisotropy) (Fig. 16)

The torque curves of the related systems can be separated in three main categories:

1. $d_{Fe} \leq 2d_{Fe\ coupl} = 1.5$ nm; which leads to torque curves with perpendicular anisotropy;

2. $d_{Fe} > 2d_{Fe\ coupl} = 1.5$ nm; which gives a superposition of the in-plane shape anisotropy and the perpendicular anisotropy;

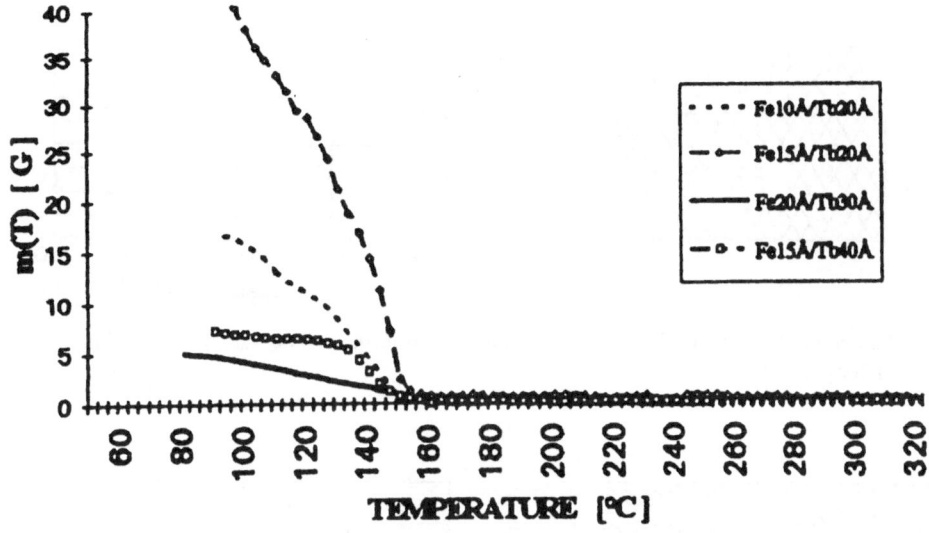

Figure 15 The temperature dependence of the perpendicular magnetic moment.

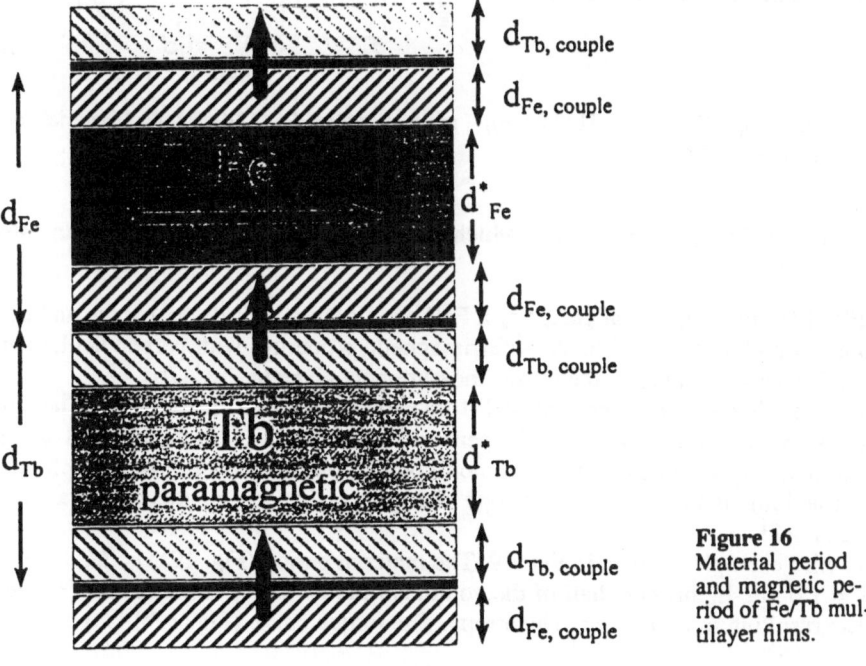

Figure 16
Material period
and magnetic pe-
riod of Fe/Tb mul-
tilayer films.

a)   b)

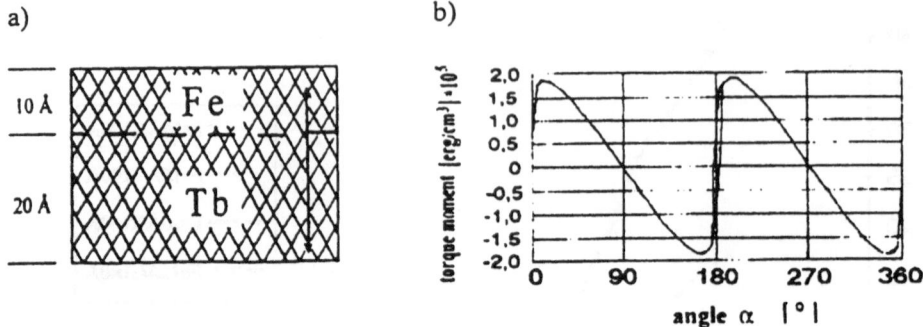

Figure 17 [Fe(1.0 nm)/Tb(2.0 nm)]$_{12}$ multilayer film, $d_{Fe} < 2d_{Fe\ coupl}$, a) Material period, b) Torque curve.

a)   b)

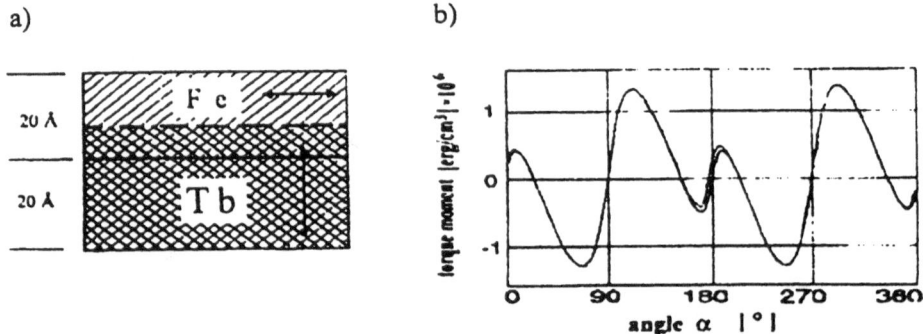

Figure 18 [Fe(2.0 nm)/Tb(2.0 nm)]$_{12}$ multilayer film, $d_{Fe} > 2d_{Fe\ coupl}$, a) Material period, b) Torque curve.

3. $d_{Fe} \gg 2d_{Fe\ coupl} = 1.5$ nm; which is expected to show mainly in-plane shape anisotropy.

Fig. 17a gives a material period of a Fe/Tb multilayer with $d_{Fe} = 1.0$ nm and $d_{Tb} = 2.0$ nm, which is expected to show ferrimagnetic properties with perpendicular anisotropy. This is confirmed by the torque measurement given in Fig. 17b.

Fig. 18a shows the combination of 2.0 nm Fe layers and 2.0 nm Tb layers. The torque curve shows the superposition of the perpendicular anisotropy in the coupled area and the in-plane shape anisotropy of the non coupled part of the Fe layers. Such curves result as long as the field is not strong enough to align the magnetization along the applied field.

A third example is given in Fig. 19. The material period (Fig. 19a) is $d_{Fe} = 3.0$ nm, $d_{Tb} = 1.0$ nm. In this case half of the volume of the Fe-layers is not coupled to the Tb layers resulting in an in-plane anisotropy. The torque curve (Fig. 19b) is typical for a

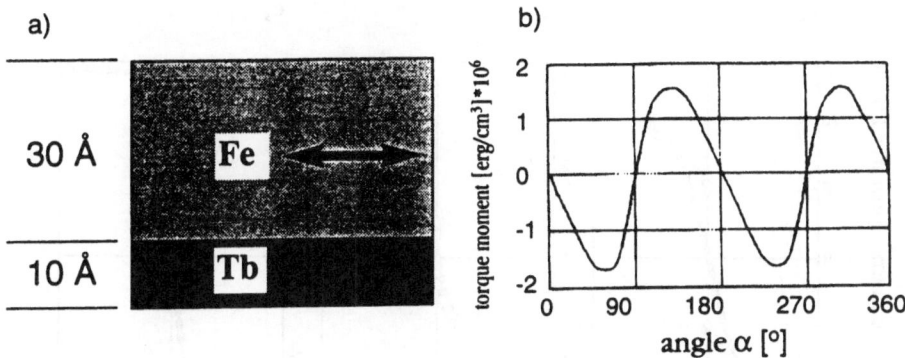

Figure 19 [Fe(3.0 nm)/Tb(1.0 nm)]$_{12}$ multilayer film, $d_{Fe} \gg 2d_{Fe\ coupl}$, a) Material period, b) Torque curve.

strong in-plane (shape) anisotropy. The torque curves confirm qualitatively the simple model of Fig. 16.

4.3 Fe/Tb Multilayers, Separated by Non Magnetic Metal Interlayers (Pt, Ta, Cu, Au)

Due to the prediction of the pair ordering model a separation of neighbouring Fe-Tb pairs at the interface by non magnetic interlayers is expected to destroy the perpendicular anisotropy. Multilayer films (Fe/Pt/Tb/Pt)$_n$ were sputtered. The thicknesses of the Fe-layers ($d_{Fe} = 1.5$ nm) and Tb-layers ($d_{Tb} = 2.2$ nm) were held at a constant value, while the thickness of the Pt interlayer was increased in steps of 0.1 nm. As soon as the thickness of the Pt interlayer was about one monolayer, the perpendicular anisotropy vanished completely. The magnetic anisotropy of the multilayer, determined from the torque curves, now was only given by the shape anisotropy.

But the system remains in a ferrimagnetic state. Even if the thickness of the interlayers is increased the magnetization loop is like that of ferrimagnetic material as shown in Fig. 20, for Pt (Fig. 20a) and Ta (Fig. 20b) interlayers. The magnetization loops with fields parallel and perpendicular to the film plane show a constant increase of the net magnetic moment with increasing field and do not saturate at the largest available field of 6 T. Despite of the non magnetic metal interlayers Pt and Ta, respectively, at zero field the magnetic moments of Fe and Tb films are antiparallel coupled to each other by a long range indirect exchange interaction via the metal interlayer. The net magnetic moment at zero field, m_0, and the susceptibility depend on the material of the non magnetic interlayers.

Introducing Au interlayers leads to a ferromagnetic parallel coupling of the magnetic moments as to be seen from Fig. 20c. The magnetization loops show a large magnetic moment and saturate at rather low fields.

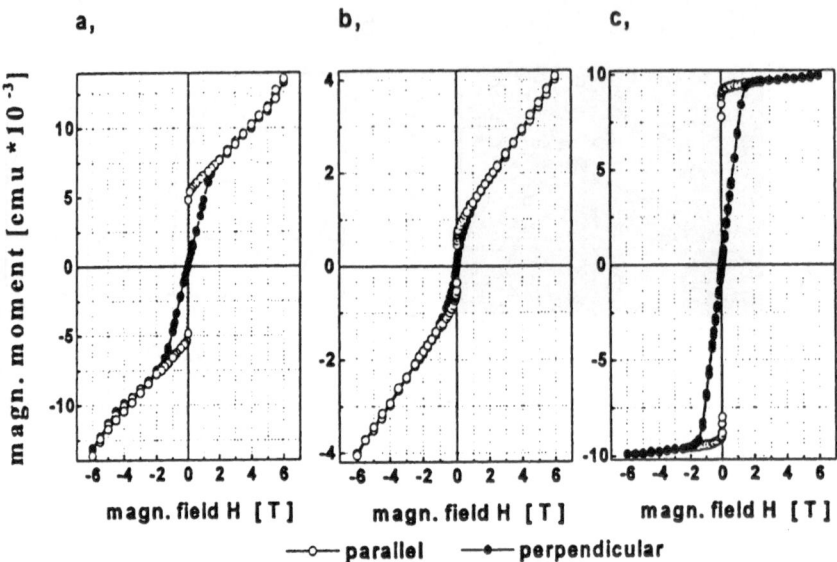

Figure 20 The magnetization loops of [Fe/(Pt,Ta,Au)/Tb/(Pt,Ta,Au)]$_{12}$ multilayer films: a) Pt, b) Ta, c) Au – interlayers, respectively.

The indirect exchange coupling between Fe- and Tb-layers depends in strength and sign on the interlayer material.

4.4   The Oscillation of the Net Magnetic Moment with Increasing Thickness of the Interlayers

Fe/Tb systems with Pt, Ta, Au interlayers of increasing thickness were investigated. The thickness of the interlayers was changed in steps of 0.1 nm. The thickness intervals of the interlayers were selected 0.5 nm $\leq d_{Pt} \leq$ 2.5 nm, 1.0 nm $\leq d_{Ta} \leq$ 3.0 nm, and 2.0 nm $\leq d_{Au} \leq$ 4.5 nm, respectively, to take care of continuous interlayers.

Investigating the film structure by transmission electron microscopy of cross-section (Fig. 21) showed continuous layers. The same result was found by atomic force microscopy, in which the topography of the surface after each respective sputtering step was investigated. In addition XPS depth profiles showed the separation of the single layers.

Evaluating the net magnetic moment of the multilayer systems m_0 at zero field from extrapolating the in-plane magnetization loops (Fig. 20) to zero field, an oscillation of m_0 with increasing interlayer thickness was found (Fig. 22). Independent experiments, evaluating the net magnetic moment from torque measurements, not reported here, led to the same result. As given in Fig. 22 the length of the oscillation period depends on the material of the interlayers:

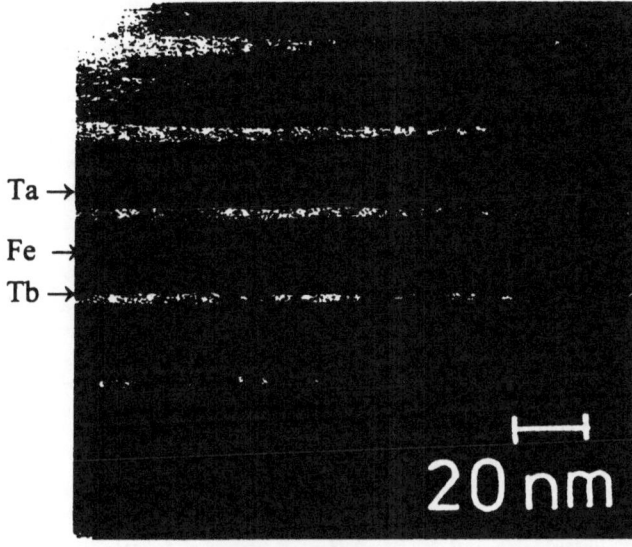

Ta →

Fe →

Tb →

Figure 21
TEM image of a crossection of a [Fe(1.5 nm)/ Ta(4.0 nm)/ Tb(2.2 nm)/ Ta(4.0 nm)]$_{12}$ multilayer film.

$$(Fe/Pt/Tb/Pt)_n \; : \quad d_{Pt \, osc} = 0.5 \text{ nm}$$
$$(Fe/Ta/Tb/Ta)_n \; : \quad d_{Ta \, osc} = 0.8 \text{ nm}$$
$$(Fe/Au/Tb/Au)_n : \quad d_{Au \, osc} = 0.7 \text{ nm}$$

In all cases $n = 12$, and the thicknesses of the Fe-layers, $d_{Fe} = 1.5$ nm, and of the Tb-layers $d_{Tb} = 2.2$ nm was held at constant values.

As to be seen from Fig. 22 the mean net moments, m_0, depend on the material of the interlayers, but in all cases an oscillation is observed. The amplitude of the oscillation does not decrease in the thickness interval of the present investigations. The sign of the indirect exchange coupling does not change: positive (parallel) for Au interlayers, negative (antiparallel) for Pt and Ta, interlayers. The $(Fe/Ta/Tb/Ta)_n$ multilayers may include more than one oscillation periods.

5 Discussion

5.1 Perpendicular Anisotropy

Indirect exchange coupling between ferromagnetic films is widely investigated and reported in many papers in the literature. Indirect exchange coupling between ferromagnetic (Fe) layers and paramagnetic (Tb) layers has been discussed in the literature concerning the perpendicular anisotropy. The long range coupling between layers, separated by non magnetic metal interlayers observed in the above reported experiments is an effect which has not yet been reported in the literature.

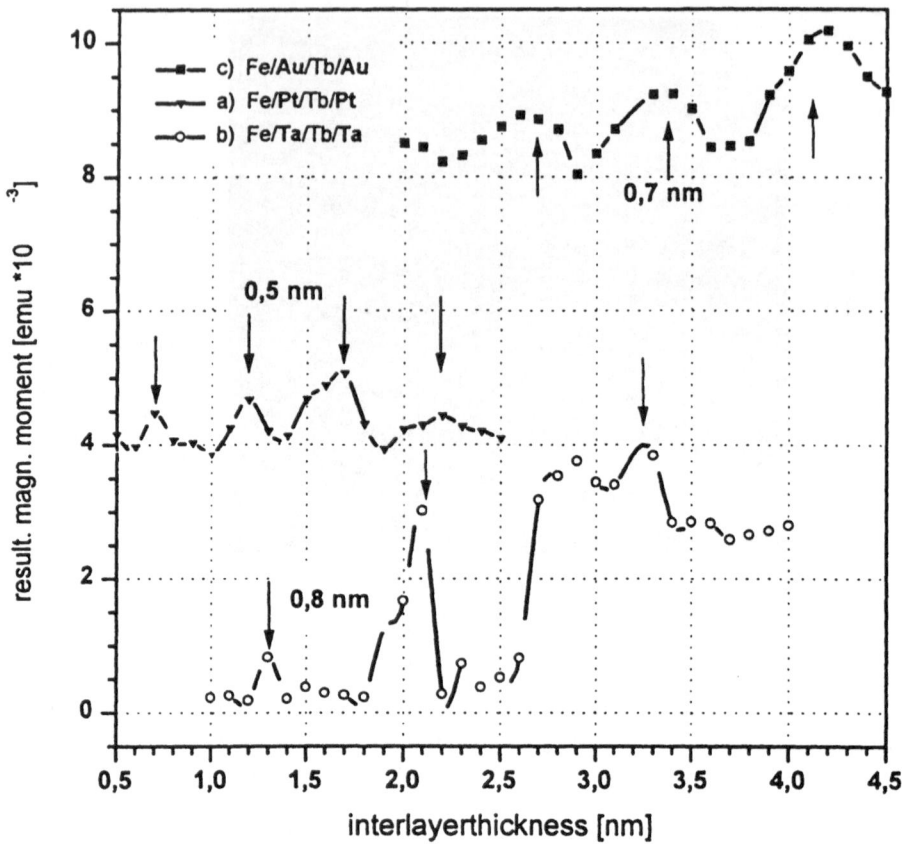

Figure 22 Oscillation of the net magnetic moment m_0 at zero field, a) Pt, b) Ta, c) Au interlayers.

The perpendicular anisotropy due to Fe-Tb pair ordering at the interface results from a short range indirect exchange coupling between nearest neighbours Fe and Tb. From the torque curve of Fig. 17b one obtains a perpendicular anisotropy constant $K_u = 2 \cdot 10^5$ ergs/cm³. The true perpendicular anisotropy $K_\perp$ is given by

$$K_\perp = K_u + 2\pi M_S^2$$

From the respective magnetization loop follows $M_S = 187$ G. Then $K_\perp = 4.2 \cdot 10^5$ ergs/cm³. This is the true anisotropy volume energy density due to the Fe-Tb pairs at the interface. The anisotropy carying volume V_a at the interface is made up by one monolayer Fe and Tb, respectively, which is only a fraction of the total volume V.

$$\frac{V_a}{V} = \frac{r_{FeTb}}{d_{Fe} + d_{Tb}} \cdot \frac{2n-1}{n} = 0.195$$

with $r_{FeTb} = 0.305$ nm, the Fe-Tb pair separation, following from the pair distribution function and d_{Fe} and d_{Tb} the thicknesses of the Fe and Tb layers, respective and $n = 12$, the number of Fe/Tb periods of the multilayer film.

The true anisotropy due to this volume fraction V_a/V is

$$K_{\perp a} = K_{\perp} \frac{V}{V_a} = 21.1 \cdot 10^5 \text{ ergs/cm}^3$$

which leads to an interface anisotropy

$$K_{\perp in} = K_{\perp a} \cdot r_{FeTb} = 6.4 \cdot 10^{-2} \text{ ergs/cm}^2$$

Finally the anisotropy energy, contributed by each Fe-Tb pair is roughly estimated by

$$k_{\perp} = K_{\perp a} \cdot r_{Tb}^2 r_{FeTb} = 8.1 \cdot 10^{-17} \text{ ergs/pair}$$
$$k_{\perp} = 5 \cdot 10^{-5} \text{ eV/pair}$$

with $r_{Tb} = 0.355$ nm from the pair distribution function.

5.2 Oscillation of the Net Magnetic Moment

As mentioned before, the indirect exchange coupling between layers could be explained by quantum well states of spin polarized electrons in the interlayer. From the experimental observation follows that one has to discuss two effects of coupling:

1. Coupling results in a magnetic moment of the originally paramagnetic Tb layers. At least three monolayers Tb at the Tb/Pt interface exhibit a ferromagnetic order. This coupling is not restricted to the first Tb monolayer.

2. The induced magnetic moments of the Tb layers couple via the non magnetic metal interlayers with the Fe layers.

The magnetic moment of the Tb layer is caused by an exchange field H_{ex}. Microscopically, the moments of the first Tb monolayer at the interface is polarized by the spin polarized quantum well states of the interlayer. The polarization of the first monolayers leads to a weaker polarized the second monolayer and so forth. The decrease of the Tb polarization with distance from the interface is approached by an exchange field H_{ex}, decreasing exponentially with distance from the first Tb monolayer at the interface.

$$H_{ex} = H_0 \exp\left(-\frac{d_{Tb} - d_0}{\ell}\right)$$

with d_0 = thickness of the Tb monolayer, ℓ the coupling length, and H_0 the exchange field at the first Tb monolayer. Now the Tb layer can be treated like paramagnetic material in an external field, given by the exchange field.

Following Curie's law the magnetic moment of the Tb layer than is given by

$$m_{Tb} = \chi \cdot H_0 \ell \left(1 - \exp\left\{ -\frac{d_{Tb} - d_0}{\ell} \right\} \right)$$
$$= m_{Tb}^* \left(1 - \exp\left\{ -\frac{d_{Tb} - d_0}{\ell} \right\} \right) \tag{5.1}$$

with m_{Tb}^* the homogeneous magnetic moment of a layer with thickness ℓ in a homogeneous magnetic field H_0.

The net magnetic moment of the layers at zero external field is given by

$$m_0 = |m_{Tb} - m_{Fe}|.$$

Oscillation of m_0 with the interlayer thickness (Fig. 22) than can be explained by an oscillation of the amplitude of the exchange field H_0 due to the quantum well states of the spin polarized electrons at the Fermi surface of the interlayer, causing an oscillation of m_0. The above given discussion, does not explain the sign of the exchange field. Due to spin polarized quantum well states of the interlayer at the Fermi surface one should expect an oscillation of the sign with increasing interlayer thickness. This was not observed in our experiment. Always one polarization is favoured. Assuming that spin polarized quantum wells couple antiparallel like Fe-Tb atoms, then for:

Pt and Ta interlayers: spin polarization of the quantum wells is parallel to the magnetic moment of the ferromagnetic layer of the quantum wells

Au interlayers: spin polarization is antiparallel to the moment of the ferromagnetic layer.

So far this problem remains unsolved.

5.3 The Magnetization Loop

The magnetization loops of Tb/(Pt,Ta)/Fe/(Pt,Ta) multilayers are those of ferrimagnetic materials. In zero applied field the moment $\vec{m}_{Tb}$ of the Tb layers couples antiparallel to the moment $\vec{m}_{Fe}$ of the Fe layers.

From the experiments (magnetization loops) the net magnetization in zero field was determined

$$m_0 = |m_{Tb} - m_{Fe}|$$

Therefore, from the amount alone one cannot decide which moment dominates. In the following common discussion we introduce $\vec{m}_1$ and $\vec{m}_2$ with $m_1 > m_2$. Furthermore we observed only a very small coercive force H_C. The magnetization loop above $H_C (H \geq H_C)$ is free of hysteresis. We assume coherent rotation of the moments $\vec{m}_1$ and $\vec{m}_2$ at increasing applied field $\vec{H}$.

The total energy of the system is given by

$$E = -\vec{m}_1\vec{H} - \vec{m}_2\vec{H} - 2A\frac{\vec{m}_1\vec{m}_2}{m_1m_2}$$

$$E = -m_1H\cos\phi_1 - m_2H\cos\phi_2 - 2A\cos(\phi_1 + \phi_2)$$

Equilibrium demands simultaneously

$$\frac{\partial E}{\partial\phi_1} = m_1H\sin\phi_1 + 2A\sin(\phi_1 + \phi_2) = 0$$

$$\frac{\partial E}{\partial\phi_2} = m_2H\sin\phi_2 + 2A\sin(\phi_1 + \phi_2) = 0$$

From these equation follows

$$m_1\sin\phi_1 = m_2\sin\phi_2 \tag{5.2}$$

and

$$m_1H\sin\phi_1 = -2A\sin(\phi_1 + \phi_2) \tag{5.3a}$$

or

$$m_2H\sin\phi_2 = -2A\sin(\phi_1 + \phi_2) \tag{5.3b}$$

The magnetization loop according to the experiment at multilayers with Pt interlayers is given in Fig. 23a. The related rotation of $\vec{m}_1$ and $\vec{m}_2$ at increasing field is schematically drawn in Fig. 23b–e. At $H = H_C$ (small) the layers are in a single domain state with antiparallel magnetic moments (Fig. 23b). At increasing field $\vec{m}_1$ rotates counterclockwise from the direction of the applied field, $\vec{m}_2$ rotates counterclockwise against the field direction (Fig. 23c). At a critical field, H_{cr}, $\vec{m}_2$ is in-plane perpendicular to the field ($\phi_2 = \pi/2$), $\vec{m}_1$ is at an critical angle, ϕ_{1cr}, (Fig. 23d). At fields $H > H_{cr}$, $\vec{m}_2$ rotates further counterclockwise, while $\vec{m}_1$ must rotate clockwise against the field direction.

In summary

$$
\begin{aligned}
&H_C \leq H < H_{cr}: &&\phi_1 < \tfrac{\pi}{2}; &&\tfrac{\partial\phi_1}{\partial H} > 0 \\
& &&\phi_2 > \tfrac{\pi}{2}; &&\tfrac{\partial\phi_2}{\partial H} < 0 \\
&H > H_{cr}: &&\phi_1 < \tfrac{\pi}{2}; &&\tfrac{\partial\phi_1}{\partial H} < 0 \\
& &&\phi_2 < \tfrac{\pi}{2}; &&\tfrac{\partial\phi_2}{\partial H} < 0 \\
&H = H_{cr}: &&\phi_1 < \tfrac{\pi}{2}; &&\tfrac{\partial\phi_1}{\partial H} = 0 \quad \text{because of steadiness} \\
& &&\phi_2 = \tfrac{\pi}{2}; &&\tfrac{\partial\phi_2}{\partial H} < 0
\end{aligned}
\tag{5.4}
$$

The net magnetic moment at the magnetization loop is given by

$$m(H) = m_1\cos\phi_1 + m_2\cos\phi_2$$

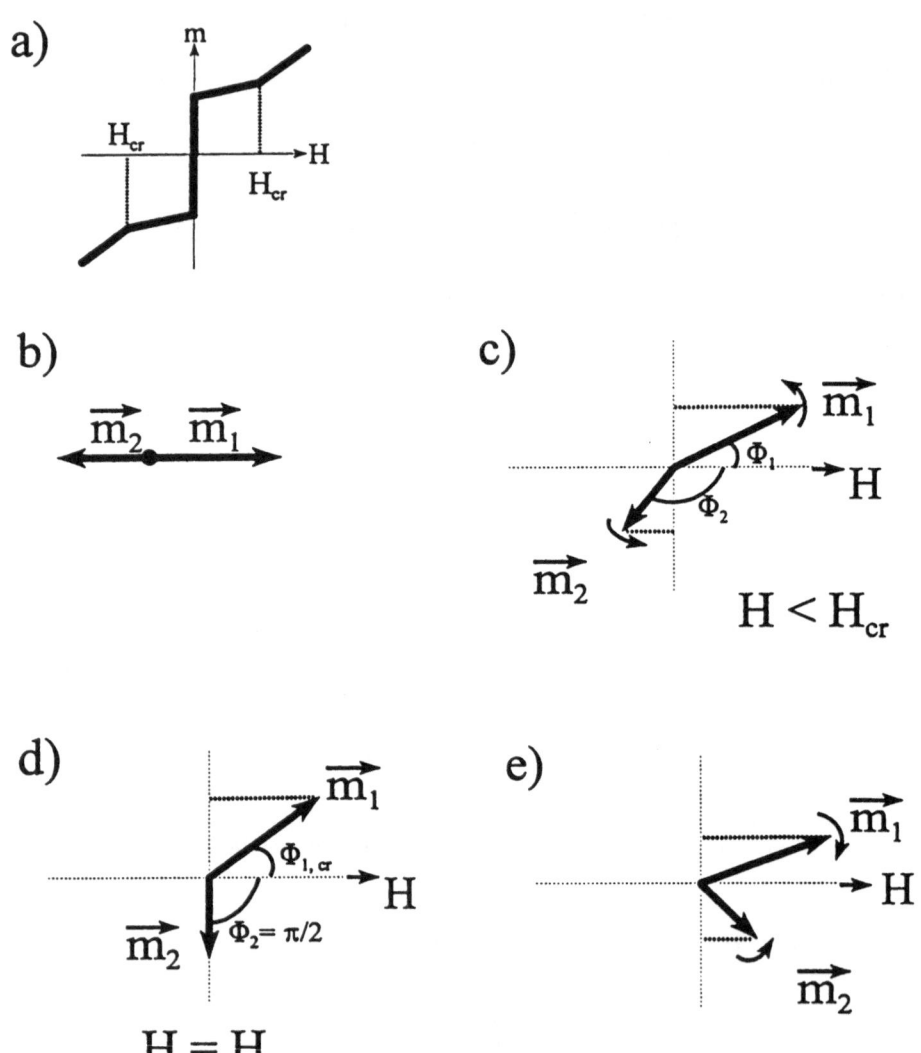

Figure 23 Model of coherent rotation of $\vec{m}_1$ and $\vec{m}_2$ at increasing field.

and the related susceptibility

$$\chi = \frac{dm}{dH} = -m_1 \sin\phi_1 \frac{\partial\phi_1}{\partial H} - m_2 \sin\phi_2 \frac{\partial\phi_2}{\partial H}$$

At $H = H_{cr}$ the susceptibility changes ($\frac{\partial\chi}{\partial H} > 0$):
Below the critical field:

$$\chi(H < H_{cr}) = -m_1 \sin\phi_1 \frac{\partial\phi_1}{\partial H} < 0$$

$$-m_2 \sin\phi_2 \frac{\partial\phi_2}{\partial H} > 0$$

Above the critical field:

$$\chi(H > H_{cr}) = -m_1 \sin\phi_1 \frac{\partial\phi_1}{\partial H} > 0$$

$$-m_2 \sin\phi_2 \frac{\partial\phi_2}{\partial H} > 0$$

At the critical field:

$$\chi(H = H_{cr}) = -m_2 \sin\phi_2 \frac{\partial\phi_2}{\partial H} > 0$$

The susceptibility increases at the critical field

$$\chi(H < H_{cr}) < \chi(H > H_{cr})$$

leading to a kink in the magnetization loop, which is to be seen in Fig. 20a for multi-layers with Pt interlayers.
The amount of the critical field can be used to discuss the coupling constant A.
At $H = H_{cr}$ we find $\phi_2 = \pi/2$ and

$$m(H_{cr}) = m_1 \cos\phi_{1cr} \tag{5.5}$$

From Eq. (5.3b) follows

$$m_2 H_{cr} = -2A \cos\phi_{1cr} \tag{5.6}$$

Introducing Eq. (5.5) into Eq. (5.6) gives

$$A = -\frac{1}{2} \frac{m_1 m_2}{m(H_{cr})} H_{cr} \tag{5.7}$$

The moments m_1 and m_2 could be determined from the saturation of the ferrimagnetic magnetization loop

$$m_s = m_1 + m_2$$

and from the zero field value

$$m_0 = m_1 - m_2$$

As can be seen from the magnetization loops, (Fig. 20a,b) the maximum field, 6 T, of our experiment is far below the saturation field. Therefore we are at present unable to separate m_1 and m_2. A rough estimation can be tried. Assuming a magnetic moment of $1.5\mu_B$ per Atom for the very thin (1.5 nm) amorphous Fe layers, the magnetic moment of one Fe layer is $m_{Fe} = 1.76 \cdot 10^{-4}$ emu/cm^2. Compared to homogeneous amorphous alloys of similar composition the magnetic moment of the Tb layers is expected to dominate the magnetic moment of the Fe layers

$$m_1 = m_{Tb} > m_2 = m_{Fe}$$

At zero field

$$m_{Tb} = m_0 + m_{Fe}$$

From measurements the net magnetic moment per layer period Fe/Tb is given by

$$m_0 = 1.84 \cdot 10^{-4} \text{ emu/cm}^2$$

which leads to $m_{Tb} = 3.6 \cdot 10^{-4}$ emu/cm^2. The critical field is $H_{cr} = 5 \cdot 10^4$ Oe and the net moment is $m(H_{cr}) = 4.04 \cdot 10^{-4}$ emu/cm^2. (From Fig. 20a)

According to eq. 6 these values result in a coupling constant between the Fe and Tb layers $A \approx 4$ erg/cm^2.

Comparing the magnetization loops of Fig. 20a and 20b one finds a smaller net magnetic moment at zero field for multilayers with Ta interlayers than with Pt interlayers. The thicknesses of the Fe layers and the Tb layers are the same in both multilayer films. A smaller net magnetic moment results from a lower moment of the Tb layers in the system with Ta interlayers. The exchange field in the Ta system is lower than in the Pt system.

Fe/Tb layers, separated by Au interlayers couple their magnetic moments parallel to each other. The magnetization saturates at low fields. The net magnetic moment is larger than that of the pure Fe layers, i.e. the parallel coupled Tb layers contribute to the magnetization. These different coupling modes for different interlayer materials need further investigation.

6 Conclusion

The properties of multilayer films Fe/Tb with layer thicknesses of three monolayers of each material are ferrimagnetic like homogeneous amorphous $Fe_{68}Tb_{32}$ films. The perpendicular magnetic anisotropy is caused by indirect exchange coupling between nearest neighbours Fe-Tb at the interface. Separating the Fe-Tb-pairs by non-magnetic metal interlayers destroys the magnetic anisotropy. The anisotropy is due to a short range interaction. Beside this interaction a long range indirect exchange via the non magnetic metal interlayer was observed, which induces a magnetic order into the originally paramagnetic Tb films. At Pt and Ta interlayers the induced magnetic moment of the Tb layers couples antiparallel to the magnetic moments of the Fe layers. The coupling is strong. Large fields are necessary to saturate the films. At Au interlayers the magnetic moments of Fe and Tb couple parallel to each other, leading to material with large magnetization. The strength and sign of indirect exchange coupling between Fe and Tb layers depend on the material of the non magnetic metal interlayer.

Bibliography

[1] P. Grünberg, R. Schreiber, Y. Pang, M.B. Brodski, H. Sowers Phys. Rev. Letters 57, 2442 (1986)

[2] M. Baibich, J. U. Bruto, A. Fert, F. Ngnyen van Dan, F. Petroff, P. Etienne, G. Creuzot, J. Chazeles Phys. Rev. Letters 61, 2472 (1988)

[3] S.S.P. Parkin, R. Badhra, K.P. Roche Phys. Rev. Letters 66, 2152 (1991)

[4] E.F. Fullerton, M.J. Conover, J.E. Mattson, C.H. Sowers, S.D. Bader Phys. Rev. B48, 15755 (1993)

[5] S.S.P. Parkin Phys. Rev. Letters 67, 3598 (1991)

[6] S.S.P. Parkin, N. More, K.P. Roche Phys. Rev. Letters 64, 2309 (1990)

[7] O. Leng, V. Cros, R. Schäfer, A. Fuss, P. Grünberg, W. Zinn Journ. Magnetism and Magnetic Materials 126, 367 (1993)

[8] J. Unguris, R.J. Celotta, D.T. Pierce Phys. Rev. Letters 67, 140 (1991)

[9] J. Unguris, R.J. Celotta, D.T. Pierce J. Appl. Phys. 75, 6437 (1994)

[10] P. Bruno, C. Chappert Phys. Rev. B 46, 261 (1992)

[11] D.M. Edwards, J. Mathon J. Magnetism. and Magnetic Materials 93, 85 (1991)

[12] P. Bruno J. Appl. Phys. 76, 6972 (1994)

[13] J.E. Ortega, F.J. Himpsel, G.J. Mankey, R.F. Willis Phys. Rev. B 47, 1540 (1993)

[14] R. Klasges, D. Schmitz, C. Carbonne, W. Eberhardt, C. Pampuch, T. Kachel, W. Gudat Spring meeting, AK Festkörperphysik, Münster 1997, AM 1.3

[15] K. Wildberger, R. Zeller, P.H. Dederichs Spring meeting, AK Festkörperphysik, Münster 1997, AM 7.3

[16] S.N. Okuno, K. Inamoto Phys. Rev. Letters 72, 1553 (1994)

[17] C. Dorner, M. Haidl, H. Hoffmann Phys. stat. sol. 145, 551 (1994)

[18] A. Bechert, R. Trautsch, H. Hoffmann, W. Andrä, Phys. stat. sol. (a) 161, 483 (1997)

Contents of Volumes 33–37